Information Communication Technologies for Enhanced Education and Learning:
Advanced Applications and Developments

Lawrence Tomei
Robert Morris University, USA

INFORMATION SCIENCE REFERENCE
Hershey · New York

Director of Editorial Content: Kristin Klinger
Director of Production: Jennifer Neidig
Managing Editor: Jamie Snavely
Assistant Managing Editor: Carole Coulson
Typesetter: Lindsay Bergman
Cover Design: Lisa Tosheff
Printed at: Yurchak Printing Inc.

Published in the United States of America by
Information Science Reference (an imprint of IGI Global)
701 E. Chocolate Avenue, Suite 200
Hershey PA 17033
Tel: 717-533-8845
Fax: 717-533-8661
E-mail: cust@igi-global.com
Web site: http://www.igi-global.com

and in the United Kingdom by
Information Science Reference (an imprint of IGI Global)
3 Henrietta Street
Covent Garden
London WC2E 8LU
Tel: 44 20 7240 0856
Fax: 44 20 7379 0609
Web site: http://www.eurospanbookstore.com

Library of Congress Cataloging-in-Publication Data

Information communication technologies for enhanced education and learning : advanced applications and developments / Lawrence Tomei, editor.

p. cm.

Includes bibliographical references and index.

Summary: "This book offers an examination of technology-based design, development, and collaborative tools for the classroom"--Provided by publisher.

ISBN 978-1-60566-150-6 (hardcover) -- ISBN 978-1-60566-151-3 (ebook)

1. Educational technology. 2. Information technology. I. Tomei, Lawrence A.

LB1028.3.I519346 2009

371.33--dc22

2008013146

British Cataloguing in Publication Data
A Cataloguing in Publication record for this book is available from the British Library.

Information Communication Technologies for Enhanced Education and Learning: Advanced Applications and Developments is part of the *IGI Global series named Advances in Information and Communication Technology Education (AICTE)* Series, ISBN: 1935-3340

Table of Contents

Preface xviii

Section I
Design Tools

Section I.a
Theory

Chapter I
Media and Women in Technology 1
Mara H. Wasburn, Purdue University, USA

Chapter II
The Gender Communication Gap in Online Threaded Discussions 15
David Gefen, Drexel University, USA
Nitza Geri, The Open University of Israel, Israel
Narasimha Paravastu, Metropolitan State University, USA

Chapter III
The Technology Acceptance Model (TAM) and the Continuance Intention of Using WebCT: A Case of College Students in Estonia 29
Princely Ifinedo, Cape Breton University, Canada

Chapter IV
The Influence of Constructivist E-Learning System on Student Learning Outcomes 45
Thanakorn Wangpipatwong, King Mongkut's University of Technology Thonburi, Thailand
Borworn Papasratorn, King Mongkut's University of Technology Thonburi, Thailand

Section I.b
Practice

Chapter V
The Didactical Agency of Information Communication Technologies for Enhanced Education and Learning 59
Andreas Wiesner-Steiner, University of Applied Sciences Bremen, Germany
Heike Wiesner, Berlin School of Economics, Germany
Heidi Schelhowe, University of Bremen, Germany
Petra Luck, Liverpool Hope University, UK

Chapter VI
Comparative Analyses of Online and Traditional Undergraduate Business Law Classes: How Effective is E-Pedagogy? 76
Daniel J. Shelley, Robert Morris University, USA
Louis B. Swartz, Robert Morris University, USA
Michele T. Cole, Robert Morris University, USA

Chapter VII
Student Perceptions of Data Flow Diagrams vs. Use Cases 94
Ido Millet, Penn State University Erie, USA
Robert Nelson, Penn State University Erie, USA

Chapter VIII
Promoting Undergraduate Education with Agent Based Laboratory 103
Hong Lin, University of Houston-Downtown, USA

Chapter IX
Supporting Arguments for Including the Teaching of Team Competency Principles in Higher Education 122
Tony Jewels, Queensland University of Technology, Australia
Rozz Albon, Curtin University of Technology, Malaysia

Section II
Development Tools

Chapter X
Creating an Interactive PowerPoint Lesson for the Classroom 135
Lawrence Tomei, Robert Morris University, USA

Chapter XI
Planning Staff Training for Virtual High Schools 142
Chris Thompson, Elmbrook Schools, USA
Zane L. Berge, University Maryland Baltimore Campus, USA

Chapter XII
Training Prospective Online Instructors: Theories Utilized by Current Online Instructors 151
MarySue Cicciarelli, Duquesne University, USA

Chapter XIII
The Impact of PowerPoint Presentations on Student Achievement and Student Attitudes 166
Michael Fedisson, Bellefonte Area Middle School, USA
Silvia Braidic, California University of Pennsylvania, USA

Chapter XIV
Teaching Java™: Managing Instructional Tactics to Optimize Student Learning............................. 185
Henry H. Emurian, University of Maryland—Baltimore, USA

Section III
Collaborative Tools

Section III.a
Asynchronous Tools

Chapter XV
Toward an Increase in Student Web Portfolios in New York Colleges and Universities.................... 204
John DiMarco, St. John's University, USA

Chapter XVI
Competent Web Dialogues: Text-Based Linking of Thoughts 219
Marianne Döös, Stockholm University, Sweden
Eva R Fåhræus, Stockholm University, Sweden
Karin Alvemark, Dalarna University, Sweden
Lena Wilhelmson, Stockholm University, Sweden

Chapter XVII
Employing Interactive Technologies for Education and Learning: Learning-Oriented
Applications of Blogs, Wikis, Podcasts, and More... 234
Jeffrey Hsu, Fairleigh Dickinson University, USA

Chapter XVIII
Assessing Online Discussion Forum Participation.. 259
Matthew Shaul, Kennesaw State University, USA

Section III.b
Synchronous Tools

Chapter XIX
Synchronous Hybrid E-Learning: Empirical Comparison with Asynchronous and Traditional Classrooms 269
Solomon Negash, Kennesaw State University, USA
Michelle Emerson, Kennesaw State University, USA
John Vandegrieft, Blackstone & Cullen, Inc., USA

Chapter XX
Understanding the Effectiveness of Collaborative Activity in Online Professional Development with Innovative Educators through Intersubjectivity 283
Diane Hui, The University of Hong Kong, Hong Kong
Donna L. Russell, University of Missouri-Kansas City, USA

Chapter XXI
Effective Questioning to Facilitate Dynamic Online Learning 303
Silvia Braidic, California University of Pennsylvania, USA

Chapter XXII
Transitioning from Face-to-Face to Online Instruction: How to Increase Presence and Cognitive/Social Interaction in an Online Information Security Risk Assessment Class 313
Cindy S. York, Purdue University, USA
Dazhi Yang, Purdue University, USA
Melissa Dark, Purdue University, USA

Compilation of References 324

About the Contributors 358

Index 366

Detailed Table of Contents

Preface xviii

Section I
Design Tools

Section I.a
Theory

Chapter I
Media and Women in Technology 1
Mara H. Wasburn, Purdue University, USA

Many Western nations face a critical shortage of skilled professionals in science, technology, engineering, and mathematics (STEM). However, despite abundant opportunities, few women prepare themselves for careers in these fields. Several of those concerned with the problem have proposed that new media programming, such as television dramas with women engineers, computer professionals, and/or engineers in leading roles, might help attract more women to STEM fields. This paper identifies a theoretical rationale for a media centered strategy, and describes a pilot study whose data suggest that a media-centered approach might have some success in producing greater interest among women in pursuing STEM careers, particularly information technology careers.

Chapter II
The Gender Communication Gap in Online Threaded Discussions 15
David Gefen, Drexel University, USA
Nitza Geri, The Open University of Israel, Israel
Narasimha Paravastu, Metropolitan State University, USA

Threaded discussions are one of the central tools of online education. These tools enhance student learning and compensate for the lack of social interaction. This study examines whether these social interactions are affected by some typical gender related conversational behaviors, despite the fact that these threaded discussion are designed to operate in a seemingly gender neutral online environment. That men and women communicate differently in open conversation is at the core of sociolinguistic theory. A direct result of these differences is a tendency toward same-gender oral conversations. This study analyzes data from 233 students in 27 online courses and examines students based on whom they reference in the threaded discussion and the way they reference others. Theoretical and practical implications on managing threaded discussions are discussed along with directions for further research.

Chapter III
The Technology Acceptance Model (TAM) and the Continuance Intention of Using WebCT: A Case of College Students in Estonia 29
Princely Ifinedo, Cape Breton University, Canada

The authors investigate the technology acceptance model (TAM) and the continued use of a popular course management system for teaching. The study investigates a sample of 72 students with experience using the software from four higher education institutions. In order to study the nature of the relationships among the constructs, eight hypotheses were formulated and tested using the structural equation modeling technique, Partial Least Squares. The predictive power of the model was adequate and the study found support for seven of eight hypotheses. The study also found that when computer anxiety is low, students are able to use and continue to use the system without much difficulty. The data did not support the relationship between perceived usefulness and usage. The study's implications for research and practice are succinctly outlined.

Chapter IV
The Influence of Constructivist E-Learning System on Student Learning Outcomes 45
Thanakorn Wangpipatwong, King Mongkut's University of Technology Thonburi, Thailand
Borworn Papasratorn, King Mongkut's University of Technology Thonburi, Thailand

In this article, the study of how a constructivist e-learning system affects students' learning outcomes was explored and a two-phase study was designed. The first study sought to create a constructivist e-learning environment (CEE) and discover how students expected their learning outcomes under CEE. CEE is composed of three constructs, which are exploration, collaboration, and construction. The statistical results showed the high level of student expectation on every construct. Consequently, constructivist e-learning system (CES) was developed. In the second study, CES was used in the actual classroom environment. The purpose was to compare the learning outcomes and knowledge development of students who studied the course using CES with those of students who learned it under a traditional learning environment. A T-test method was used to analyze the learning outcomes. The results showed that students who used CES had better learning outcomes and knowledge development than students who did not use CES.

Section I.b
Practice

Chapter V
The Didactical Agency of Information Communication Technologies for Enhanced Education and Learning 59
Andreas Wiesner-Steiner, University of Applied Sciences Bremen, Germany
Heike Wiesner, Berlin School of Economics, Germany
Heidi Schelhowe, University of Bremen, Germany
Petra Luck, Liverpool Hope University, UK

The authors present two projects that deal with teaching and learning using digital media in basic and higher education and offer new perspectives on the active role of technology in learning processes. Their first case draws on a project that aims to promote girls' interest in sciences, mathematics and technology. It suggests a new pedagogical approach towards the use of robotics in education and discusses how didactics and technology interact and how the character of robotics itself plays an important role in gender determination. The second case focuses on distance education teaching methods in childcare management. The options remaining for practitioners in higher education are either to embrace the new media or to watch its inevitable unfolding.

Chapter VI
Comparative Analyses of Online and Traditional Undergraduate Business Law Classes:
How Effective is E-Pedagogy? 76
Daniel J. Shelley, Robert Morris University, USA
Louis B. Swartz, Robert Morris University, USA
Michele T. Cole, Robert Morris University, USA

E-learning and e-pedagogy continue to grow in importance in higher education due in large measure to cost, changing student profiles, scarcity of traditional classroom space, and the recognition that distance learning has created a genuinely new paradigm for instruction. To respond to the changing demographics, working adults, military students, and residents of rural and even international communities, universities are adding a considerable array of online courses. As they do, the question arises whether online instruction is, or can be, as effective as the traditional classroom. Investigating the question is the focus this study that compares students enrolled in both online and traditional classroom versions of one business law course where all elements were the same except for the instructional format. The first study found no significant difference between the two formats with regard to student satisfaction and student learning. However, the second study did find statistically significant differences between the online and the traditional course formats with regard to student satisfaction with the instructor, and student satisfaction with the course structure.

Chapter VII
Student Perceptions of Data Flow Diagrams vs. Use Cases 94
Ido Millet, Penn State University Erie, USA
Robert Nelson, Penn State University Erie, USA

Data flow diagrams and use cases are two popular methods for teaching as well as practice. For the last four years, the authors have been using both methodologies in a systems analysis course. Questionnaire results indicate that students find the use cases methodology slightly easier to understand. However, students believe that data flow diagrams are significantly better at communicating with users and programmers.

Chapter VIII
Promoting Undergraduate Education with Agent Based Laboratory 103
Hong Lin, University of Houston-Downtown, USA

Agent-oriented design has become one of the most active areas in the field of software engineering, serving as a focal point for accountability and responsibility for coping with the complexity of software systems both during design and execution. Research has found that software engineering challenges in developing large scale distributed systems can be overcome by an agent-based approach. In this chapter, this author discusses how a distributed system can be modeled as a set of autonomous, cooperating agents that communicate intelligently with one another, automate or semi-automate functional operations, and interact with human users at the right time with the right information.

Chapter IX
Supporting Arguments for Including the Teaching of Team Competency Principles in Higher Education 122
Tony Jewels, Queensland University of Technology, Australia
Rozz Albon, Curtin University of Technology, Malaysia

For optimum workplace effectiveness in knowledge-intensive industries in which principles of knowledge management need to be applied, it is necessary to take into account not only the competencies of individuals themselves but also the competencies of the teams in which they must operate. Although the incorporation of various types of group work into pedagogies is already widespread within institutes of higher education, many examples fail to embrace a rationale for, or the potential benefits of, multiple contributor environments. We present in this chapter arguments for including the teaching of team competency principles in higher education, supported by an original multi-dimensional team competency teaching model, a taxonomy for assessing team competency levels and an example of the implementation of these principles.

Section II
Development Tools

Chapter X
Creating an Interactive PowerPoint Lesson for the Classroom 135
Lawrence Tomei, Robert Morris University, USA

Power Point users will find this chapter invaluable when creating an "Interactive Lesson," a self-paced, student-controlled, individualized learning opportunity embedded with assessments. Interactive lessons are offered to learners who need individualized instruction in the form of remedial instruction, additional practice, or enrichment activities. Interactive lessons may not be new; however, the practical, sequential methodology offered herein provides an innovative design model for creating and presenting self-paced, personalized lesson content. The resulting presentation can be captured to a floppy diskette, burned onto a CDROM, or sent as an email attachment to students in a classroom, computer lab or at home. The interactive lesson has many practical applications for students needing remedial attention or those attending cyber schools or home-bound students.

Chapter XI
Planning Staff Training for Virtual High Schools.. 142
Chris Thompson, Elmbrook Schools, USA
Zane L. Berge, University Maryland Baltimore Campus, USA

This chapter briefly profiles three virtual schools, each at a different stage of development, yet each dependent upon a successful and sustained distance education program for its professional staff in order to remain viable into the future. As virtual schools become more acknowledged by the public and the attention given to the online schools shifts from their sources of funding to their standardized test scores, a model for sustained distance training and education must be in place to deliver quality professional development that can positively impact students' achievement scores on standardized tests for each school's online student population.

Chapter XII
Training Prospective Online Instructors: Theories Utilized by Current Online Instructors 151
MarySue Cicciarelli, Duquesne University, USA

Research shows that training prospective online instructors in an online learning environment has its advantages. One very effective training topic concerns the use of theory when designing curriculum. This study reports on the empirical research about online instructors and their use of different design theories. It identifies design theories that have not been researched in regard to online instructor utilization of theory, and it illustrates how frequently online instructors use nine of the design theories.

Chapter XIII
The Impact of PowerPoint Presentations on Student Achievement and Student Attitudes 166
Michael Fedisson, Bellefonte Area Middle School, USA
Silvia Braidic, California University of Pennsylvania, USA

The study discussed in this chapter was conducted over a two-year time frame with classes grouped heterogeneously. Seventh grade students were tested on their knowledge of sentences and nouns in a language arts classroom. During instruction, classes were taught using traditional book work and handouts for one unit and technological enhancement for the second unit. When test results were compared, the data indicated the use of technological aids as teaching tools increased student test grades in year one. The increases were especially note for low-achieving students and for those with identified learning disabilities. However, in year two, those same results were not achieved. A technology survey was also used to establish each student's comfort level with technology and their attitudes towards the use of technological aids in the classroom.

Chapter XIV
Teaching Java™: Managing Instructional Tactics to Optimize Student Learning............................. 185
Henry H. Emurian, University of Maryland—Baltimore, USA

Information systems students in a graduate section and an undergraduate section of an introductory Java graphical user interface course completed the following initial assignments to learn a simple program: (1) automated programmed instruction tutoring, (2) hands-on learning with a lecture, and (3) collaborative peer tutoring. Tests of knowledge transfer and software self-efficacy were administered before students began the first assignment and following completion of each one. The results showed progressive improvement in rule test performance and software self-efficacy across the several instructional events. Taken together, the results of these classroom observations extend the generality of previous work to an updated set of instructional materials and assignments, and that outcome shows the reliability of the learning processes with new groups of students. Students who are new to Java had the privilege of exposure to an initial repertoire of teaching tactics that are synergistic and cumulative.

Section III
Collaborative Tools

Section III.a
Asynchronous Tools

Chapter XV
Toward an Increase in Student Web Portfolios in New York Colleges and Universities.................... 204
John DiMarco, St. John's University, USA

This research project investigated the existence of web portfolios on academic websites in New York State. The goal of the project was to promote web portfolios, become acquainted with the current level of student web portfolio use, and suggest a sample syllabus to build web portfolios into a curriculum. The chapter cites disappointing results when surveying websites looking for web-based portfolios. Recognizing this shortfall in the use of web portfolios, this chapter offers a syllabus sample that can be used in technology- based classroom environments across disciplines to integrate portfolios into curriculums. Major findings were that there is a low quantity of web portfolios in relationship to overall student enrollment, thus providing impetus to study a new phenomenon, lack of web portfolios. The study yielded data providing a breakdown of where and how many web portfolios were found. This study provides a basis for further research by scholars into web portfolios within academic settings.

Chapter XVI
Competent Web Dialogues: Text-Based Linking of Thoughts 219
Marianne Döös, Stockholm University, Sweden
Eva R Fåhræus, Stockholm University, Sweden
Karin Alvemark, Dalarna University, Sweden
Lena Wilhelmson, Stockholm University, Sweden

Conducting a dialogue on the Web is a process of linking thoughts in virtual conversations. A dialogue differs from a discussion; a dialog shares ideas whereas a discussion seeks to convince other participants in the conversation. The chapter highlights group dialogues as conversations in which people learn with and from each other. Learning dialogues have the potential of developing the learners' capacities for critical thinking and complex problem solving. A model of a competent dialogue is offered to help improve the linking of thoughts in web dialogues. The chapter concludes with considerations when developing dialogue-based communication forms for learning purposes and contributes to teachers' demand for more support in pedagogic and educational issues.

Chapter XVII
Employing Interactive Technologies for Education and Learning: Learning-Oriented Applications of Blogs, Wikis, Podcasts, and More 234
Jeffrey Hsu, Fairleigh Dickinson University, USA

A number of new communications technologies have emerged in recent years which were originally used primarily for personal and recreational purposes. The weight of these tools is now on social networking and communications. However, these "conversational, constructivist Web 2.0 learning tools," coupled with the power and reach of the Internet, have been employed effectively in both educational learning and knowledge-oriented applications. In particular, the technologies attended to in this chapter include Instant Messaging (IM), weblogs (blogs), wikis, and podcasts. A discussion of these technologies and their uses, underlying educational and cognitive psychology foundations, and applications for education and the management of knowledge are examined in detail. The implications for education, as well as areas for future research are also explored.

Chapter XVIII
Assessing Online Discussion Forum Participation 259
Matthew Shaul, Kennesaw State University, USA

As a socially constructive learning tool, discussion forums remain central to online education. They have continued to evolve in functionality, acquiring ever-increasing usability features. However, development has lagged in providing instructors the means to assess student work in forums. The author submits an overview of his software program that provides instructors with the means to evaluate forum work quickly, easily, and repeatedly. The software accomplishes this by accessing the forums' underlying database, searching for manifest and latent data, and calculating data associated with an array of metrics. This is a Web-based tool built on Open Source and standards-based languages, providing opportunities to port the program to numerous Learning Management Systems. It is the intention of this author to provide this tool, when completed, for such use as a free, Open Source tool. Interested parties may e-mail the author for progress updates. Currently, however, further work on the project must await the completion of another project, the author's dissertation.

Section III.b
Synchronous Tools

Chapter XIX
Synchronous Hybrid E-Learning: Empirical Comparison with Asynchronous and Traditional Classrooms 269
Solomon Negash, Kennesaw State University, USA
Michelle Emerson, Kennesaw State University, USA
John Vandegrieft, Blackstone & Cullen, Inc., USA

This chapter relates an empirical analysis conducted to compare synchronous hybrid e-learning environments with traditional classrooms. The study of 165 students from eight colleges at a large public university produced results that show contrary to prior research, students taking unfamiliar subjects online in synchronous format were overall satisfied with the results of their learning. No statistical difference was found in student satisfaction between synchronous online and traditional face-to-face formats, Also, overall satisfaction, as measured by intent to use the same format again, found no statistical difference between the two formats.

Chapter XX
Understanding the Effectiveness of Collaborative Activity in Online Professional Development with Innovative Educators through Intersubjectivity 283
Diane Hui, The University of Hong Kong, Hong Kong
Donna L. Russell, University of Missouri-Kansas City, USA

Effectiveness of professional development is affected by the quality of social interaction. This chapter examines how online collaborative dialogues might influence teachers' decisions in their classrooms. The study extends principal sociocultural approaches to cognitive concepts of intersubjectivity and activity through an examination of empirical data. Part of a larger innovative professional development program involving four classrooms, the investigation examined synchronous chatroom dialogues between teachers and researchers, and utilized pre- and post-unit interviews using qualitative discourse and focused microanalyses techniques. The results argue that teachers purposefully used their dynamic intersubjective spaces and strategies in the management of meaning-making negotiations within an online interactive environment. The findings reveal two novel variable forms of intersubjectivity: temporary suspension, and resistance and disagreement. These findings provide useful implications for advanced applications and developments with information communication technology in innovations for enhanced learning and teaching as they relate to the evaluation of teacher effectiveness in implementing collaborative online problem-based activities.

Chapter XXI
Effective Questioning to Facilitate Dynamic Online Learning 303
Silvia Braidic, California University of Pennsylvania, USA

Teaching is a complex activity that involves careful preparation, delivery and reflection. As an educator, it is essential to create a sense of community in which students feel significant and are truly engaged as learners. A central focus of the educator is to maximize the capacity of each learner. How this happens in an online learning environment is the thrust of this chapter that addresses the need for learning communities that promotes effective discussion. Specifically, the practice of questioning that lies at the heart of classroom practice is examined. Similar to the traditional classroom, questioning occurs in a variety of ways for online learners. The article shares ideas for effective questioning strategies in an online environment.

Chapter XXII
Transitioning from Face-to-Face to Online Instruction: How to Increase Presence and Cognitive/Social Interaction in an Online Information Security Risk Assessment Class 313
Cindy S. York, Purdue University, USA
Dazhi Yang, Purdue University, USA
Melissa Dark, Purdue University, USA

This article briefly reviews two important goals in online education: interaction and presence. These are important goals in online education because they are linked to learning and motivation to learn. The article provides guidelines and an extended example of how to design an online course in information security in a manner that will enhance interaction and presence. This article's contribution is to provide guidelines with a corresponding extended and concrete example for those who are tasked with designing and delivering online courses. Although the guidelines and example were targeted to the field of information security, they can be readily adopted by other disciplines.

Compilation of References 324

About the Contributors 358

Index 366

Preface

INTRODUCTION

The International Journal of Information Communication and Technology Education (IJICTE) published a striking series of manuscripts pertaining to teaching and learning with technology in its publication year 2007. The articles contained in this Volume 3 of the *Advances in Information and Communication Technology Education Series* are the best of the best in the areas of design, development, and collaborative tools for addressing technology for the classroom.

Design tools offered in Section I have been subdivided into *Theory* and *Practice*. The theory-based tools discuss gender bias in science, technology, engineering, and mathematics as well as the technology acceptance model. The importance of understanding why, after nearly five decades of progress in information technology, women are still underrepresented in the field is critical to the future of the discipline. Too, computer anxiety caused by low acceptance of technology as a viable educational instrument for learning is another cause for concern by leaders in the IT community. The influence of constructivist e-learning system on student learning outcomes rounds out our look at design tool theories. From a .Practice-based perspective, five chapters argue issues of didactic teaching, e-pedagogy, teaching practices, and agent-oriented design – all with a bent toward best practices of teaching and technology. Designing courses rich in digital media offers new hope in furthering distance education. And, further research in designing traditional versus online courses is always welcomed by IT advocates.

In Section II of *Information Communication Technologies for Enhanced Education and Learning: Advanced Applications and Development*, several very interesting chapters introduce advanced applications of PowerPoint for classroom and staff training. Graphics presentation software is now capable of advanced features providing innovative models for creating self-paced lessons. As a result, these technology-based lessons are producing increased student attention, comprehension, and, most of all, achievement. To successfully implement technologies into the classroom, increased notice is being taken of training programs for faculty and instructors. In this section of the book are several professional development theories and models that have been proven effective for designing staff and faculty training environments. The final chapter in the section describes the techniques used by the author to optimize student learning while teaching Java programming language.

Collaborative Tools (Section III) are on the rise, both in the classroom and throughout society in general. Asynchronous tools support any learning event where interaction occurs intermittently with a time delay. Learners participate according to their own individual schedule and are typically separated geographically from the instructor. The text offers an examination of Web-based portfolios, Web-based dialogues, plus an array of interactive asynchronous technologies such as blogs, wikis, podcasts, etc. as well as procedures for assessing online discussion forum participation. Synchronous tools are becoming even more plentiful with the rise in popularity of learning management systems. Several chapters compare the various formats for synchronous communication and discuss how to sustain online collaboration and successfully transition from face-to-face to online instruction.

DESIGN TOOLS

In the first chapter, Wasburn shares her investigation into the critical shortfall of skilled professional in the science, technology, engineering, and mathematics (STEM) disciplines. Part of the solution, the author posits, is to attract more women to careers in these areas. In Chapter I, "*Media and Women in Technology*," several pertinent questions explore the possibilities of using a media-centered approach to achieve the goal if increased participation by women. The chapter examines the theoretical presumption that exposure to positive television images of women as technology professionals will attract more of them to STEM careers. Also studied are the causal factors as well as an understanding of the dynamics taken to produce desired results. The chapter finishes with a look at what the empirical data suggest about the viability of the hypothesis and the anecdotal evidence that supports the hypothesis.

Chapter II, "*The Gender Communication Gap in Online Threaded Discussions*," by Gefen, Geri, and Paravastu focuses on threaded discussions as key tools of online education. The implications of this study to practice cannot be underrated. For example, the investigation found that men and women communicate differently online; a finding that was probably intuitively assumed in the past now comes with an impact on what that means in the distance education environment. The chapter discusses how discussion sub-groups form and how they encourage reticent students to actively participate in the course discussions. Gender stereotypes are presented—even those precipitated by the computer. The paper concludes with recommendations for controlling online conversations, discreetly but directly, focusing the positive discussion on learners who might otherwise be ignored because of gender preferences.

Chapter III, "*The Technology Acceptance Model (TAM) and the Continuance Intention of Using WebCT: A Case of College Students in Estonia*," by Ifinedo investigates the influence of *ease of finding* and *Computer anxiety* on the technology acceptance model and a popular course management system, WebCT. Eight hypotheses were developed to test the structural model and the data supported all but one of the eight hypotheses. The study's implications for research offer the reader many opportunities to expand upon Ifinedo's investigation beyond the initial usage phase and offers insights about adopting content management systems for teaching and learning. While the scope of the study is not sufficient to generalize the findings, the author suggests that future studies should increase the sample size as well as incorporate the impacts of other relevant variables, including peer-pressure, age, gender, and facilitating conditions.

Chapter IV establishes the *Influence of Constructivist E-Learning System on Student Learning Outcomes*. The two-phased study begins by examining the process of creating a constructivist e-learning *environment*; phase two expands the investigation to constructivist e-learning *systems* in actual classroom environments. Student learning outcomes are compared between students who used constructivist e-learning with those who used a traditional learning environment. CES-trained students did better than traditional.

Chapter V presents substantial results from two projects that deal with teaching and learning with digital media in basic and higher education. The first project studied electronic learning tools perceived as "didactical actors" and uncovered new relations between learners and didactical technology. The second project found that linking evaluation and technology increased the learner's commitment to e-learning modules in higher education. Both projects in "*The Didactical Agency of Information Communication Technologies for Enhanced Education and Learning*," offer a new perspective on the active role of technology in learning processes. Wiesner-Steiner, Wiesner, Schelhowe and Luck advocate that these cases clearly imply both a social and technology sensitivity to the didactical approach and its key role for learning with information communication technologies.

Shelley, Swartz and Cole propose that distance learning is the new paradigm of instruction in their Chapter VI, "*Comparative Analyses of Online and Traditional Undergraduate Business Law Classes: How Effective is E-Pedagogy?*" In their study of e-learning and e-pedagogy growth in importance in the delivery of higher education, they investigate the cost of higher education, changing student profiles, and scarcity of traditional classroom space. They examine changing student demographics, working adults, students in the military, and residents of rural communities as well as of other countries. Their original study (IJICTE, 2007) found no statistically significant difference between the online and traditional instructional/learning formats with regard to any of the four research questions on student satisfaction and student learning. The results from the second study presented here had more mixed results. There was a significant difference found in student satisfaction with the instructor and with the course structure. Also, student learning, as measured by final course grades, was higher for the online course students. Read more about this study and the similarities and differences it found between studies barely two years apart.

Continuing the theme of Design Tools, Chapter VII, "*Student Perceptions of Data Flow Diagrams vs. Use Cases,*" by Millet and Nelson presented their investigation into data flow diagrams and use cases, two popular methodologies in teaching as well as in practice. Fifteen sections of the author's systems analysis course were introduced to structured analysis techniques as well as object-oriented methodologies. Results indicate that, while students find the use cases methodology slightly easier to understand, they believe that data flow diagrams are significantly better at communicating with users and programmers. Exposing students to one methodology before the other apparently did not lead to significant changes in student perceptions of these methodologies, so the authors posited that future systems analysis courses are free to cover these two methodologies without concern for their sequence in the course.

Chapter VIII, "*Promoting Undergraduate Education with Agent Based Laboratory,*" is presented by Hong Lin. In the field of software engineering, agent-oriented design provides for accountability and responsibility for complex software systems during design and execution. The research presented was partially supported by NSF grant, "Acquisition of a Computational Cluster Grid for Research and Education in Science and Mathematics." Student research projects were supported by U.S. Army Research Office Award through Scholars Academy of the University of Houston-Downtown. The goal of the project was to integrate various networking technologies into one client/server model to provide a uniform lab environment for different lab activities. Read how they accomplished this objective by recognizing, considering, and adding/deleting services or features in a top-down strategy.

The final manuscript dealing with Design Models, "*Supporting Arguments for Including the Teaching of Team Competency Principles in Higher Education,*" examines optimum workplace effectiveness in knowledge intensive industries. Chapter VIII takes into account not only the competencies of individuals but also those that comprise the teams within which they must operate. This study finds that although the incorporation of various types of group work into pedagogies is already fairly common within institutes of higher education, such incidents fail to embrace a rationale for, or the potential benefits of, multiple contributor environments. It continues to argue for including the teaching of team competency principles in higher education and a competency teaching model is introduced for consideration by the reader.

DEVELOPMENT TOOLS

PowerPoint continues to play a primary role in adding technology to classroom learning. Whether it is used for formal classroom presentations or individualized training scenarios, graphics presentation supports visual learners. In Chapter X by Tomei, "*Creating an Interactive PowerPoint Lesson for the Classroom,*" examines many features of PowerPoint not usually considered and even less often implemented into classroom presentations. The interactive lesson is a self-paced, student-controlled, individualized learning opportunity embedded with assessments and offered to augment individualized instruction; corrective instruction, additional practice, or enrichment activities. Learn all about action buttons, hidden slides, and the kiosk browser and follow the step-by-step instructions on how to construct assessment slides in this chapter that walks the reader through the steps needed to create a lesson suitable for either a formal multimedia classroom presentation, an individualized lesson, or a self-taught enrichment experience on home computers.

Chapter XI profiles three virtual schools, each at a different stage of development and each employing a successful distance education program to develop its professional staff. Several innovative professional development environments are discussed, including the Electronic Classroom of Tomorrow, iQ Academies, and Virtual I.D.E.A.L. school as well as barriers to sustaining distance education. "*Planning Staff Training for Virtual High Schools*," by Thompson and Berge conclude that many of the factors they studied to address the issue of virtual schools and online education are really not much different than the standards of success identified by brick and mortar institutions.

"*Training Prospective Online Instructors: Theories Utilized by Current Online Instructors*," by Cicciarelli reports on empirical research about online instructor use of different design theories. The review of the literature does an excellent job of familiarizing the reader with the three widely recognized schools of educational thought: behaviorism, cognitivism, and humanism. The, Chapter XII takes the reader beyond this discussion to a look at the empirical research describing theories preferred by online instructors. Mastery learning, simulations, multiple intelligences, transactional distance, and social and cooperative learning theories are some of the top 15 most common applications mentioned. The study found nine of the 15 theories were in widespread use in online courses. The reader is encouraged to read the results of this paper to determine the reasons why.

A second chapter focusing on graphics presentation in general and PowerPoint specifically is offered by Fedisson and Braidic in their Chapter XIII manuscript, "*The Impact of PowerPoint Presentations on Student Achievement and Student Attitudes*." The research study was grounded in an examination of seventh grade students tested on their knowledge of sentences and nouns in a language arts classroom and conducted over a two-year period. Students were asked questions regarding the use of the projector and PowerPoint presentations, the factors that helped them achieve a better grade on classroom tests, and their recommendations/ preferences for using graphics packages in the future for teaching writing, spelling, and grammar. The use of technology to motivate students achieve a higher mastery of skills is well documented in this paper.

Chapter XIV, "*Teaching Java™: Managing Instructional Tactics to Optimize Student Learning*," portrays the results of a study targeting information systems students in a graduate section and an undergraduate section of an introductory Java graphical user interface course. Knowledge transfer and software self-efficacy were the targeted criteria of the study and the results showed progressive improvement in rule test performance and software self-efficacy across the several instructional events. These results extend previous work that the author shares with the reader in an early issue of the *International Journal of Information Communication and Technology Education*.

COLLABORATIVE TOOLS

Chapter XV, "*Toward an Increase in Student Web Portfolios in New York Colleges and Universities*," investigated the existence of Web portfolios on academic Websites citing disappointing results when surveying New York State colleges and universities for these tools of authentic assessment. DiMarco's goal for this project was to promote Web portfolios by offering the current level of student Web portfolio usage and activity within New York colleges and universities and suggesting a sample syllabus to build Web portfolios into curriculums. He found a low number of portfolios (a mere .39 percent) of the enrollment population and yielded some interesting data for further investigation. Two facts seemed to evolve from this study. The first fact was few Web portfolios are readily available; the second was that many academic Websites posted documents regarding the virtues and involvement of Web portfolios, yet these institution's Websites showed no tangible implementation of Web portfolios by students.

Döös, Fåhræus, Alvemark, and Wilhelmson offer their investigation into group, Web-based dialogues as conversations that *link ideas via digital conversations.* The introductory remarks of Chapter XVI suggest a number of factors influencing the development of group discussions on the Web and their potential value to participants. Their study, "*Competent Web Dialogues: Text-Based Linking of Thoughts,*" examines experience-based learning, collective learning, dialogue competence, synchronous or asynchronous text meetings, and other considerations for teachers and students. The conclusions center on how the experience of distance education programs noted in this paper using technology to supplement learning platforms produced several positive benefits for consideration by the reader.

"*Employing Interactive Technologies for Education and Learning: Learning-Oriented Applications of Blogs, Wikis, Podcasts, and More,*" by Hsu discusses several interactive technologies and their uses, the underlying educational psychology that governs their uses and some possible applications for education in general and the management of knowledge specifically. For readers who have not explored blogs, podcasts, wikis, and the like, Chapter XVII defines these "conversational technologies" along with the characteristics and suitable applications most appropriate to course-related activities. For those readers inclined to research, the author suggest some of the broader research issues that should be examined include measuring the quality and quantity of learning that occurs when employing these specific technologies and tools.

Social constructivist learning tools, in the form of online discussion forums, remain central to online education as the modality continues to evolve in functionality. Chapter XVIII, "*Assessing Online Discussion Forum Participation,*" by Shaul, examines how the development of student assessment has caused social constructivist theory to lag behind other schools of educational psychology. The author introduces a software program for instructors to help them evaluate online discussion forums quickly, easily, and consistently. Then, he updates the reader on the latest status of the project before making the software available to users.

This next study classified students in both traditional and e-learning (i.e., synchronous) classrooms. Traditional classroom students (64%) attended all classes in a face-to-face format while the e-learning students (36%) attended some of their classes face-to-face and some classes via the synchronous format. "*Synchronous Hybrid E-Learning: Empirical Comparison with Asynchronous and Traditional Classrooms,*" examined numerous hypotheses. The first investigated whether students were less satisfied with the synchronous learning environment when learning unfamiliar courses. The second hypothesis evaluated overall student satisfaction with the synchronous and traditional learning formats. The third and final hypothesis measured overall student satisfaction by evaluating student intent to enroll in future courses. While the results offered in Chapter XIX offered by Negash, Emerson, and Vandegrieft may be limited to the specific courses examined in this study, they do provide important new information in the assessment of online learning.

Hui and Russell explore the dynamics of intersubjectivity on online professional development and reveals new evidence for the management of two variable forms of intersubjectivity, temporary suspension and resistance and disagreement. Findings from Chapter XX, "*Understanding the Effectiveness of Collaborative Activity in Online Professional Development with Innovative Educators through Inter-*

subjectivity," provide useful implications for advanced applications and developments with information communication technology in innovations for enhanced learning and teaching as they relate to the evaluation of teacher effectiveness in implementing collaborative online problem-based activities.

"*Effective Questioning to Facilitate Dynamic Online Learning*," addresses the need for a learning community to promote effective discussion through the practice of questioning. Braidic shares ideas for effective questioning strategies in an online environment in Chapter XXI that can help instructors achieve well defined goals. Whether in a traditional classroom or in an online learning environment, instructors must develop a place where students feel comfortable with questions. Readers will become familiar with I.Q. (I Question), an extension of Bloom's prompts that infuses questions-asking techniques into student assignments via article readings, cases, and the like to engage the learners in various levels of questioning.

The final Chapter XXII in the book, *Transitioning from Face-to-Face to Online Instruction: How to Increase Presence and Cognitive/Social Interaction in an Online Information Security Risk Assessment Class*," explores interaction and presence as two of the most important goals of online education. The authors provide guidelines and examples of how to design an online course in information security in a manner that will enhance interaction and presence and are readily adopted by other disciplines.

Information Communication Technologies for Enhanced Education and Learning: Advanced Applications and Development represents a unique examination of technology-based design, development, and collaborative tools for the classroom. Theory is mixed with practice and asynchronous is combined with synchronous apparatus with the expressed purpose to foster teaching and learning with technology. Enjoy the latest installment of the *Advances in Information and Communication Technology Education Series – Volume 3.*

Section I
Design Tools

Section I.a
Theory

Chapter I
Media and Women in Technology

Mara H. Wasburn
Purdue University, USA

ABSTRACT

Many Western nations face a critical shortage of skilled professionals in science, technology, engineering, and mathematics (STEM). However, despite abundant opportunities, few women prepare themselves for careers in these fields. Several of those concerned with the problem have proposed that new media programming, such as television dramas with women engineers, computer professionals, and/or engineers in leading roles, might help attract more women to STEM fields. This paper identifies a theoretical rationale for a media centered strategy, and describes a pilot study whose data suggest that a media-centered approach might have some success in producing greater interest among women in pursuing STEM careers, particularly information technology careers.

INTRODUCTION

"It is still news whenever women tackle any job American society traditionally has seen as male" (Vavrus, 2002, p. 11). In July 2005, fifteen major American business groups, led by the Business Roundtable, issued a joint statement decrying the declining prominence of the United States in science, technology, engineering, and mathematics (STEM), and calling for the nation to double the number of college graduates in those fields by 2015. The statement cited data indicating that more than 50 percent of the current United States science and engineering workforce is approaching retirement age and that by 2010, if present trends continue, the vast majority of all scientists and engineers in the world will be living in Asia. The report claimed that the scientific and technical capacity of the United States has already begun to atrophy, threatening America's standard of living at home and leadership in the world (Business Roundtable, 2005). Within the engineering community in particular, concerns about a shortfall of qualified professionals have been voiced for over a decade (Heckel, 1996; National Science Board, 2000). Corresponding concerns for their nation's welfare and standing in the global political economy have been expressed in many countries throughout Western Europe (Femtec, 2002).

It is widely understood that part of the solution to the escalating problem of the shortage of well trained technical personnel in all advanced industrial nations involves attracting considerably more women to careers in STEM disciplines. In the United States, there is substantial occupational segregation by sex. Although women constitute 46 percent of the labor force, less than a quarter of the scientists and engineers in the country are women (Mervis, 2000). Precise international comparisons of occupational segregation are difficult because nations seldom use comparable detailed occupational coding systems (Jacobs, 1993, p. 133). However, available data do indicate not only the existence of such a gendered division of labor throughout Western Europe, but also the likelihood of its persistence. For example, while half of all university students in Germany are women, women represent only 34 percent of all students in the natural sciences and 19 percent of all students in engineering (Femtec, 2002, p. 2). Similarly, men were found to be over represented among computer science graduates in all 21 industrial nations considered in a recent study. In the United States, the "male over representation factor" is 2.10, in the United Kingdom 3.10, in France 4.57, and in Germany 5.58 (Charles & Bradley, 2005).

Approximately half the potential STEM talent pool consists of women. Therefore, in 2000, a United States government commission was charged with developing strategies to attract more women and minorities in STEM careers. The commission reported to the Committee on Science of the House of Representatives that significant barriers to these goals persist (Committee on Science, 2000). Such deterrents range from differing male/female attitudes toward science and technology that begin to diverge as early as elementary and middle school, to the absence of women faculty, mentors, and fellow students in college and university classrooms that create a "chilly climate for women" in these areas (AAUW, 2000; Seymour, 1999).

A recent report by the Committee on Maximizing the Potential of Women in Academic Science and Engineering, created by the National Academies (2007) affirmed that women have the ability and drive to succeed in science and engineering, but they face persistent structural barriers and personal bias. As the result, they continue to be lost throughout every phase of their education. The report concludes that failure to act will be detrimental to our nation's competitiveness.

In the field of information technology, career opportunities for women abound. Yet despite the obvious advantage of entering this area, there has been a steady decline in the number of computer science bachelor's degrees awarded, particularly to women (Camp, 1997). In 1983-84, more than 37 percent of the bachelor's degrees in computer science were awarded to women. Ten years later, the percentage had fallen to 28 percent, and it has held relatively steady through the new millennium (Camp, 2002).

An examination of research on women in computer science revealed that the emphasis at the post-secondary level is on the social psychological factors that prevent women's inclusion (Dryburgh, 2000). Margolis and Fisher (2002) used the metaphors of a "clubhouse" to describe the extent to which women are excluded from the male purview of computing, and "dreaming in code" as "emblematic of a male standard of behavior in this computer-oriented world." The authors no longer want to try to fit women into this male culture. They issued a call to arms for a revolution in the culture and curriculum of computer science that will encompass and respect the contributions that women can make to the discipline.

As young women grow older, fewer of them express interest in studying STEM subjects. One factor cited is social identity threat, the concern that one's identity may be at risk in certain contexts (Abrams & Hogg, 1999; Major & O'Brien, 2005). The literature refers "leaky" pipeline of women from elementary school through graduate studies and employment, eventually leading to their under-representation in the STEM professions. (Freeman, 2004; Jones, Howe, & Rua, 2000.)

A 2003 United States National Science Foundation publication described 211 ongoing projects in the country designed to attract and retain women and in STEM courses. More than $90 million had already been poured into these projects. Given the proliferation of such efforts, some measurable effect on the entry and persistence of women in these professions should be expected. However, studies indicate no substantial gains (Freeman, 2004; Huang, Tadolese, & Walter, 2000). In fact, much of the progress that women have made in these areas has stalled or eroded (National Council for Research on Women, 2001). Such findings indicate the importance of developing additional new strategies for attracting more women into STEM programs. One such approach, which is the focus of this paper, involves using the mass media to create a more positive understanding of women in these professions.

The approach was the topic of a seminar entitled "Women in Science and Engineering, and TV Drama: Sex, Lives, and Videotape" held in November 2004 in London's Institution of Electrical Engineers. The event was organized by the Public Awareness of Science and Engineering (PAWS) Drama Fund and was supported by six of the United Kingdom's leading science, engineering, and technology organizations. The seminar brought together scientists, engineers, and television drama producers and writers. Its goal was to offer recommendations for helping and encouraging the media to present more well-rounded, up-to-date, and attractive images of women in STEM careers through the development of new programming such as science-based television dramas with women in leading roles. Some of the research presented centered on strategies for ensuring that a such media messages are heard and then propagate (Gladwell, 2000).

THEORETICAL RATIONALE FOR A MEDIA STRATEGY

In the 1960s, the international feminist movement helped advance the idea that cultural understandings of gender roles are socially constructed and have to do with ideology and power rather than being "natural." Feminist scholars began directing attention to the media's role in making women's minority status experienced as part of the "natural order of things." Numerous content analyses found that women were under-represented in the media and portrayed in ways that tended to sexualize, commodify, and trivialize them. Such presentation supported an inequalitarian status quo in which women played marginal roles in political, economic, and intellectual life (Brunsdon et al., 1995; Gunter, 1995; McQueen, 1998; Tuchman, 1978).

The basic theoretical insight that our understanding and experience of the world of everyday life is socially constructed was first fully articulated in the early 1930s by German philosopher Alfred Schutz (1932) who sought to develop a sociological variant of phenomenology. The work gained considerable influence in the United States when it appeared in English in 1967. This was one year after Peter Berger and Thomas Luckmann's theoretically similar study "The Social Construction of Reality" (1966) had gained the attention of American social scientists. The publication of the two studies corresponded with the height of feminist activity. For example, the National Organization of Women was founded in 1966.

In Schutz's view, all of us carry in our minds a "stock of knowledge of physical things and fellow citizens, of social collections and artifacts, including cultural objects" (Schutz, 1932/1967, p. 81). This stock of knowledge provides a frame of reference or orientation with which we can interpret objects and events as we conduct our everyday lives. Moreover, the objects and events of the world have no inherent or universal meaning apart from this imposed framework.

For Schutz, our stock of knowledge *is* our reality. It is experienced as the objective world existing "out there," independent of our will and confronting us as fact. This stock of knowledge has a taken-for-granted character and is seldom the object of conscious reflection. It is understood by us in a common sense fashion as reality itself. Although we can doubt this reality, we rarely do so, and we cannot do so when we are engaged in our routine activities. This perspective suggests that most of us might feel too busy to attend seriously to the fact that boys monopolize classroom computers or to the low probability that one of the attractive female characters in the enormously popular American television series *Sex and the City* might be a physical scientist or an engineer or a mathematician or a systems analyst. We would be unlikely to react to the fact that the brilliant, crime-solving mathematician on the series *Numb3rs* could just have easily been cast as a woman without disturbing the plot, or to the fact that *Friends*, one of all-time most popular television programs of all time, had three prominent female characters: a masseuse, a restaurant owner, and a member of the fashion industry. Not one of them was an engineer or a computer scientist. Even the female leads on *CSI, Crime Scene Investigation,* who are forensic scientists using the most advanced scientific and technical methods to apprehend criminals, came to their jobs by chance, rather than by completing formal programs of scientific education.

Schutz contends that we assume other members of our society generally share our stock of knowledge and will experience the world in the same way we do. We assume that others will see the world as being made up of the same types of objects and events, that these objects and events will have the same meaning for them, and that they will respond to them in ways they themselves have learned are appropriate. After all, even today, how many people really believe that being a chemical engineer is just as suitable a career for a woman as being a teacher or a nurse?

According to Schutz, we rely on typifications or "recipes" for action that exist in our culture. These typifications, which are part of our stock knowledge, provide us with ready-made courses of action, solutions to problems, and interpretations of the social world. Although the typifications constitute a cultural framework that is experienced as requiring no further analysis, problematic situations can arise that call the typifications into question. For example, frequently encountering mass media images of reasonably attractive women successful doing "men's work" might, over time, encourage people to reconsider their views about how "natural" is the traditional gendered division of labor.

In a manner similar to that of Schutz, Peter Berger and Thomas Luckmann (1966) discuss the process by which we create the realities of our everyday lives. They observed that social institutions appear to have an objective reality of their own as given, self-evident aspects of the world. The social world, which is a human product, confronts its producer as an external reality, as something other than a human product. New generations learn about this reality through the process of socialization, just as they learn about other things that make up the world they encounter.

New generations also learn meanings of the social order, which bestows on that order not only cognitive validity but normative legitimacy as well. Socialization involves the simultaneous transmission of knowledge and values. All understandings of the social world carry with them evaluations. Berger and Luckmann's position rejects the standard distinction between the explanation and evaluation of the social world.

Presentation of traditional images, such as those of a gendered division of labor, legitimates that institutional order.

By the time Schutz's study was published, the power of radio and motion pictures as agents of socialization that could be used to legitimate or challenge social institutions was well recognized worldwide by those in the media industry, as well as by social scientists and national governments (see Blumer, 1933; Cantril & Allport, 1935; Furhammer & Isaksson, 1971; Lacey, 1996). Nevertheless, Schutz fails to discuss the media's role as a major source of our "stock of knowledge" and as a creator of the typifications on which we rely as we go about our everyday lives. More than three decades later, Berger and Luckmann also ignore the importance of the mass media in legitimating or changing social orders. This is remarkable in light of the vast literature on the influence of television that was produced in the 1960s by cultural critics and feminist theorists as well as social scientists.

A major constructionist theory focusing on the influence of mediated reality on social behavior was introduced by George Gerbner and his associates in the 1970s and subsequently elaborated (Genuter, 1995; Gerbner, 1976; Gerbner, et al., 1994; Signorielli & Morgan, 1990). Initial concern was with how the vast amount of violence portrayed on American television exaggerated the fears people have about encountering violence in their own neighborhoods. Later developed as cultivation theory, the approach asserts that, at least among heavy users, television produces a "mainstreaming" effect whereby differences in beliefs in otherwise heterogeneous populations are muted. Heavy television viewers internalize many of the perspectives on the social world presented by television. Such influence occurs as a result of continual and lengthy exposure to television in general, not just exposure to individual programs or genres.

In terms of the works of Schutz (1932/1967), Berger, and Luckmann (1966), discussed earlier, television presents typifications, which, after prolonged repetitious exposure, the viewing public accepts as accurate representations of social reality. Ubiquitous images come to represent not only the social order but the normative order as well. Viewers use typifications to negotiate the social world by understanding "the natural order of things." If girls, compared to boys, are never or almost never represented as interested in STEM disciplines in youth-oriented television programming, or seen as enrolled in advanced physics, chemistry, calculus, or computers classes, that is just the way things are—just natural. Change in the social order is not called for. If women, compared to men, are never or almost never portrayed as scientists, technical experts, or engineers on television programs, that, too, is a reflection of "the way things are." The likely viewer reaction to such under-representation is not dissatisfaction with the apparent inequality, but simple acceptance of the consequences of how "natural" interests and abilities are distributed by sex.

Taken together, social constructionism and cultivation theory clearly suggest an approach to attracting more women to STEM careers. The strategy is to vastly increase media representation of women in these occupations. This should be undertaken in all varieties of programming including children's shows, dramas, situation comedies, talk shows, soap operas, and even commercials. The goal is to cultivate a social understanding of middle and high school girls enrolled in STEM classes, of women scientists, engineers, and technical experts as simply part of the natural social order—as nothing unusual. STEM education and careers can be presented as legitimate spheres of participation for women—areas of professional activity in which women not only *do* but *should* participate.

Positive images of women technology professionals are not necessarily images of bright, articulate, personable, physically attractive people. While, quite obviously, they should not be less attractive than the other characters with whom they interact or than others who appear regularly on other programs, the important point is that women technology professionals appear frequently in the full variety of television programs. Young women should be commonly encountered as characters enrolled in technology-rich classes rebuilding computers.

It is important to note that women are not a monolithic group and, consequently, no one approach will work for attracting women to STEM disciplines. In addition to gender, other factors such as race, class, ethnicity, and sexual orientation influence education and career choices (Rosser, 1998). This suggests that if a media strategy is to be helpful, it must involve diverse programming appealing to audiences composed of women with diverse demographic characteristics.

Women should be commonly encountered as characters competently doing technologically sophisticated work that is just as legitimately "woman's work" as it is "man's work." This proposal is consistent with research concluding that, at least in the United States, media routinely ignore and/or trivialize women's participation in STEM, and thereby discourage their career aspirations in scientific and technical fields (Potts & Martinez, 1994; Steinke, 1997).

THE PILOT STUDY

The preceding suggests the hypothesis that television viewers who have encountered images of women technical professionals are more likely to believe that STEM careers are acceptable (legitimate) careers for women than are those who have not seen such presentations. An opportunity presented itself to conduct a pilot study that, though limited in several ways, does shed some new light on the tenability of this suggestion.

Between April 1 and May 15, 2004, the Survey Research Institute at Purdue University conducted as a graduate student training exercise one of its periodic social surveys of the entire continental United States via a computer assisted telephone-interviewing (CATI) system. In the system, telephone numbers are selected randomly from a list of random digit dialing telephone numbers that include all area codes and telephone prefixes throughout the United States. The CATI system allows graduate student interviewers to record responses into a database while conducting the interview. CATI software prevents a researcher from calling anywhere in the United States before 9:00 a.m. or after 9:00 p.m. in each time zone.

For each potential interview, the respondent is first asked if he or she is 18 or not. If the respondent is not 18 or older, the researcher asks if anyone in the house is at least 18 years of age. Once an 18 year old is contacted, the respondent is asked if he or she has enough time to answer the survey. As long as the respondent agrees, the survey is administered. Should a respondent show an interest in taking the survey but state that he or she does not have the time to take the survey, a better time is scheduled in the CATI system and the respondent is called back at the rescheduled time. Upon completion of the interview, the researcher records the sex of each respondent. In the case that a respondent prematurely terminates the interview, all of the responses up to that point are saved and a callback is scheduled in an attempt to complete the interview.

Four hundred interviews were initiated of which 284 were completed for a response rate of 71 percent. The social survey allocated 30 minutes for questions of the pilot study investigating media images of women technology professionals. Three categories of STEM careers were selected to represent a broad range of specific professions: engineer, research scientist, and computer technician. The analysis dealt with the proportion of the sample that had seen actresses playing these occupational roles, an indirect measure of the relative attractiveness of those occupations, and the differences in attitudes toward the acceptability (legitimacy) of various careers for women between those who had and those who had not seen actresses playing those occupational roles on television.

Limitations

Although the sample did randomly include individuals representing a wide range of demographic characteristics, it cannot be considered representative of the entire population of the continental United States. Women, in particular, were over represented: 201 (70.7 percent) as opposed to 83 (29.2 percent) male respondents. Since use of the CATI system prevented interviewing young women and men below the age of 18, students are not represented in the sample. Also, due to the necessary brevity and other characteristics of telephone surveys, respondents were asked short questions, some of which were only proxy measures of central concepts, such as the relative attractiveness of women in various STEM occupations, and the terms for those occupations such as "computer technician," which was used to represent a variety of computing-focused careers. The cumulative effect of encountering numerous positive images over time, central to cultivation theory, could not be explored.

Findings

Data in Table 1 show that over 90 percent of the sample had seen actresses portray nurses, medical doctors, lawyers, and secretaries. This is not surprising in light of the long-term popularity of medical and legal dramas on United States prime time television. Female secretaries are likely to appear in work contexts in most varieties of television programming.

There is a considerable gap between the frequency with which respondents report seeing actresses portraying those roles and the frequency with which they report seeing actresses portraying other occupational roles, including three technology roles selected to represent a large cluster of related occupations: research scientist, computer technician, and engineer. Research scientists are seen much more frequently than are computer technicians and engineers. This probably reflects their appearances on several types of television dramas including crime, law, mystery, and science fiction. The only two occupational roles in which the majority of respondents had not seen actresses were computer technician and engineer. This is certainly due, at least in part,

Table 1.

Question: Would you tell me if you have ever seen actresses on television playing characters with the following professions? (In percents)

	Yes	No
Nurse	96.1	3.9
Medical Doctor	93.0	7.0
Lawyer	91.2	8.8
Secretary	90.1	9.9
Research Scientist	78.2	21.8
Factory Worker	65.8	34.2
Elementary School Teacher	63.7	36.3
Engineer	50.6	49.4
Computer Technician	44.7	55.3

n=284

to the comparative rarity that such roles appear in any variety of programming. When decisions are being made as to the careers to assign female characters in television comedies, dramas, soap operas, and even commercials, having those characters portrayed as engineers or computer technicians would take advantage of a particular opportunity to establish women in technology as a part of the natural order of things.

How attractive are women technology professionals compared to women in other occupations? Elementary school teacher was selected as the profession for comparison. This is one of the most traditional middle-class occupations, and has had a long history of being gender stereotyped as appropriate for women. Data in Table 2 below indicate that the relative attractiveness of women in technology generally is a reflection of the standing of their occupation in the occupational prestige hierarchy. However, there are exceptions. When comparing respondent perceptions of the attractiveness of a woman who is an engineer or a research scientist with that of a woman who is an elementary school teacher, the latter is more frequently judged more attractive. Generally, engineers and research scientists have more education, higher income, and higher occupational prestige than elementary school teachers. Yet, data show that respondents believed most men would prefer an elementary school teacher as a spouse or partner. Such a finding indicates the need to improve the image of women in technology-rich professions.

Are those who have seen actress in a STEM occupation significantly more likely than others to believe it is an acceptable (legitimate) occupation for a woman? The answer to this question bears directly on the tenability of the theoretical assumption central to this paper: Media images of social reality come to be regarded not only as the empirical but also as the normative "natural order of things."

Respondents were asked the extent to which they agreed with the statement that each of several STEM careers (research scientist, engineer, computer technician) was an "acceptable career for a woman." The theoretical expectation is that those respondents who had seen an actress on television playing the role of a technology professional would more frequently report that the role is legitimate for a woman than those who indicated they had never seen such a representation. Because attitudes of men and women toward women in technology might be quite different, their responses also were analyzed separately.

In the case of each of the three careers, five comparisons were made: overall between the attitudes of those who had and those who had not

Table 2.

Question: I would like you to tell me which of two women, described by their careers, do you think most men would prefer as a spouse or partner? (In percents)

		Elementary School Teacher
Medical Doctor	64.4	35.6
Nurse	61.3	38.7
Lawyer	49.6	50.4
Engineer	44.2	55.8
Research Scientist	39.1	60.9
Computer Technician	28.2	71.8
Secretary	24.6	75.3
Factory Worker	14.1	85.9

n=284

seen a representation of a woman in that occupational role on television, between men who had seen and men who had not seen such a portrayal, between women who had seen and women who had not seen such a portrayal, between men who had and women who had seen such a portrayal, and between men who had not seen and women who had not seen such a portrayal.

Data in Table 3 were used to calculate the magnitude of difference in the distribution of attitudes for the overall sample. Chi square tests were used to assess the probability that the magnitude of each of the observed differences was due to chance. Due to space limitations, data used to calculate additional chi square values are not presented here. However, they are available from the author. Chi square tests require row and column totals greater than zero. Consequently, in some tables, strongly disagree and even disagree responses were eliminated in the calculation.

None of the comparisons using the data in Table 3 reveals a statistically significant difference

Table 3.

Question: Do you agree or disagree: Is it acceptable for a woman to be a research scientist?

Saw	Strongly Disagree	Disagree	Uncertain	Agree	Strongly Agree	Total
Yes	0	13	20	92	97	222
No	0	0	4	29	29	62

n=284

Chi Square Values for Comparisons

		P value (.05 or less)
a. Overall yes vs. overall no	0.25	ns
b. Men yes vs. men no	0.74	ns
c. Women yes vs. women no	0.34	ns
d. Men yes vs. women yes	0.14	ns
Men no vs. women no	0.70	ns

Table 4.

Question: Do you agree or disagree: Is it acceptable for a woman to be an engineer?

Saw	Strongly Disagree	Disagree	Uncertain	Agree	Strongly Agree	Total
Yes	0	2	3	51	66	122
No	0	5	14	72	71	162

n=284

Chi Square Values for Comparisons

		P value (.05 or less)
a. Overall yes vs. overall no	0.20	ns
b. Men yes vs. men no	0.21	ns
c. Women yes vs. women no	0.31	ns
d. Men yes vs. women yes	0.43	ns
Men no vs. women no	0.37	ns

between the attitudes of those who had and those who had not seen an actress portray a research scientist on television. In the case of research scientists, data do not conform to the theoretical expectation.

Data in Table 4 present similar results. None of the five comparisons reveals a statistically significant difference. In the case of engineers, as in the case of research scientists, data do not conform to the theoretical expectations.

Data in Table 5, however, tell a different story. Table 1 showed that fewer viewers had seen an actress on television playing a computer technician than playing any one of the eight other professional roles considered in this study. Table 2 indicated that overall, women who are computer technicians are much less frequently viewed as attractive than are women who are engineers or research scientists. Table 5 shows that overall, of the 284 respondents, the majority (53.2 percent) disagreed with the statement that it is acceptable for a woman to be a computer technician. This is an impressive statistic when contrasted with the corresponding 4.6 percent for research scientist and 2.5 percent for engineer. If there is a need to create a more positive view of women technology professionals, computer technicians would appear to be among those in greatest need.

Consistent with theoretical expectations, overall differences in the attitudes of those who saw an actress portray a computer technician and those who did not are statistically significant. Corresponding differences also were found for women but not for men. Significant differences were found between the attitudes of men and women all of whom had seen a portrayal and between men and women, all of whom had not seen a portrayal. These results indicate the existence of gender differences in attitudes toward computer technicians (and perhaps toward other STEM occupations s well) and in the apparent ability of media representations to influence perceptions of and attitudes toward women in certain STEM careers. If more women are to be attracted to STEM occupations, it would seem important to influence the perceptions and attitudes of men as well as those of women. It is primarily men who teach classes in STEM disciplines in high schools and colleges, make admission decisions to college and university science, engineering, and

Table 5.

Question: Do you agree or disagree: Is it acceptable for a woman to be a computer technician?

Saw	Strongly Disagree	Disagree	Uncertain	Agree	Strongly Agree	Total
Yes	25	32	6	37	27	127
No	54	40	12	33	18	157

n=284

Chi Square Values for Comparisons

		P value (.05 or less)
a. Overall yes vs. overall no	9.90 E-22	<.001
b. Men yes vs. men no	0.43	ns
c. Women yes vs. women no	0.05	<.05
d. Men yes vs. women yes	1.42 E-10	<.001
Men no vs. women no	5.18 E-14	<.001

technology programs, hire scientists, engineers, and other technology professionals, and constitute the majority of colleagues with whom women work in these professions.

DISCUSSION AND CONCLUSION

Part of the solution to the shortage of trained scientists, engineers, and computer professionals in advanced industrial societies is to attract more women to careers in these areas. One widely discussed strategy for accomplishing this goal is to make such careers more attractive through the use of the media, particularly television. While this proposal makes common sense, several questions have yet to be addressed. These are the concerns of this paper. Are there sound reasons to believe that a media-centered approach will achieve some success? That is, what is the theoretical basis for the hypothesis that exposure to positive television images of women as technology professionals will attract more of them to STEM careers? What causal mechanism is involved? Understanding causal dynamics can inform actions taken to produce desired results. What will empirical data suggest about the tenability of the hypothesis? Can we move beyond common sense and anecdotal evidence in evaluating the hypothesis?

The works of social theorists Alfred Schutz, Peter Berger, and communication researcher George Gerbner provide an explanation for our understanding, evaluation, and reaction to the social world. Commonly encountered representations of actors, conditions, and events in the "real world" come to be understood, correctly or not, as the nature of reality itself. Furthermore, this understanding of the "natural order of things" comes to be accepted as the proper or legitimate structure. Such an understanding provides a guide for social behavior. The theory suggests that if children seldom or never encounter, directly or indirectly through the media, girls in laboratory classes or solving difficult mathematics problems, they are likely to believe that such educational pursuits are naturally the purview of boys. Similarly, if adults seldom or never encounter, directly or indirectly through the media, women scientists, engineers, or technicians, they are likely to believe that such careers are "naturally" careers for men.

The social construction of reality perspective and cultivation theory suggest a strategy for attracting more women to STEM careers: use media to present the public continuously with images of women in a wide variety of technology-rich educational programs and occupations. The object is to cultivate the view of women scientists, engineers, mathematicians, and technology professionals, and of young women preparing for those careers, as nothing exceptional. The goal is to construct the socially shared perspective that it is just as "natural" for a woman to be a STEM professional as it is for her to be a medical doctor or a lawyer.

Social constructionism and cultivation theory call attention to the importance of the frequency with which audiences encounter positive media images women technology professionals. To reach large and diverse audiences, representations must appear in a wide range of television program formats including soap operas, situation comedies, talk shows, dramas, commercials, and arguably most important of all, programs that appeal to teenage girls. It is helpful to produce programs that feature strong, competent and otherwise attractive female characters in the role of technology professionals. However, for the most part, positive images of women in technology need not be glamorous images. Primarily producing such images might actually discourage women whose self-assessments are not nearly so glamorous from pursuing STEM careers.

This pilot study investigated the tenability of one of two hypotheses derived from social theory and communication research. The first hypothesis states that those who have been exposed to positive media images of women technology professionals are more likely to believe that STEM careers are legitimate careers for women than are those who have not been exposed to such images. The second hypothesis states that widespread belief that STEM careers are legitimate careers for women will actually move more women to those careers. Investigating this second hypothesis is beyond the scope of this pilot study.

As noted earlier, although data were drawn randomly from the entire population of the continental United States, the sample was too small to be considered representative of the entire country. They certainly cannot be taken as representative of the views of those in other nations as well. Also previously noted, the necessary brevity and other characteristics of telephone surveys imposed further limitations. The cumulative effect of encountering numerous positive images over time, central to cultivation theory, could not be explored. However, while findings are tentative, they are suggestive. Among these are:

1. There may be vast differences in the frequency with which television audiences have encountered representations of women in different STEM occupations.
2. There may be vast differences in public perceptions of the relative attractiveness of various technology-rich occupations.
3. In each of the cases of research scientist, engineer, and computer technician, men were more likely than women to agree that the profession is an acceptable career for a woman. This is encouraging, since, as noted, those hiring women into these positions and serving as faculty in their university courses are more likely to be men than women.
4. Exposure to positive images of women in technology may increase the likelihood that viewers will believe these are acceptable careers for women in the case of some STEM careers (e.g., computer technician) but not others (e.g., research scientist and engineer.)
5. Significantly fewer women than men believe that the computing category of STEM career represented by computer technician is acceptable for women. It might be more profitable to invest more effort in creating positive media images of women in computer-focused careers.

Computer technician emerged from the pilot study as the career category deemed least acceptable to both men and women as being appropriate for a women. Fewer viewers surveyed had seen an actress on television playing a computer technician than playing any one of the eight other professional roles considered in this study. Women who are computer technicians were viewed as less attractive than women in other STEM careers.

Additional research, using a representative sample of the United States, more rigorously defined concepts, and more sensitive measures is needed to determine whether or not these findings of the pilot study, tentatively suggesting that the media strategy most likely to be effective is one targeting young women likely to be interested in becoming computer technicians, are, in fact, valid. Future research should also include respondents under the age of 18, because it can be useful to know the views of students preparing to select their future professions. The impact of variables such as age, race, and class would also be instructive.

Similar international data are needed to determine whether or not the findings would apply in other countries as well. The findings can be understood as suggesting patterns of beliefs, values, and their sources likely to be found in the country as a whole. If the findings hold, it would appear that the majority both men and women find other STEM careers appropriate for women, suggesting that a media strategy would be less effective in raising those numbers, and that other explanations and strategies should be explored.

REFERENCES

AAUW. (2000). *Tech-savvy: Educating girls in the new computer age*. Washington, DC: AAUW Educational Foundation.

Abrams, D., & Hogg, M. A. (1999). *Social identity and social cognition*. Malden, MA: Blackwell.

Berger, P. L., & Luckmann, T. (1966). The social construction of reality: A treatise in the sociology of knowledge. Garden City, NY: Anchor Books.

Blumer, H. 1933. *Movies and conduct*. New York: MacMillan.

Brunsdon, C., D'Acci, J., & Spigel, L. (Eds.). (1997). *Feminist television criticism*. Oxford: Oxford University Press.

Business Roundtable. (2005). *Tapping America's potential: The education for innovation initiative*. Washington, DC: The Business Roundtable.

Camp, T. (2002). Forward to women and computing, *SIGSE Bulletin, 34* (2). 1-6.

Camp, T. (1997). The incredible shrinking pipeline. *Communications of the Association of Computing Machinery, 40 (10)*, 103-110.

Cantril, H., & Allport, G. (1935). *The psychology of radio*. New York: Harper & Bros.

Charles, M., & Bradley, K. (2005). *Women and information technology: Research on the reasons for under representation*. Paper presented at the 100th Annual Meeting of the American Sociological Association, Philadelphia, PA, August 13.

Committee on Maximizing the Potential of Women in Academic Science and Engineering. (2007). *Beyond bias and barriers: Fulfilling the potential of women in academic science and engineering*. Washington, DC: The National Academies Press.

Committee on Science, U. S. House of Representatives. (2002). *A review of the Morella Commission report on recommendations to attract more women and minorities in science and engineering*. Serial No. 106-82. Education 6 (10): 1-2. Washington, DC: U. S. Government Printing Office.

Dryburgh, H. (2000). Underrepresentation of girls and women in computer science: Classification of 1990s research. *Journal of Educational Computing Research, 23*, 181-202.

Femtec. (2002). *Introduction to Femtec: University-based career center for women Berlin, Inc.* Berlin, Germany: Femtec.

Freeman, C. E. (2004). *Trends in educational equity of girls & women: 2004* (No. NCES 2005-016). Washington, DC: U.S. Government Printing Office: U.S. Department of Education, National Center for Education Statistics.

Furhammer, I., & Isaksson, F. (1971). *Politics and film*. New York: Praeger..

Genuter, B. (1995). *Television and gender representation*. London: John Libbey.

Gerbner, G. (1976). *Television and its viewers: What social science sees*. Santa Monica, CA: Rand Corporation.

Gerbner, G., Gross, L., Morgan, M., & Signorielli, N. (1994). Growing up with television: The cultivation perspective. In Bryant, J., & D. Zillman (Eds.). *The cultivation perspective* (pp. 17-41). Hillsdale, N. J.: Erlbaum.

Gladwell, M. (2000). The tipping point: How little things can make a big difference. London: Little Brown.

Gunter, B. (1995). *Television and gender representation*. London: John Libbey

Heckel, R. W. (1996). Engineering freshman enrollments: Critical and non-critical factors. *Journal of Engineering Education, 85* (1), 15-21.

Huang, G., Taddese, N., & Walter, E. (2000). *Entry and persistence of women and minorities in college science and engineering.* (No. NCES 2001-601). U.S. Department of Education National Center for Education Statistics. Washington, DC: U. S. Government Printing Office.

Jacobs, J. A. (1999). The sex segregation of occupations." In Powell, G. N. (Ed.). *Handbook of gender and work*, (pp. 125-139). Thousand Oaks, CA: Sage Publications.

Jones, M. G., Howe, A., & Rua, M. J. (2000). Gender differences in students' experiences, interests and attitudes toward science and scientists. *Science Education, 84*(2), 180-192.

Lacey, K. (1996). *Feminine frequencies: Gender, German radio, and the public sphere: 1923-1945.* Ann Arbor: University of Michigan Press.

Major, B., & O'Brien, L. T. (2005). The social psychology of stigma. *Annual Review of Psychology, 56,* 393-421.

Margolis, J. & Fisher, A. (2002). *Unlocking the clubhouse: Women in computing.* Cambridge, MA: MIT Press, pp. 3-5).

McQueen, D. (1998). *Television: A media student's guide.* London: Arnold Press.

Mervis, J. (2000). Diversity: Easier said than done. *Science, 289* (5478), 378-379.

National Center for Research on Women. (2001). *Balancing the equation: Where are the women and girls in science, engineering, and technology?* New York: National Council for Research on Women.

National Science Board. (2000). *Science and engineering indicators, 2000.* Arlington, VA: National Science Foundation.

Potts, R., & Martinez, I. (1994). Television viewing and children's beliefs about scientists. *Journal of Applied Development Psychology, 15,* 287-300.

Rosser, S. V. (1998). Applying feminist theories to women in science programs. *Signs, 24,* 174-200.

Schutz, A. (1932/1967). *The phenomenology of the social world.* Evanston, IL: Northwestern University Press.

Seymour, E. (1999). Therole of socialization in shaping the career-related choices of undergraduate women in science, mathematics, and engineering majors. In C.C. Selby (Ed.), *Women in science and engineering: Choices for success* (pp. 118-126). New York: The New York Academy of Sciences.

Signorielli, N., & Morgan, M. (1990). Cultivation analysis: New directions in media effects research. Newbury Park, CA: Sage Publications.

Steinke, J. (1997). A portrait of a woman as a scientist: Breaking down barriers erected by gender role stereotypes. *Public Understanding of Science, 6,* 409-428.

Tuchman, G. (1978). The symbolic annihilation of women by the mass media. In Tuchman, G., A. Kaplan Daniels, & J. Benet (Eds.). pp. 3-17. *Hearth and home: Images of women and the media.* New York: Oxford University Press.

Vavrus, M. D. (2002). *Postfeminist News: Political Women in Media Culture.* Albany: State University of New York Press, p.13).

Chapter II
The Gender Communication Gap in Online Threaded Discussions

David Gefen
Drexel University, USA

Nitza Geri
The Open University of Israel, Israel

Narasimha Paravastu
Metropolitan State University, USA

ABSTRACT

Threaded discussions are one of the central tools of online education. These tools enhance student learning and compensate for the lack of social interaction. This study examines whether these social interactions are affected by some typical gender related conversational behaviors, despite the fact that these threaded discussion are designed to operate in a seemingly gender neutral online environment. That men and women communicate differently in open conversation due to their different respective social objectives in communication is at the core of sociolinguistic theory. A direct result of these differences is a tendency toward same-gender oral conversations. To some extent, according to sociolinguists, cross-gender communication resembles cross cultural conversations. This study analyzes threaded discussions in online courses through the lens of sociolinguistic theory, and conjectures that these gender differences should be reflected in mild gender segregation in the threaded discussions as well as men showing a greater inclination to dominate the discussion. Data from 233 students in 27 online courses support these hypotheses and enable a significant identification of the gender of the student based on whom they reference in the threaded discussion and the way they reference others. Theoretical and practical implications on managing threaded discussions are discussed along with directions for further research.

INTRODUCTION

Conversation, as we all know, is more than a mere exchange of words and the meaning these words convey. Language, being a central aspect of culture and social interaction, also carries a social message and the resulting social segregation and hierarchy such a message creates. This unconscious inclusion of a rich social message is common in conversations by both men and women, albeit each inserts a different social meaning. The problem is that men and women communicate with very different social objectives. So different, in fact, that men and women may totally misunderstand the underlying meaning an opposite gender member is making. Think of shopping as an example. When a woman discusses her shopping it is often with the intent of including the other in the conversation, nothing to do with asking permission, but too often men understand this communiqué as a request of approval. This is because men, more than women, typically communicate with an objective of establishing and maintaining their social status. Commenting on and approving a communiqué establishes their importance. On the other hand, women, more than men, communicate to broadcast rapport. Sharing their shopping excursion story is a good opportunity to involve others or be involved oneself in a conversation. The opposite approach to communication often results a cross cultural misunderstanding (Tannen, 1994). A direct consequence of these differing social objectives and cross cultural misunderstanding is the emergence of gender segregated discussions, as evidenced in many cocktail parties. Men prefer to talk to other men, and women prefer to talk to other women. This is the basic premise of sociolinguistics (Yates, 2001).

Although sociolinguistics research has dealt mainly with the context rich scenarios of oral discourse, the applicability of this idea to the Internet with its more lean social context has received some verification in recent years (Gefen & Ridings, 2005). Virtual communities are online meeting places in which people freely interact as though they were interacting in a face to face manner in a social club. Virtual communities apparently exhibit much of the same gender related behavior predicted by sociolinguistics. Men join these communities to gather and share information, women join to give and share social support. Moreover, although many virtual communities are voluntarily mostly single gender communities, when men seek social support in virtual communities they go to mixed gender communities, supporting the typically stereotyped tendency of women to center their communication on the social side of things. Likewise, when women seek information they go to mixed gender communities, supporting the typically stereotyped tendency of men to center their communication on information exchange (Gefen & Ridings, 2005). These cross gender boundary preferences portray the characteristic gender behavior observed in oral communication (Hannah & Murachver, 1999). And, across cultures, business related email messages, although generally not there to serve a social purpose, are perceived differently by men than by women, with women significantly sensing more social presence in these emails and as a result perceiving them as a more useful medium in their work (Gefen & Straub, 1997). Similar results were reported about the differences of reaction to online purchases by men and women, men being more impulsive online shoppers than women (Zhang, Prybutok, & Strutton, 2007).

But whether and how this applies to online class settings remain open questions. These are important questions to answer because threaded discussions are a among the most valuable activities in online classes (Levy, 2006). If gender is a consideration in how students interact online, then teachers should be aware of this. On the face of it, the controlled social environment of a threaded discussion in an online class and the limited power play available in these settings should make these gender tendencies, especially the social dominance claimed to be needed by men and the resulting voluntary gender segregation, rather mute. On the other hand, if these are indeed ingrained gender based characteristics, as opposed to being socially and context oriented, then these gender tendencies should come through even in these very lean social settings. Moreover, and relating to the second part of the research question, the very controlled online class environment with its typically dictated rules of conduct makes many of the typical gender related behaviors inapplicable. How these behaviors may nonetheless come through is the other open ended question. The objective of this study was to empirically examine this and in doing so to raise the need to consider these gender differences in threaded discussion in online courses.

These issues are crucial since online learning has gained considerable growth in recent years, replacing face-to-face instruction (Hiltz & Turoff, 2005). However, not enough consideration has been given to the implications of this major change (Hirschheim, 2005). One of the main problems of online learning is the high student dropout rates, which make student retention a major concern (Levy, 2007; A.P. Rovai, 2002; A. P. Rovai, 2003; Simpson, 2003; Tinto, 1998; Woodley, 2004). Online e-learning services have been found to reduce MBA students inclination to withdraw (Geri, Mendelson, & Gefen, 2007), and online threaded discussions are among the main components of such services (Levy, 2006). One facet of the online threaded discussions is the replacement of face-to-face class discussions, and as such, they are supposed to enhance learning. The other facet is overcoming the "loneliness of the long-distance learner" (Eastmond, 1995). Online discussions are aimed at solving this problem and increasing retention (Guri-Rosenblit, 2005). Hence, it is crucial to conduct these discussions effectively by creating the appropriate social atmosphere to support the online learning process.

The data, examining some prominent gender differences in communication style embedded in the online discussions, show that men and women do generally communicate differently and there is some preference for same gender communication within the shared class threaded discussion even in the socially lean and rigid environment of online course discussions. While there was no support to the hypothesis that women would show more empathy than men would, there was support for the hypothesis that men would show more socially dominating behavior. These effects while weak in the entire data became strong when examining only students who took advantage of the online conversations to engage with other students.

The contributions of this study are twofold, practically and theoretically. Practically, the study highlights the different conversational behavior men and women have also in online courses, giving instructors some idea of what to expect and hence how to better manage online courses. Theoretically, the study introduces sociolinguistics to the hitherto unstudied context of conversations in online courses with their relative lean and controlled social environment, showing that while the empathy of women may not have extended to this context, men's tendency to control the conversation as a way of showing social standing does. The paper discusses the implications of these findings on online learning and suggests directions for further research.

THEORY AND HYPOTHESES

Gender Communication Style Differences

Although communication is about the exchange of information, there almost always is also a strong social aspect which permeates conversation and that carries meaning way beyond the actual words spoken and the direct meaning they convey. The social meaning embedded in conversation and the way it is understood, according to sociolinguists, are both to a large extent gender dependent. Men and women may speak what on the surface may be the same language, but the underlying social message is often very different. So different in fact that some sociolinguistics claim that cross gender conversations are almost bound to be misunderstood. For example, when a woman says she wants to buy something, there often is a social message of inclusion and rapport permeating this statement. This social message may actually be at the core of the message, being more important than the information conveyed itself. It is a message of come share this idea with me and talk to me about it. But, many men might understand this message either as a matter of informing them of this purchase intention or as requesting permission, although neither of which were initially intended. Tannen exasperatedly and famously caught this cross gender misunderstanding it in her best selling book with the all telling title *You Just Don't Understand* (Tannen, 1994).

According to sociolinguistic theory, the basic gender difference in communication is that men communicate with the social objective of attaining and maintaining social status while women communicate more to create rapport. Consequently, men try to control the conversation and are more critical of others while women try to be inclusive and supportive (Guillier & Drndell, 2006; Kilbourne & Weeks, 1997; Mulac, Erlandson, Farrar, & Hallett, 1998; Tannen, 1994, 1995) . This holds true across cultures (Costa, Terracciano, & McCrae, 2001; Hofstede, 1980). These differences originally observed in oral conversations apply also to the Internet and to listservs (Herring, 1996b; Stewart, Shields, & Sen, 2001) as well as to ecommerce (Gefen, 2000), online purchase behaviors (Zhang et al., 2007), the reason people join virtual communities (Gefen & Ridings, 2005), their assessment of email (Gefen & Straub, 1997), why people use the Internet (Fallows, 2005), how they take computer training (Venkatesh & Morris, 2000), and trying to innovate in IT (Ahuja & Thatcher, 2005), or adopt a new technology such as multipurpose information appliances (Hong & Tam, 2006). Supporting these conclusions, women, more than men, utilize email and the Internet to maintain social ties (Boneva, Kraut, & Frohlich, 2001; Parks & Floyd, 1995).

A second consequence of these gender differences is the preference in some cases to have same gender conversations. Men generally prefer to talk to other men and women to other women. Considering the cross cultural communication aspects of cross gender communications this is no surprise (Tannen, 1994). This preference can easily be seen in cocktail parties but it also applies to virtual communities. This applicability to virtual communities is important because people can hide or masquerade their gender in these communities (Caspi & Gorsky, 2006) and nonetheless some expected gender differences and the preference for same gender conversations come through. Perhaps even more telling is that when people prefer cross gender communications it is in accordance with the stereotypical gender behavior. Men go to mixed communities because they want social support, a known female communication attribute, while women go to mixed communities when they want to concentrate on obtaining information, a known male communication attribute (Gefen & Ridings, 2005).

Online Course Discussions

These social segregations and misunderstandings in communication may seem rather amusing, and indeed Tannen's book (1994) was a best seller, but their repercussions are far reaching. Tannen (1995) demonstrates how differences in conversational style may undermine women in the workplace by making them seem less competent and confident. A similar phenomenon was observed in classroom discussions (Tannen, 1991). These gender related discourse differences also affect some aspects of the way people learn online, notably that men use online resources more to obtain information while women do so more to communicate personal issues (Herring, 1993, 1996a; Yates, 2001). Unbeknown to them and with no bad intentions, students may bring these gender related social messages into their class conversations, as people generally do into many of their conversations. The result of all this may be misunderstandings and with these an impairment of the learning process. That is why it is imperative upon online instructors to recognize these differences and realize their cross gender differences.

One place where these misunderstandings may come about and have unintended consequences, is in online courses. Online courses are courses taught through the Internet where students download course materials online, send in their assignments electronically, and even take quizzes and exams online. An integral part of many of these online courses is the threaded discussion section. In the threaded discussion the professor posts a question or topic of discussion and the students are then expected, and are often graded on, to take part in an asynchronous discussion of this topic. As an integral part of this discussion the students are expected not only to bring up their own ideas but also to discuss the ideas brought up in the postings of the other students.

Hypotheses

If these gender differences and segregation tendencies in communication, so highlighted in oral discourse (Crawford, 1995; Gray, 1992), are a matter of gender differences rather than dependent on the type of media involved (Gefen & Straub, 1997), then some of these differences should be evident also in online courses. Specifically, in the case of the type of online course discussions that goes on in a threaded discussion we would expect, extrapolating from sociolinguistics, to find that women will be more supportive of other threaded discussion participants while men will be more critical. This is in accordance with women's reputed tendency to be inclusive and men's tendency to be controlling in conversation. Moreover, men do tend to try and control the conversation more than women do (Edelsky, 1993), and are more prone to try to create their superior social standing through the conversation (Tannen, 1994, 1995). Women, on the other hand, tend to encourage more participation by all involved and are less forceful toward other participants (Weatherall, 1998), encourage cooperation (Coates, 1986), and are more complementary (Coates, 1986; Yates, 2001). Moreover, men tend to be more aggressive and competitive in their speech (Kilbourne & Weeks, 1997) and to interrupt others more (Anderson & Leaper, 1998; West & Zimmerman, 1983; Zimmerman & West, 1975) in an attempt to be dominant in the conversation (Herring, 1993; Holmes, 1992).

Taken together, this should translate to women being more supportive of others in the conversation as a way of being inclusive, paying complements, and being encouraging, while men should be more critical of others as a way of showing their domination, higher social standing, and generally being more competitive, especially as this could downplay the importance of others (Guillier & Drndell, 2006; Tannen, 1994) .

H1: Women will be more supportive of others in the threaded discussions

H2: Men will be more critical of others in the threaded discussions

In the context of these gender related communication characteristics, the reputed same gender congregation tendency should also be evident in threaded discussions. In oral conversations cross gender conversations are akin to cross cultural conversations and hence there is a tendency toward same gender conversations (Tannen, 1994). If this applies also to online courses then this tendency should carry over also to threaded discussions. Practically this means that there should be more references by men to previous postings by men and more references by women to previous postings by women. Indeed, in virtual communities there is such a preference with many communities being almost entirely all men or all women (Gefen & Ridings, 2005).

H3: There will be more references to previous postings by same gender students in the threaded discussion than to posting by members of the other gender.

If these tendencies are as pronounced in threaded discussions as they are in oral discourse then just as in oral discourse where conversation styles could be a predictor of the person's gender (Hannah & Murachver, 1999), styles should be a predictor of gender in a threaded discussion too.

H4: Student gender can be identified by the supportiveness and criticism in the posting.

Research Methodology

Examining threaded discussion postings in online courses this study answers these two questions. In the university where the data were collected these conversations were weekly units with a new topic started once a week. These weekly topics would evolve during the week, much as a guided discussion in a face to face setting would, with new questions being posted to the students as older ones were discussed in full. Typically, there were around 20 students participating in each online class. Participation was graded. A large number of courses and their online discussion components were examined. The content of the threaded discussion in these courses was copied and then classified. In all, 1335 postings in 27 online courses by 233 individuals were classified. Each posting was initially classified as whether it referred to previous postings by other students. If the posting did refer to a previous posting, then it was recorded whether the posting agreed or disagreed with the postings it referred to and whether it related to the person who posted the referred to posting by name. The data were classified by two raters with a 100% agreement between them during the training period on the actual data.

DATA ANALYSIS

Hypotheses H1, H2 and H3 were examined with T tests. These are shown in Table 1 together with general statistics. There is no significant difference in the percent of supportive, agreeing, or disagreeing messages. Neither H1 nor H2 are supported, although men did post more and longer messages, a trend that has been reported in previous computer-mediated communication research (Prinsen, Volman, & Terwel, 2007) and is possibly related to an attempt attributed to men in the literature to try and dominate the conversation (Edelsky, 1993).

A more detailed analysis of Table 1 however shows a more complex picture. Men refer back more to men and women more to other women, and in doing so women agree more with other women and men are more supportive of other men. This supports H3. Practically then, while the anticipated typical gender behaviors were not evident in the data, what was evident was a gender oriented group boundary of the kind one typically comes across in a cocktail party. Women communicate more with other women and men more with other men. Interestingly, Gefen and Ridings (2005) came to much the same conclusion when analyzing voluntary participation in virtual communities.

Hypothesis 4 was examined by verifying whether the gender of the poster could be significantly classified based on the nature of the reference to previous posters. Based on these 1335 postings the logistic regression did significantly (χ^2_8=8.951, p=.346)[1], albeit weakly (Nagelkerke R^2= 3.8%) identify 56% of the student gender correctly. This weak result is not surprising. Many students in online courses do not refer back to

Table 1. Group statistics

		All the Data (1335 postings)			Only those who referred back to others at least twice (162 postings)		
		Mean	Std.	T-statistic	Mean	Std.	T-statistic
Total length in words	Women	292.90	175.05	-4.36**	400.71	184.01	-2.48*
	Men	343.54	205.07		480.84	202.18	
Number of Postings	Women	2.58	1.71	-2.70**	4.05	2.02	-2.06*
	Men	2.92	2.32		4.82	2.41	
Percent of Supportive Messages	Women	9.1%	.29	.463	.37	.49	1.68
	Men	8.4%	.28		.25	.43	
Percent of Disagreeing Messages	Women	5.4%	.23	.246	.28	.50	.88
	Men	5.1%	.22		.22	.49	
Percent of Agreeing Messages	Women	19%	.40	1.89	.49	.50	1.43
	Men	15%	.36		.38	.49	
How many of these postings referred explicitly to postings by others	Women	.54	.96	-0.16	2.58	.98	-1.74
	Men	.55	1.07		2.91	1.17	
How many postings relate to postings by Men by name	Women	.31	.65	-1.91	1.42	.96	-3.74**
	Men	.40	.86		2.12	1.24	
How many postings relate to postings by Women by name	Women	.21	.58	2.08*	1.02	1.11	1.78
	Men	.15	.47		.73	.94	
How many of these postings to Men are supportive in tone	Women	.08	.31	-.198*	.42	.65	-2.33*
	Men	.13	.47		.76	1.00	
How many of these postings to Women are supportive in tone	Women	.05	.22	1.26	.23	.46	2.42*
	Men	.03	.21		.08	.33	
How many of these postings to Men agree with others	Women	.20	.56	0.13	1.02	1.08	.20
	Men	.20	.55		.98	1.08	
How many of these postings to Women agree with others	Women	.13	.40	1.97*	.51	.76	.27
	Men	.09	.37		.47	.84	
How many of these postings disagree with Men	Women	.04	.19	-1.32	.21	.41	-1.20
	Men	.06	.30		.33	.71	
How many of these postings disagree with Women	Women	.02	.14	-0.99	.11	.31	-1.24
	Men	.03	.21		.19	.50	

postings by other students and so identifying the gender of the student by typical gender communication style with others should be mostly weak. Nonetheless, the data did show some characteristic communication behavior even in these data. Men were significantly identified by having longer postings (β=.002, $p<.001$) and referring to women by name significantly less than women did (β=-.372, p=.002). Trying to control the conversations by speaking longer is a typical male trait (Herring, 1993, 1996b) and men did tend to interrupt others (Anderson & Leaper, 1998; Coates, 1986), in this case since the conversation is asynchronous the equivalent of interruption is downplaying the contribution of others.

The data were then reanalyzed but only including those postings where students referred back to postings by others. Since this type of posting corresponds more readily to the equivalent of a discussion, rather than a monolog, we expected the results to be much stronger in this case, as indeed they were. Based on the 401 postings of students who did refer back to other students at least once in their posting, the results of the logistic regression did significantly (χ^2_8=9.043, p=.339), albeit still rather weakly (Nagelkerke R^2= 7.4%) identify 58% of the respondent gender correctly. Again, the data showed some characteristic communication behavior, but in a more pronounced manner. Men significantly referred more to postings by other men by name (β=.550, $p<.001$) and agreed less with them (β=-.362, p=.032). This tendency to prefer to communicate with same gender others resembles findings about virtual communities where men and women tend mostly to congregate in same gender communities (Gefen & Ridings, 2005), as in H3. This tendency of men to agree less with other men is also in agreement with theory about oral discourse where men are supposed to be more motivated to compete with other men over social dominance (Tannen, 1994, 1995) and downplay the contribution of others in an attempt to bolster their own social standing (Kilbourne & Weeks, 1997).

This pattern is even stronger once the analysis was limited to only those who referred back to others at least twice. Here the results became much stronger. Limiting the analysis to these 162 postings alone, the logistic regression did significantly (χ^2_8=6.819, p=.556) and rather strongly (Nagelkerke R^2=25%) identify 66% of the respondent gender correctly. As before, the data show some characteristic communication behavior, but in a more pronounced manner. Men significantly referred more to postings by other men by name (β=.837, $p<.001$) and agreed less with them (β=-.470, p=.022). H4 was supported.

DISCUSSION

Summary of Results

Men and women do communicate differently also in threaded discussions. Apparently, the dictated nature of the conversation in an online course and the limitations technology puts on how people can interact with each other do much to repress the typical gender type behaviors which are so evident in oral discourse and virtual communities. Nonetheless, the conversational behavior of men and women is still sufficiently different, primarily in preferring same gender references, that the gender of a student can be significantly identified based on this behavior. Specifically, although the specific typical gender related communication patterns do not carry over, as shown by women not being more supportive and men not being more critical, the impact of the typical gender communication pattern does carry over in the type of company people prefer, namely same gender groupings. As the data show, men tend to refer more to postings by other men and are more supportive of other men while women refer more to postings by other women and are more supportive of other women. This tendency is evident especially when the analysis focuses on those students who refer more to others. Although it would be an exaggeration to say there are two separate discussions going on, men to men and women to women, there are some signs of a slight tendency in this direction. This is interesting because overall men and women are on average about as supportive, agreeing, and disagreeing with other students if the gender of the other student is not included in the analyses. In other words, even in the pedagogically controlled setting of online courses there are gender effects.

Limitations

Online courses and the way they are taught are not cast in a mold. As courses generally are, also online courses differ in the way they are taught and are influenced by the personality of the online instructors. Thus, the results and conclusions of this exploratory study with its convenience sample should be understood as such. The data raise the need to consider these interesting implications and warrant additional research. Whether the results can be generalized to other online course environments depends on additional research and many more samples.

Implications to Theory

The theoretical contribution of this study is in its initial validation of the need to consider the gender mix of students, even when they are attending online courses, and by brining sociolinguistics theory aspects into this context of online course threaded discussion. Much more research is needed to establish exactly how sociolinguistics applies to this setting and to better understand how the gender mix affects online classes, and, hence, exactly what online instructors and course designers need to do to make the online threaded discussion more of a success. But this study does verify this need. Gender differences also apply in an online course environment, even though the circumstances of these courses might seem to prevent the typical gender related conversational behavior. In a threaded discussion men cannot dominate by controlling airtime or by interrupting and women cannot show compassion through their tone of voice. Moreover in a threaded discussion, because the teacher holds the leads on the discussion, students cannot choose to discuss the type of topics typically related to their gender, but are led to discuss gender neutral topics. Despite all of this, gender preferences still come through, and men still prefer to converse with men and women with women.

This suggests some directions for further research. The first is the question whether gender mix in an online class matters? More specifically, now that this study has shown that the gender mix does make a difference, at least in the sample examined, in how students participate in online discussions, the question is does this gender mix effect also affect student learning quality and student satisfaction with the online courses? If men prefer to communicate with other men and women with other women, then being a minority might have a negative impact on how students participate and are accepted by the majority gender group in the threaded discussion. This question then ties into the next question of what impact does this have on student retention. Considering what the gender mix does in regular classroom settings (Felder, Felder, Mauney, Hamrin, & Dietz, 1995), it is quite possible that this gender mix effect might negatively affect students and, therefore, should be at least recognized. This study takes the first step in that direction by showing that there is such a gender effect and by suggesting a possible theory base which, in part, explains it.

The role of the course instructor in shaping the social atmosphere in online discussions and its influence on the conversational style of students is another important question which requires further research. Women and men have different interaction styles and that may have implications for pertinent online tutoring support (Price, 2006). As Salmon (2004, p.4) explains, students' experience is central to knowledge construction and the e-moderator is essential in promoting constructive communication. She also encourages e-moderators to pay attention to cultural differences regarding styles of address, hierarchy and authority, attitudes towards gender, criticism, the proper ways of asking and answering questions, personal disclosure and even the names that students use (Salmon, 2004, p.157-159). It is still an open question exactly how this can be done in an online course, but some ideas can be borrowed from other settings. As Tannen (1991) reports regarding non online class settings, women respond differently to challenges during class discussions and are relatively silent, but can be encouraged to participate with open-ended questions. Whether this applies also to online discussions requires more research. Indeed, our findings show that men posted more messages than women, and similar results were reported by Lawlor (2006) regarding graduate student participation in online discussions. However, further research is needed to find effective ways for encouraging women participation in online class discussions.

IMPLICATIONS TO PRACTICE

Men and women do communicate differently and therefore in certain cases they prefer gender segregated communications (Tannen, 1994). The same applies with online virtual communities (Gefen & Ridings, 2005). As this study shows, to some extent this tendency also applies to online course threaded discussions. Being aware of this could help instructors facilitate better threaded discussions. While this preference by students to relate more to same gender students may not interfere with their or with others' learning experience, it might spoil the online course social atmosphere if a student feels relatively ignored or not supported enough by his or her cohorts. After all, communication is of paramount importance in learning. If online teachers are aware of these preferences to communicate within the same gender group, then they can take action. This preference can be controlled, especially if, as we suspect, it is unintentional by the students.

Understanding this student proclivity to cluster in their threaded discussions in same-gender subgroups should be considered when managing these discussions especially because one of the main advantages of online courses is in the way they allow also reticent students to actively participate in the course discussions. However, as this study findings show, even in the lean environment of threaded discussions, there are conversational style gender differences which may imperil this advantage.

This actually is not so surprising, apparently, even the perceived gender of the computer itself, manifested by its sound being based on a male or a female narrator, is enough to elicit typical responses associated with this artificial gender manipulation (Nass, Moon, & Green, 1997). Likewise, associating a computer generated message with a cartoon representation of a man or a women elicited the same type of gender response. People are more inclined to accept the answer given by the computer when the gender of the computer generated cartoon giving this answer matches gender stereotypes: people were more accepting of answers about sports when given by a male cartoon and more accepting of answers about fashion when given by a female cartoon (Lee, 2003).

Being aware of the student and his or her unique needs is a hallmark of good teaching. Although arguably this is much harder in an online course than in a regular classroom, nonetheless, at least this gender aspect is something online instructors can and should consider. Controlling the online conversation and discreetly but directly focusing the positive discussion on a student who might otherwise have been relatively ignored because of these same gender preferences is one way of doing so. Creating smaller and mixed gender teams is another method instructors in online courses can apply. Such a method directly utilizes one of the great advantages of online courses, and may thus make the online environment more conductive to overcome these same gender preferences than other teaching environments.

CONCLUSION

Socio-linguists typify cross gender communication as almost a cross cultural experience. Our data do not quite support such a view but do show certain patterns of online communication in threaded discussion which instructors should be aware of. Birds of a feather stick together, as the expression goes, also when it concerns the gender of online students. Students prefer to relate to other students of their own gender. Instructors should seek ways to use the flexibility and unique capabilities of online courses to better serve their students, recognizing these gender preferences.

REFERENCES

Ahuja, M. K., & Thatcher, J. B. (2005). Moving Beyond Intentions and Toward the Theory of Trying: Effects of Work Environment and Gender on Postadoption Information Technology Use. *MIS Quarterly, 29*(3), 427–459.

Anderson, K. J., & Leaper, C. (1998). Meta-analyses of Gender Effects on Conversational Interruption: Who, What, When, Where, and How. *Sex Roles, 39*(3/4), 225-252.

Boneva, B., Kraut, R., & Frohlich, D. (2001). Using e-Mail for Personal Relationships: The Difference Gender Makes. *American Behavioral Scientist, 45*(3), 530-549.

Caspi, A., & Gorsky, P. (2006). Online Deception: Prevalence, Motivation and Emotion. *Cyberpsychology and Behavior, 9*(1), 54-59.

Coates, J. (1986). *Women, Men and Languages: Studies in Language and Linguistics*. London, UK: Longman.

Costa, P. T. J., Terracciano, A., & McCrae, R. R. (2001). Gender Differences in Personality Traits across Cultures: Robust and Surprising Findings. *Journal of Personality and Social Psychology, 81*(2), 322-331.

Crawford, M. (1995). *Talking Difference. On Gender and Language*. London, UK.: Sage.

Eastmond, D. V. (1995). *Alone but Together: Adult Distance Study through Computer Conferencing*. Cresskill, NJ: Hampton Press.

Edelsky, C. (1993). Who's Got the Floor? In D. Tannen (Ed.), *Gender and Conversational Interaction* (pp. 189-227). New York, NY: Oxford University Press.

Fallows, D. (2005, 12/28). How Women and Men Use the Internet. Retrieved December 28, 2005, from http://www.pewinternet.org/PPF/r/171/report_display.asp

Felder, R. M., Felder, G. N., Mauney, M., Hamrin, C. E. J., & Dietz, E. J. (1995). A Longitudinal Study of Engineering Student Performance and Retention. III. Gender Differences in Student Performance and Attitudes. *Journal of Engineering Education, 84*(2), 151-174.

Gefen, D. (2000). Gender Differences in the Perception and Adoption of E-mail and Computer-Mediated Communication Media: A Sociolinguistics Approach. In A. Kent (Ed.), *The Encyclopedia Of Library And Information Science (Volume 69)* (pp. 139-153). New York: Marcel Dekker, Inc.

Gefen, D., & Ridings, C. (2005). If You Spoke as She Does, Sir, Instead of the Way You Do: A Sociolinguistics Perspective of Gender Differences in Virtual Communities. *The DATA BASE for Advances in Information Systems, 36*(2), 78-92.

Gefen, D., & Straub, D. W. (1997). Gender Differences in Perception and Adoption of E-mail: An Extension to the Technology Acceptance Model. *MIS Quarterly, 21*(4), 389-400.

Geri, N., Mendelson, O., & Gefen, D. (2007, August 9-12). *On the Centrality of Social Norms in Determining MBA Program Success.* Paper presented at the Americas conference on Information Systems, Keystone, Colorado, USA.

Gray, J. (1992). *Men are From Mars, Women Are From Venus.* New York, NY: HarperCollins.

Guillier, J., & Drndell, A. (2006). I Totally Agree with You: Gender Interactions in Educational Online Discussion Groups. *Journal of Computer Assisted Learning, 22*(5), 368-381.

Guri-Rosenblit, S. (2005). Eight Paradoxes in the Implementation Process of E-learning in Higher Education. *Higher Education Policy, 18*(1), 5-29.

Hannah, A., & Murachver, T. (1999). Gender and Conversational Style as Predictors of Conversational Behavior. *Journal of Language and Social Psychology, 18*(2), 153-174.

Herring, S. C. (1993, January 20). Gender and Democracy in Computer Mediated Communication. *Electronic Journal of Communication 3(2)* Retrieved July 29, 2006, from http://www.cios.org/www/ejc/v3n293.htm

Herring, S. C. (1996a). Posting in a Different Voice: Gender and Ethics in Computer-Mediated Communication. In C. Ess (Ed.), *Philosophical Perspectives on Computer-Mediated Communication* (pp. 115-145). Albany: State University of New York Press.

Herring, S. C. (1996b). Two Variants of an Electronic Message Schema. In S. C. Herring (Ed.), *Computer-Mediated Communication Linguistic, Social and Cross-cultural Perspectives* (pp. 81-106). Philadelphia, PA: John Benjamins Publishing Company.

Hiltz, S. R., & Turoff, M. (2005). Education Goes Digital: The Evolution of Online Learning and the Revolution in Higher Education. *Communications of the ACM, 48*(10), 59-64.

Hirschheim, R. (2005). The Internet-Based Education Bandwagon: Look before You Leap. *Communications of the ACM, 48*(7), 97-101.

Hofstede, G. (1980). *Culture's Consequences: International Differences in Work Related Values.* London, UK: Sage.

Holmes, J. (1992). Women's Talk in Public Contexts. *Discourse and Society, 3*(2), 131-150.

Hong, S. J., & Tam, K. Y. (2006). Understanding the Adoption of Multipurpose Information Appliances: The Case of Mobile Data Services. *Information Systems Research, 17*(2), 162-179.

Kilbourne, W., & Weeks, S. (1997). A Socio-economic Perspective on Gender Bias in Technology. *Journal of Socio-Economics, 26*(1), 243-260.

Lawlor, C. (2006). Gendered Interactions in Computer-Mediated Computer Conferencing. *The Journal of Distance Education / Revue de l'Éducation à Distance, 21*(2), 26-43.

Lee, E.-J. (2003). Effects of "Gender" of the Computer on Informational Social Influence: The Moderating Role of Task Type. *International Journal of Human-Computer Studies, 58*, 347–362.

Levy, Y. (2006). The Top 10 Most Valuable Online Learning Activities for Graduate MIS Students. *International Journal of Information and Communication Technology Education, 2*(3), 27-44.

Levy, Y. (2007). Comparing Dropouts and Persistence in e-learning Courses. *Computers & Education, 45*(2), 185-204.

Mulac, A., Erlandson, K. T., Farrar, W. J., & Hallett, J. S. (1998). Uh-huh. What's That all about? Differing Interpretations of Conversational Backchannels and Questions as Source of Miscommunication Across Gender Boundaries. *Communication Research, 25*(6), 641-668.

Nass, C., Moon, Y., & Green, N. (1997). Are Computers Gender-neutral? Gender Stereotypic Responses to Computers. *Journal ofApplied Social Psychology, 27*, 864–876.

Parks, M. R., & Floyd, K. (1995, January 10, 2003). Making Friends in Cyberspace. *Journal of Computer Mediated Communication 1(4)* Retrieved October 3, 2007, from http://www.ascusc.org/jcmc/vol1/issue4/parks.html

Price, L. (2006). Gender Differences and Similarities in Online Courses: Challenging Stereotypical Views of Women. *Journal of Computer Assisted Learning, 22*(5), 349-359.

Prinsen, F. R., Volman, M. L. L., & Terwel, J. (2007). Gender-Related Differences in Computer-Mediated Communication and Computer-supported Collaborative Learning. *Journal of Computer Assisted Learning, 23*(5), 393-409.

Rovai, A. P. (2002). Building Sense of Community at a Distance. *International Review of Research in Open and Distance Learning, 3*(1), www.irrodl.org/index.php/irrodl/article/viewPDFInterstitial/79/153.

Rovai, A. P. (2003). In Search of Higher Persistence Rates in Distance Education Online Programs. *Internet and Higher Education, 6*(1), 1-16.

Salmon, G. (2004). *E-moderating: The Key to Teaching and Learning Online* (2nd ed.). London, UK: RoutledgeFalmer, Taylor and Francis Group.

Simpson, O. (2003). *Student Retention in Online, Open and Distance Learning.* Sterling, VA: Kogan Page.

SPSS. (2004). *SPSS 13.0 for Windows*. Chicago, IL: Apache Software Foundation.

Stewart, C. M., Shields, S. F., & Sen, N. (2001). Diversity in On-Line Discussions: A study of Cultural and Gender Differences in Listervs. In C. Ess & F. Sudweeks (Eds.), *Culture, Technology, Communication: Towards an Intercultural Global Village* (pp. 161-186). Albany, NJ: State University of New York Press.

Tannen, D. (1991). Teachers' Classroom Strategies Should Recognize that Men and Women Use Language Differently. *The Chronicle of Higher Education, June 19*, B1-B3.

Tannen, D. (1994). *You Just Don't Understand Women and Men in Conversation.* New York, NY: Ballantine Books.

Tannen, D. (1995). The Power of Talk: Who Gets Heard and Why. *Harvard Business Review, 73*(5), 138-148.

Tinto, V. (1998). Colleges as Communities: Taking Research on Student Persistence Seriously. *The Review of Higher Education, 21*(2), 167-177.

Venkatesh, V., & Morris, M. G. (2000). Why Don't Men Ever Stop to Ask for Directions? Gender, Social Influence, and their Role in Technology Acceptance and Usage Behavior. *MIS Quarterly, 24*(1), 115-139.

Weatherall, A. (1998). Re-visioning Gender and Language Research. *Women and Language, 21*(1), 1-9.

West, C., & Zimmerman, D. (1983). Small Insults: A Study of Interruptions in Cross-sex Conversations between Unacquainted Persons. In B. Thorne, H. Kramarae & N. Henley (Eds.), *Language, Gender and Society* (pp. 103-118). Rowley, MA: Newbury House.

Woodley, A. (2004). Conceptualizing Student Dropout in Part-Time Distance Education:

Pathologizing the Normal? *Open Learning, 19*(1), 47-63.

Yates, S. J. (2001). Gender, Language an CMC for Education. *Learning and Instruction, 11*(1), 21-34.

Zhang, X., Prybutok, V. R., & Strutton, D. (2007). Modeling Influences on Impulse Purchasing Behaviors During Online Marketing Transactions. *Journal of Marketing Theory & Practice, 15*(1), 79-89.

Zimmerman, D., & West, C. (1975). Sex-roles, interruptions and silences in conversation. In B. Thorne, H. Kramarae & N. Henley (Eds.), *Language and Sex: Difference and Dominance* (pp. 89-101). Rowley, MA: Newbury House.

ENDNOTE

1 The Hosmer and Lemeshow Chi Square test shows good fit when the p-value is insignificant (SPSS, 2004).

Chapter III
The Technology Acceptance Model (TAM) and the Continuance Intention of Using WebCT: A Case of College Students in Estonia

Princely Ifinedo
Cape Breton University, Canada

ABSTRACT

In this study, we investigate the influence of two external influences i.e., Ease of finding and Computer anxiety on the technology acceptance model (TAM) and the continuance intention of using a popular course management system (CMS): WebCT. The study used a sample of 72 students that have experience using the software. The students came from four local higher education institutions. In order to study nature of the relationships among the constructs, eight (8) hypotheses were formulated and tested using a structural equation modeling technique: Partial Least Squares (PLS). The predictive power of the model was adequate and the study found support for seven of eight hypotheses. Regarding the impact of the antecedents on continuance intention in the use of technology, the results offer the following insights: when computer anxiety is low, students are able to use the system without much difficulty, and are likely to continue to use CMS in the future. Similarly, students will continue the tool as long as they find it easy to navigate. Perhaps due to contextual factors, the data did not support the relationship between Perceived usefulness and Usage. This particular finding is at variance with the TAM's results and viewpoint. The study's implications for research and practice are succinctly outlined.

INTRODUCTION

Higher learning institutions across the world have started adopting a type of information and communication technology (ICT), generally referred to as course management systems (CMS) to improve pedagogy (Limayem et al.; 2003; Tavangarian et al., 2004; Ifinedo, 2006; 2007a; Ngai et al., 2007). CMS are used in the management of asynchronous academic environments (Tavangarian et al., 2004). Examples of CMS include Blackboard, Learning Space, and WebCT (the example used in this study). In brief, the technology or tools enable students to learn at their own speed, give and receive feedback from peers and instructors alike. Further, it provides a wide variety of learning and teaching opportunities, such as course content and syllabi tools, student progress tracking, group project organization, student self-evaluation, email, and on-line chat. Morss (1999) studied the relevance of WebCT in higher learning settings noting that students generally have favourable attitudes towards the tool. This is due to the fact that WebCT is easy to use and requires little or no technical background (see Ifinedo, 2006; 2007a).

Virtually, hundreds of universities around the globe have adopted WebCT to enhance their e-learning platforms (Tavangarian et al., 2004: Ifinedo, 2005b; 2006; 2007b; Ngai et al., 2007). The same is true for higher learning institutions in Estonia, where CMS, including WebCT, have been adopted to facilitate web-based learning or e-learning (Ifinedo, 2005a). Estonia is an emerging country in Eastern Europe where ICT use at all the levels of education has been supported and encouraged (the Tiger Leap Foundation, 1997, Estonian eUniversity, 2004a). To that end, Estonian colleges were chosen as a model to test the efficacy of web-based learning using CMS. Researchers (e.g., Morss, 1999; Limayem et al.; 2003; Tavangarian e et al., 2004; Ngai et al., 2007) have studied the acceptance of CMS among college students in developed countries. Results suggest that the acceptance and success with such tools are high. Unfortunately, a search of relevant literature shows little or no empirical studies exist in which the Estonian student's perspectives have been discussed. Success in the use and acceptance of these technologies among students in developed countries does not necessarily represent the attitudes of students from other regions of the world (Brown, 2007). Conflicting results could be due to cultural and socio-economic differences (Straub et al., 1995; Gefen and Straub, 2000). It is hoped that by studying the perceptions of Estonian student intent to continue the use of WebCT, policy makers and e-learning project administrators in the country will benefit from the results of this study.

This current effort complements other research in Estonia examining e-learning project success assessment. For example, Ifinedo (2005a) reports the risks of implementing e-learning projects from the information systems (IS) project managers' point of view. The Estonian eUniversity (2004b) conducted a survey to determine the needs of e-studies and e-learning environments among teachers in the country. In both studies, the views of students' were not sought. Indeed, Keller and Cernerud (2002) note that the discourse of ICT use in pedagogy tends to focus on how faculty members use such technologies, with little or no attention paid to students' perspectives on these issues. They argue that by researching students' views, we stand to increase our knowledge in the success of this learning environment. More importantly, e-learning project managers and other policy makers in Estonia, as elsewhere, are beginning to realize that as new ICT are introduced, if administrators are not educated in the success of these learning strategies, a valuable resource may be lost (Davis, 1989; **Straub et al., 1995;** Gefen and Straub, 2000; Lee et al., 2003; Estonian eUniversity, 2004b). The notion of acceptance in this chapter refers to "the demonstrable willingness within a user group to employ information technology for the tasks it is designed to support" (Dillon and Morris, 1996, p.4).

This study is motivated by the lack of empirical studies on WebCT continuance intention of use among college students. Additionally, this research aims at presenting empirical evidence from a region of the world that has not been featured prominently in the literature. Importantly, this research did not limit its scope to presenting evidence on WebCT use as the major indicator for success with the tool. Previous studies were limited by an approach using only such to measure success in the context of the Technology Acceptance Model (TAM). Such studies overlook the fact that use (or *Usage*) is the first step in achieving success with the new IS. Indeed, Bhattacherjee (2001) and Limayem et al. (2003) argue for IS continuance intention to be incorporated into studies investigating the adoption or acceptance of IS. These researchers assert that by taking this point of view into consideration, the overall levels of IS acceptance and use would be better understood. Similarly, Davis (1989) argues the predictive capability of the TAM could be improved when relevant variables or factors are considered.

In response, researchers examining the acceptance of technologies have heeded the warning made by Davis (1989) and others by incorporating the influence of a variety of external factors (see Venkatesh and Davis, 1994; Brown, 2002; Lee et al., 2003; Ifinedo, 2006; Ngai et al., 2007). Having said that, *Computer anxiety*, *Ease of finding*, *Ease of understanding*, and *Self-efficacy* are among the external influences that have been used to increase the predictive power of the TAM in the context of e-learning technologies and tools. Accordingly, the choices made for this study reflect the need to enhance the predictability of the TAM. Thus, the study's specific objectives are as follows:

- To develop a hypothetical model comprising factors such as *Ease of finding* and *Computer Anxiety* and *Continuance intention* use of WebCT among Estonian undergraduates
- To test the predictive power of the structural model
- To determine the relationships among the foregoing constructs or factors

The chapter is organized as follows. The next section presents a review of the background literature. This is followed by the development of the relevant hypotheses. Next, the research methodology is discussed. Afterwards, the data analysis is presented. The chapter ends with a discussion and conclusion section.

BACKGROUND LITERATURE

The Technology Acceptance Model (TAM) is regarded as the most widely used theoretical framework for assessing the acceptance of technologies in the literature (see e.g. Legris et al., 2003). The TAM was developed by Davis (1989) who drew from Theory of Reasoned Action (TRA) proposed by Fishbein and Ajzen (1975). The TAM proposes that users' acceptance of a new IS can be predicted by the users' perceptions. These perceptions include the ease of use and perceived usefulness of the information technology (Davis, 1989). In brief, the three core constructs in the TAM include the following: *Perceived ease of use*, *Perceived usefulness*, and *Usage*. The *Perceived ease of use* describes "The degree to which a person believes that using a particular system would be free of effort" (Davis, 1989, p. 320). *Perceived usefulness* describes the user's perceptions of the expected benefits derived from using a particular IS system (Davis, 1989). *Usage* is the dependent variable in the TAM, and it is "theorized to be influenced by perceived usefulness and perceived ease of use" (Ibid, p 320). Further, *Perceived usefulness* mediates the effect of *Perceived ease of use* on *Usage* and at the same time, both directly influence *Usage* (Davis, 1989). Although the relationships among the TAM constructs have yielded consistent results in the literature (see Szajna, 1996; Straub et al., 1995; Igbaria et al. 1997; Legris et al., 2003), some conflicting results have also been reported (see e.g. Hu et al., 1999; Anandarajan et al. 2002; Brown, 2002; Pan et al., 2003). For example, Hu et al. (1999) reported *Perceived usefulness* as a significant determinant of *Usage*, a result that is inconsistent with findings in Anandarajan et al. (2002) and Brown (2002). The latter studies demonstrated that *Perceived ease of use* was the construct that best predicts the use of ICT.

Furthermore, some commentators, including Davis (1989), have argued that for future technology acceptance to be fully appreciated, the impact of relevant variables needs to be integrated into future research. Perhaps as a result of the limitations in the TAM, other researchers (e.g., Igbaria, 1990; Venkatesh and Davis, 1994; Steer et al., 2000; Brown, 2002; Lee et al., 2003; Ngai et al., 2007) have re-modeled the TAM to include influences from external factors or variables. Steer et al. (2000) used the TAM in web-based environments to study usage behavior and concluded that a wider range of factors are needed to understand the user's adoption actions. Similarly, previous studies examining the acceptance of e-learning technologies among students have incorporated such factors as *Computer anxiety*, *Ease of finding*, *Ease of understanding*, and *Self-efficacy*, among others, to increase the predictive power of the TAM (Brown, 2002; Lee et al., 2003; Pan et al., 2003; Ifinedo, 2006; Ifinedo, 2007b). However, in this chapter we all the aforementioned influences will not be as those have been discussed in-depth elsewhere (see Ifinedo, 2007a). Having said, this current effort seeks to find answers to the following question: Can a hypothetical model that incorporates the TAM and the other relevant factors be developed to help us understand the future acceptance of IS (in this case CMS)?

The development of CMS, in general and WebCT, in particular is closely linked to the Internet (and the Web) (Morss, 1999; Lu et al., 2005; Ifinedo, 2006; 2007b; Hsu et al., 2006). The ability to navigate through web-based media directly influences how users of such facilities perceive the usefulness, ease of use and success of such applications (e.g., Hara and Kling, 1999; Lu et al., 2005; Hsu et al., 2006). It goes without saying that those who are able to successfully navigate, to find and understand such media are more satisfied than those who are unable to do so (Lederer et al., 2000; Lu et al., 2005; Hsu et al., 2006). In fact, Lederer et al. (2000) noted that *Ease of finding* is an important variable that significantly predict the use of Web-based facilities. In the context of CMS acceptance among students, Brown (2002) and Ifinedo (2006) revealed that these variables positively influence WebCT use through *Perceived usefulness* and *Perceived ease of use*. That said, *Computer anxiety* describes "the tendency of individuals to be uneasy, apprehensive, or fearful about current or future use of computers" (Igbaria and Parasuraman 1989, p. 375). The literature shows that *Computer anxiety* influences IS acceptance (Igbaria and Parasuraman 1989; Igbaria, 1990; Compeau and Higgins, 1995).

As was briefly discussed above, use only of new IS, though vitally important in enhancing acceptance, may be insufficient in ensuring the overall success of the IS. Bhattacherjee (2001) and Limayem et al. (2003) assert that long term success in enhanced IS continuance intention should not be overlooked. Bhattacherjee (2001) proposes the post acceptance model (PAM), which borrows from the Expectation-Confirmation Theory in consumer behaviour. The PAM suggests that the user forms an initial expectation of an IS prior to its use, then he or she accepts and uses the IS or rejects and does not use the IS. Afterwards, he or she develops perceptions about the IS (i.e. perceived usefulness). The user then assesses his or her original expectations, from which they determine a level of satisfaction.

Finally, a satisfied user forms an IS *continuance intention*, while a dissatisfied user may discontinue the use of the IS. In the development of a hypothetical model that incorporates the concepts discussed above, this chapter draws from the work of Pan et al. (2003). A framework that included a set of factors divided into exogenous, endogenous, and dependent variables were used in the discourse of students' attitudes towards web-based learning environments. The model includes *Computer anxiety*, *Ease of Finding*, and TAM, with the dependent variable being *Usage*. Thus, a hypothetical model is developed to include all the relevant concepts discussed herein (Figure 1).

Hypotheses Formulation

Eight hypothesized paths are evaluated in this chapter and illustrated in Figure 2. The statements of hypotheses are presented below.

The literature shows that the ease of finding information on a website is strongly related to the perceived ease of use of such websites (Ledera et al., 2000). The same is true for e-learning platforms (Lu et al., 2005) and CMS (e.g., Brown, 2002; Ifinedo, 2006). Further, students who are capable of finding information using such technologies do tend to have higher perception of the usefulness of such software (Ifinedo, 2007a). Thus, it can be hypothesized that:

H1: Ease of finding is positively related to perceived ease of use of WebCT.

H2: Ease of finding is positively related to perceived usefulness of WebCT.

Igbaria (1990) and Igbaria and Parasuraman (1989) and found that IS acceptance is influenced by computer anxiety. This is congruent with the findings in Compeau and Higgins (1995). In the context of CMS acceptance among students, Brown (2002) and Ifinedo (2006) reported that a strong relationship exists between computer anxiety, on the one hand, and perceived usefulness and perceived ease of use, on the other. Thus, it can be hypothesized that:

Figure 1. The hypothetical model comprising relevant components

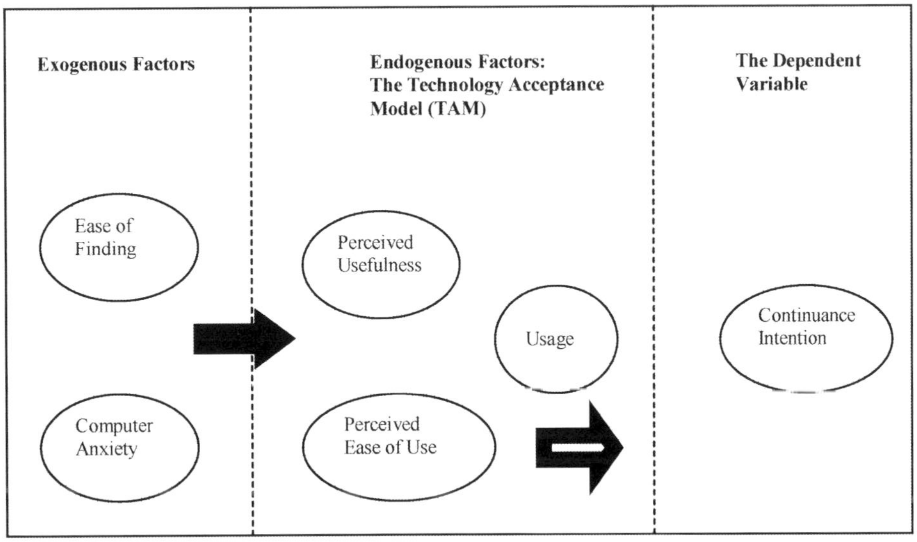

Figure 2. The research model with the eight hypotheses

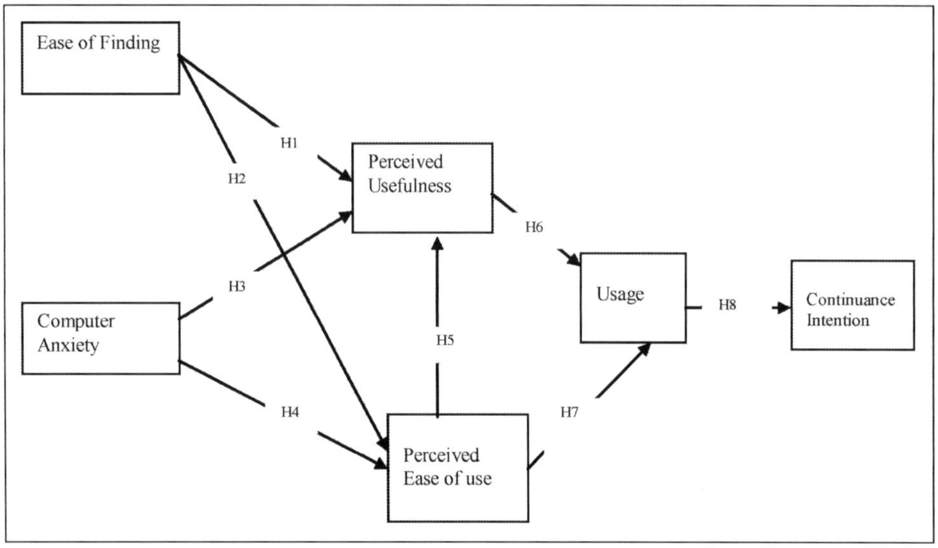

H3: Computer anxiety has a positive effect on perceived usefulness of WebCT.

H4: Computer anxiety has a positive effect on perceived ease of use of WebCT.

With regard to IS acceptance, *Perceived usefulness* mediates the effect of *Perceived ease of use* on *Usage*. In fact, Davis (1989) demonstrated that *Perceived ease of use* and *Perceive usefulness* have positive effects on use of an IS. Evidence from differing sources have supported the TAM (e.g., Igbaria, 1990; Venkatesh and Davis, 1994; Straub et al., 1995). It is also important to mention here that conflicting results have surfaced in some studies using the TAM model to understand technology acceptance (e.g. Straub et al., 1995; Hu et al., 1999; Brown, 2002; Pan et al., 2003). Nonetheless, in the face of the overwhelming evidence providing support for the nature of the relationships in the TAM, this study proposes the following set of hypotheses:

H5: Perceived ease of use has a positive effect on perceived usefulness of WebCT.

H6: Perceived ease of use has a positive effect WebCT usage.

H7: Perceived usefulness of has a positive effect on WebCT usage.

The attitude of an IS user towards the systems impact his or her continuance intention (Bhattacherjee, 2001). The results from studies by Limayem et al. (2003), Sørebø (2004), Ifinedo (2006), and Roca et al. (2006) have shown that favourable perceptions of the ease of use and usefulness of CMS influence the continuance intention among users. Thus, it can be hypothesized that:

H8: WebCT usage has a positive effect on continuance intention.

RESEARCH METHODOLOGY

This study utilized a convenient sample size of 72 students to obtain data from four tertiary institutions in Estonia including The Estonian Business School, Tallinn University of Technology, Tartu University and Estonian IT College. The four universities are among the well-attended schools in the country, and have students that have had experience with CMS, including WebCT (Ifinedo, 2005b). This study employed the judgmental sampling technique (Neuman, 1997). Following guidelines from the approach, a self-administered two-page questionnaire (please see the Appendix) was provided to students who indicated they had experience with WebCT. Participation was voluntary.

The questionnaire was translated into Estonian in accordance with Brislin's (1986) suggestions for research conducted in a different culture. The questionnaire was test-piloted by four students whose comments helped to improve the quality of the final administered instrument. Students from diverse academic backgrounds were enlisted with the hope that such considerations would permit deeper insights. Accordingly, the study's participants included students from the sciences, social sciences, and the arts/humanities. Their demographic profile is shown in Table 1.

The questionnaire contained measures that had previously been validated in the literature. The scale for *Ease of finding* (EAF) was comprised of three (3) items. The measures were taken from the work of Lederer et al. (2000) and Brown (2002). *Computer anxiety* (CAX) had three (3) items. These measures were adapted from Compeau and Higgins (1995) and Brown (2002). Four (4) and three (3) items from Davis (1989) were used to measure *Perceived ease of use* (PEOU) and *Perceived usefulness* (PUS), respectively. The *Usage* (USG) construct was represented with two (2) measures, which were taken from Davis (1989). Finally, *Continuance intention* (CIX) was comprised of two (2) items that originated with Bhattacherjee (2001) and Sørebø's (2004). All items were operationalized using a Likert-type scale ranging from 1 (strongly disagree) to 7 (strongly agree) with the exception of *Usage*, which was assessed differently (see the Appendix). The composite reliabilities (similar to

Table 1. Demographic profile of the respondents

Variable		Number	Percent (%)
Gender	Male Female	32 40	44.4 55.6
Age	Less than 25 years 26-39 years	63 9	87.5 12.5
Education (level)	First year student Second year student Third year student Fourth year student	13 22 13 24	18.1 30.6 18.1 33.3
Study programme	Business / Economics studies Information Technology Mechanical Engineering Philosophy Electrical Engineering	36 16 9 6 5	50 22.2 12.5 8.3 6.9

the Cronbach alphas) of the measures obtained in the data analysis are adequate and are consistently above the minimum value of 0.70 recommended by Nunnally (1978). Please see Table 2 below.

DATA ANALYSIS

A structural equation modeling (SEM) technique was used to examine the causal relationships among the constructs. SEM is a multivariate data analysis technique that contains mechanisms that eliminate measurement errors in the observed variables. There are two main approaches: PLS (Partial Least Squares) and covariance-based SEM. The PLS approach is chosen for its capability to accommodate small-sized samples (Chin, 1998). Additionally, PLS recognizes two components of a casual model: the measurement model and the structural model. The measurement model consists of relationships among the factors of interest (i.e., the observed variables) and the measures underlying each construct. PLS demonstrates the construct validity of the research instrument (i.e. how well the instrument measures what it purports to measure). The two main dimensions are the convergent validity and the discriminant validity. The convergent validity (composite reliability) assesses the extent to which items on a scale are theoretically related; the loadings of variables are also noted.

On the other hand, the structural model provides information on how well the hypothesized relationships predict the theoretical model. PLS software e.g. PLS Graph 3.0, provides the squared multiple correlations (R^2) for each endogenous

Table 2. Psychometric properties of measures and constructs

Construct	Item	Item loading	Composite reliability
Ease of finding (AVE = 0.848)	EAF1	0.8790	0.943
	EAF2	0.9503	
	EAF3	0.9140	
Computer anxiety (AVE = 0.776)	CAX1	0.7868	0.932
	CAX2	0.9291	
	CAX3	0.9263	
	CAX4	0.8735	
Perceived ease of use (AVE = 0.733)	PEOU1	0.8029	0.916
	PEOU2	0.8640	
	PEOU3	0.9013	
	PEOU4	0.8521	
Perceived usefulness (AVE = 0.815)	PUS1	0.9024	0.930
	PUS2	0.9359	
	PUS3	0.8695	
Usage (AVE = 0.914)	USG1	0.9513	0.955
	USG2	0.9611	
Continuance intention (AVE = 0.907)	CIX1	0.9547	0.952
	CIX2	0.9510	

Table 3. Correlations of latent constructs and AVE

	AVE	EAF	CAX	PEOU	PUS	USG	CIX
EAF	0.848	**0.921**					
CAX	0.776	0.635	**0.881**				
PEOU	0.733	0.509	0.426	**0.856**			
PUS	0.815	0.654	0.700	0.651	**0.903**		
USG	0.914	0.470	0.409	0.711	0.503	**0.956**	
CIX	0.908	0.553	0.488	0.546	0.555	0.742	**0.952**

construct in the model and the path coefficients. The R^2 indicates the percentage of a construct's variance in the model while the path coefficients (β) indicate the strengths of relationships between constructs (Chin, 1998). Unlike other structural modeling software (e.g. LISREL), PLS Graph 3.0 does not generate a single goodness-of-fit metric for the entire model. According to Chin (1998), both the β and the R^2 are sufficient for analysis, and β values between 0.20 and 0.30 yield meaningful interpretations.

Assessing the Measurement Model

Table 2 presents the Cronbach alphas, item loadings and composite reliabilities. Chin (1998) recommends item loadings of greater than 0.70. To determine if the measures are distinct and unidimensional, the discriminant validity is used. The square root of the average variance extracted (AVE), which provides a measure of the variance shared between a construct and its indicators for each construct is evaluated (Fornell and Larcker, 1981; Chin, 1998). These foregoing authors recommend AVE values of at least 0.50 and that the square root of AVE should be larger than off-diagonal elements (i.e., load highly on the measure it is intended to measure). The results in Table 3 indicate that in no case was any correlation between the constructs greater than the squared root of AVE (the leading diagonal). This suggests that the measures used in this study are distinct and unidimensional. Clearly, the convergent and discriminant validity of the study's data are psychometrically adequate.

Assessing the Structural Model

The paths coefficients (β) and the R^2 were generated by PLSGraph 3.0 and are shown in Figure 2. The constructs in the research model accounted for 34% of the variation in the model. Of note, the R^2 results compare with that of similar studies (see e.g., Gefen and Straub, 2000). The test of significance for all the paths was conducted using the bootstrap resampling procedure with 200 resamples. The results reveal no significant relationship exists between *Ease of finding* and *Perceived usefulness* (β = 0.205) (Figure 2). On the other hand, there was a strong relationship between the same construct, *Ease of finding* and *Perceived ease of use* (β = 0.173). The results reveal that *Computer anxiety* is positively related to both *Perceived usefulness* (**β = 0.399**). The data also show that *Perceived ease of use* has a positive effect on *Perceived usefulness* (**β = 0.368). All the exogenous factors account for 28**% of the variation of *Perceived ease of use*. Similarly, 66% of the variation in the *Perceived usefulness* construct is accounted for by **the exogenous factors and** *Perceived ease of use.*

Further, PLS Graph 3.0 shows that *Perceived usefulness* has no effect on *Usage* (**β = 0.060**); on the contrary, the other main variable in the TAM,

Figure 3. The Results of PLS graph 3.0 analysis

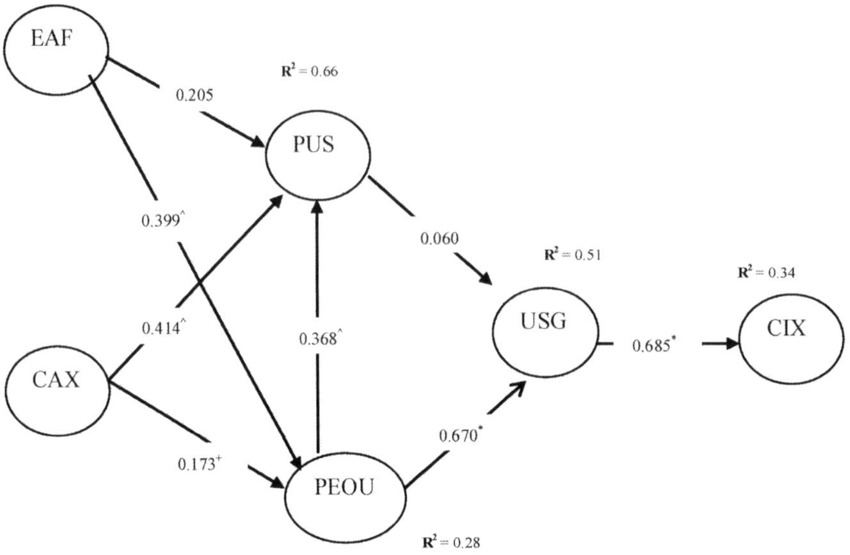

Legend:

*1) + = significant at 0.10 level, ^ significant at 0.05 level, * significant at 0.001 level.*

2) Bold arrows show the significant paths originating from the specified antecedents and terminating in the dependent variable

Perceived ease of use is shown to have a strong effect on *Usage* (**β = 0.670). Furthermore,** *Usage* has a strong positive effect on *Continuance intention* (**β = 0.685). The preceding constructs, including the exogenous factors,** *Perceived ease of use*, and *Perceived usefulness* **explain 51% of the variation in WebCT** *Usage*. Together, the exogenous and endogenous variables account for 34% of the variation in the hypothetical model.

DISCUSSIONS AND CONCLUSION

In this research, a hypothetical model was developed to examine the influence of two external; factors on the continuance intention of using a popular CMS: WebCT. The study was conducted in a part of world - Estonian Europe - where studies of this nature have not been featured in the IS literature. Essentially, the study aims to examine causality among the factors in the proposed model. The choice of the selected factors included *Ease of finding*, *Computer anxiety* and *Perceived usefulness* as informed by their relevance and prominence in the IS acceptance literature. It is acknowledged that the chosen factors are offered as illustrative rather than exhaustive examples. Eight hypotheses were developed to test the structural model. The data provides support for all the eight hypotheses, but one (i.e. H6).

As predicted, the results showed that *Ease of finding* is positively related to *Perceived ease of use.* This suggests that when students are able to navigate through WebCT's contents, their perception with regard to the ease of using such a system increases. From the results, it can be seen that when students have a high understanding of WebCT's contents, their perception of the system's usefulness tends to be high as well. These findings are consistent with others who have examined CMS acceptance among students (Brown, 2002; Ifinedo, 2006; 2007a). Further, the data also show that when students' uneasiness and fear of using ICT, including computers and CMS is low, they tend to have higher perceptions regarding the ease of using and the usefulness of such systems. Brown (2002) supports this viewpoint whereas Pan et al. (2003) do not.

Consistent with the majority of results in the literature about the causal relationships between the *Perceived ease of use* and *Perceived usefulness*, this study demonstrated that *Perceived ease of use* is a mediating variable between *Perceived usefulness* and *Usage.* This can be interpreted to mean that when students are upbeat about using WebCT and similar technologies, they discover such systems are easy to use and subsequently the students can be expected to derive benefits from their efforts. However, this interpretation is open to debate. One may argue that students do not have a choice in how they adopt new technologies. The use of ICT, including e-learning tools, is usually not offered as a voluntary opportunity. More often than not, students are told how they will learn. Nevertheless, this result supports this aspect of the TAM, and adds to the body of knowledge in this field. In other words, with respect to IS acceptance to research elements in Estonia (an emerging economy in Eastern Europe), the relationship between these two constructs is maintained. With regard to the core of the TAM, this study suggests the *Perceived ease of use* of WebCT among Estonian college students is strongly related to *Usage,* whereas the effect of *Perceived usefulness* on *Usage* is unsupported. One possible explanation for this could be attributable to the non-voluntary nature of ICT adoption and acceptance in college settings. Another plausible explanation relates to the impact of contextual influences. Straub et al. (2000) suggested that some aspects of the TAM, i.e., *Perceived usefulness,* may be more important for IS acceptance in the developed West. Along a similar line of reasoning, other researchers, including Anandarajan et al (2002) and Brown (2002), have found evidence in support of *Perceived ease of use* as a significant predictor of use (usage) for ICT products in developing countries. (Recall Estonia is an emerging economy in Eastern Europe). Rather than being taken as an inconsistent result, the analysis might be affirming a contextual reality. Further, the empirical data may be suggesting that when the use of WebCT is high, the continuance intention of use will be high. This information corroborates the findings in Limayem et al. (2003), Sørebø (2004), and Ifinedo (2006; 2007b).

The findings in the study also suggests to university administrators in the context of this research to procure funding for e-learning tools or technologies that students find easy to navigate (i.e., find information). When this is possible, students would use the system without effort and may continue to use it in the future. By the same token, administrators should guarantee that the implementation of such systems is done in manners to ensure that anxieties and fears among students are alleviated. The data analysis indicates the continuance intention of WebCT will be higher in such instances.

The study also has implications for research: 1) This endeavour is among the few to propose a model of IS acceptance that goes beyond the initial usage phase, 2) The proposed framework could stimulate future inquiries as well as generate new leads for IS acceptance studies, 3) This research offers insights about CMS adoption and use from the perspective of students in a region that has not featured prominently in the literature, 4) The findings of this study lend credence to other observations in the literature.

This study has its share of limitations. The research used a convenient sample size of 72; a larger sample size would yield a more robust data analysis and consequently more insightful results. This, however is not a major concern as the study used the PLS approach for data analysis.

The selection of the research elements could be improved. For example, the study did not separate the students into broad categories; those from the "sciences" and "arts/humanities" separated into different groups. More useful insights may have emerged from individuals in different disciplines. For example, it is possible that views espoused by Engineering students are different those studying Philosophy. Importantly, the sample may not be representative of all college students in the country. The measurement of WebCT usage was self-reported; and this might limit results. Actual usage of the system may offer better information (see Legris et al., 2003). This study is a cross-sectional study; a longitudinal study may be illuminating. As a result, the interpretations of the study's findings should be evaluated with care. To improve the generalizability of this effort, future studies could increase the sample size as well as incorporate the impacts of other relevant variables, including peer-pressure, age, gender, and facilitating conditions.

REFERENCES

Anandarajan, M., Igbaria, M. and Anakwe, U. (2002). IT Acceptance in a Less-developed Country: A Motivational Factor Perspective, *International Journal of Information Management, 22,* 47-65.

Bandura, A. (1997). Self-efficacy: The exercise of control. New York: W.H. Freeman.

Bhattacherjee, A. (2001) Understanding Information Systems Continuance: An Expectation-Confirmation Model, *MIS Quarterly, 25*(3), 351-370.

Brown, I. (2002). Individual and Technological Factors Affecting Perceived Ease of Use of Web-based Learning Technologies in a Developing Country, *Electronic Journal of Information Systems in Developing Countries, 9*(5), 1-15.

Brislin, R. (1986). "The Wording and Translation of Research Instruments." In W.J. Lonner and J.W. Berry (eds.), *Fields Methods in Cross-Cultural Research.* Newbury Park, CA: Sage.

Chin, W. (1998). Issues and opinion on Structural Equation Modeling, *MIS Quarterly, 22*(1), vii-xvi.

Compeau, D. and Higgins, C. A. (1995). Computer Self-efficacy: Development of a Measure and Initial Test, *MIS Quarterly, 19*(2), 189-212.

Davis, F. D. (1989). Perceived Usefulness, Perceived Ease of Use, and User Acceptance of Information T Technology, *MIS Quarterly, 13*(3), 319-339.

Dillon, A., and Morris, M. G. (1996). User Acceptance of Information Technology: Theories and Models, *Annual Review of Information Science and Technology, 31,* 3-32.

Estonian eUniversity (2004a). The UNIVe Project. Retrieved September 10, 2004, from http://www.e-uni.ee/Minerva/

Estonian eUniversity (2004b). Estonian e-University NEEDS Analysis. Retrieved May 5, 2004, from http://www.e-uni.ee/doc/uuring/E-ylikool_vajaduste_analyys_mp2.pdf

Fishbein, M. and Ajzen, I. (1975). Belief, Attitude, Intention, and Behavior: An Introduction to Theory and Research. Reading, Mass: Addison-Wesley.

Fornell, C. and Larcker, D. F. (1981). Evaluating Structural Equations Models with Unobservable Variables and Measurement Error, *Journal of Marketing Research, 18*, 39-50.

Gefen, D. and Straub, D. (2000). The Relative Importance of Perceived Ease of Use in the IS Adoption: A Study of E-Commerce Adoption, *Journal of Association for Information Systems, 1*(8), 1-28.

Hara, N. and Kling, R. (1999). Students' Frustrations with a Web-Based Distance Education, *First Monday, 4*(12). Retrieved March 3, 2004, from http://www.firstmonday.org/issues/issue4_12/hara/index.html

Hsu, M-H, Yen, C-H, Chiu, C-M and Chang, C-M (2006). A Longitudinal Investigation of Continued Online Shopping Behavior: An Extension of the Theory of Planned Behavior, *International Journal of Human-Computer Studies, 64*(9), 889-904.

Hu, P. J., Chau, P. Y. K., Sheng, O. R. L. and Tam, K. Y. (1999). Examining the Technology Acceptance Model Using Physician Acceptance of Telemedicine Technology, *Journal of Management Information Systems, 16*(2), 91-112.

Ifinedo, P. (2005a, September 12-15). E-learning Technology Adoption Factors in an Eastern European Country: An Exploratory Study, *Proceedings of the 9th East-European Conference on Advances in Databases and Information Systems*. Tallinn, Estonia.

Ifinedo, P. (2005b). Uncertainties and Risks in the Implementation of e-learning Information Systems Project in a Higher Learning Environment: Viewpoints from Estonia, *Journal of Information and Knowledge Management, 4*(1), 37-46.

Ifinedo, P. (2006). Acceptance and Continuance Intention of Web-based Learning Technologies (WLT) Use Among University Students in a Baltic Country, *Electronic Journal of Information Systems in Developing Countries, 23*(6), 1-20.

Ifinedo, P. (2007a). Investigating the Antecedents of Continuance Intention of Course Management Systems Use among Estonian Undergraduates, *International Journal of Information and Communication Technology Education, 3*(4), 76-92.

Ifinedo, P. (2007b). E-learning in the Nigerian Higher Education Sector: Opportunities and Challenges, *The African Symposium: An Online Journal of African Educational Research Network, 7*(2), 48-54.

Igbaria, M. and Parasuraman, S. (1989). A Path Analytic Study of Individual Characteristics, Computer Anxiety, and Attitudes toward Microcomputers, *Journal of Management, 15*(3), 373-388.

Igbaria, M. (1990). End-User Computing Effectiveness: A Structural Equation Model, *Omega, 18*(6), 637-652.

Keller, C. and Cernerud, L. (2002). Students' Perceptions of E-learning in University Education, *Journal of Educational Media, 27*(1/2), 55-67.

Lee, C., Witta, E. L. (2001). Online Students' Perceived self-efficacy: Does It Change? *Annual meeting of the Association for Educational Communications and Technology (AECT)*, Atlanta, GA.

Lee, J, Cho, H., Gay, G., Davidson, B. and Ingraffea, A. (2003). Technology Acceptance and Social Networking in Distance Learning, *Educational Technology & Society, 6*(2), 50-61.

Lederer, A., Maupin, D., Sena, M. and Zhuang, Y. (2000). The Technology Acceptance Model and the World Wide Web, *Decision Support Systems, 29*(3), 269-282.

Legris, P., Ingham, J. and Collerette, P. (2003). Why Do People Use Information Technology?: A Critical Review of the Technology Acceptance Model, *Information and Management, 40*(3), 191-204.

Leidner, D. E. and Jarvenpaa, S. L. (1993). The Information Age Confronts Education: Case Studies on Electronic Classrooms, *Information Systems Research, 4*(1), 24-54.

Limayem, M., Cheung, C. M. K. and Chan, G. W. W. (2003). Explaining Information Systems

Adoption and Post-Adoption: Towards An Integrative Model. In *Proceedings of 24th International Conference on Information Systems.*

Liu, C., Lu, J. and Yu, C. (2003). Learning Style, Learning Patterns, and Learning Performance in a WebCT-based MIS course, *Information and Management, 40*(6), 497-507.

Lu, H-P, Liu, S-H. and Liao, H-L. (2005). Factors Influencing the Adoption of E-learning Websites: An Empirical Study, *Issues in Information, 6*(1), 190-196.

Morris, M.G. and Turner, J. M. (2001). Assessing Users' Subjective Quality of Experience with the World Wide Web: An Exploratory Examination of Temporal Changes in Technology Acceptance, *International Journal of Human-Computer Studies, 54*(6), 877-901.

Morss, D. A. (1999). A Study of Student Perspectives on Web-based Learning: WebCT in the Classroom, *Internet Research: Electronic Network Applications and Policy, 9*(5), 393-408.

Neuman, W. L. (1997). *Social Research Method.* London: Allyn and Bacon.

Ngai, E. W. T., Poon, J. K. L. and Chan, Y. H. C. (2006). Empirical Examination of the Adoption of WebCT using TAM, *Computer and Education, 48*(2), 250-267.

Nunnally, J. C. (1978). *Psychometric Theory.* New York, NY: McGraw-Hill.

Pan, C., Siva, S. and Brophy, J. (2003). Students' Attitude in a Web-enhanced Hybrid Course: A Structural Equation Modeling Inquiry, *Journal of Educational Media and Library Sciences, 41*(2), 181-194.

Roca, J. C., Chiu, C-M. and Martínez, F. J. (2006). Understanding e-learning Continuance Intention: An extension of the Technology Acceptance Model, *International Journal of Human-Computer Studies, 64*(8), 683-696.

Straub, D., Limayem, M. and Karahanna, E. (1995). Measuring System Usage: Implications for IS Theory Testing, *Management Science, 41*(8), 1328-1342.

Szajna, B. (1996). Empirical Evaluation of the Revised Technology Acceptance Model, *Management Science, 42*(1), 85-92.

Sørebø, A. M. (2004). A replication of the Post-Acceptance Model in the context of E-Learning, *Proceedings of 2004 IRMA International Conference.* Louisiana, USA.

Steer, D., Turner, P. and Spencer, S. (2000). Issues in Adapting the Technology Acceptance Model (TAM) to Investigate Non-workplace Usage Behavior on the World-Wide-Web, School of Information Systems, Working Chapter, University of Tasmania.

Tavangarian, D. Leypold, M. E., Nölting, K., Röser, K. M. and Voigt, D. (2004). Is E-Learning the Solution for Individual Learning?, *Electronic Journal of E-Learning, 2*(2), 273-280.

Tiger Leap Foundation (1997) Retrieved March 9, 2007, from http://www.tiigrihype.ee/?op=&id=

Venkatesh, V. and. Davis, F. D (2000) A Theoretical Extension of the Technology Acceptance Model: Four Longitudinal Field Studies. *Management Science, 46*(2), 186-204.

APPENDIX: THE RESEARCH QUESTION

Perceived ease of use (PEOU)	Strongly Disagree 1	Disagree 2	Somewhat disagree 3	Neutral 4	Somewhat Agree 5	Agree 6	Strongly Agree 7
1. WebCT is easy to use 2. WebCT is easy to learn 3. WebCT is user friendly 4. WebCT is easy to master							
Perceived usefulness (PUS)	Strongly Disagree 1	Disagree 2	Somewhat disagree 3	Neutral 4	Somewhat Agree 5	Agree 6	Strongly Agree 7
1. WebCT is useful for my studies 2. WebCT usages improves my academic performance 3. WebCT makes my studying easier.							
Ease of finding (EAF)	Strongly Disagree 1	Disagree 2	Somewhat disagree 3	Neutral 4	Somewhat Agree 5	Agree 6	Strongly Agree 7
1. WebCT allows easy return to previous display pages. 2. I can determine my position within the WebCT program. 3. WebCT is easy to navigate.							
Computer Anxiety (CAX) (Reversed coding)	Strongly Disagree 1	Disagree 2	Somewhat disagree 3	Neutral 4	Somewhat Agree 5	Agree 6	Strongly Agree 7
1. Working with a computer makes me nervous. 2. Computers make me feel uncomfortable. 3. Computers make me feel uneasy. 4. Computers scare me.							

Please answer the following with regard to your WebCT use.

Intention to use (USG1)	1	2	3	4	5	6
On an average working day that you use WebCT, how much time do you spend on the system?	almost never	< ½ hr	½ - 1 hr	1-2 hrs	2-3 hrs	> 3 hrs
(USG2) On average (for the period that you were using WebCT), how frequently do you use it?	1 once a month	2 a few times a month	3 a few times a week	4 about once a day	5 several times a day	

Continuance Intention (CIX)	Strongly Disagree 1	Disagree 2	Somewhat disagree 3	Neutral 4	Somewhat Agree 5	Agree 6	Strongly Agree 7
1. I intend to continue to using WebCT rather than discontinue its use							
2. My intentions are to continue my use of WebCT rather than use alternative means							

Demographic information

Faculty/department of study:______________________________

Year of study:____

Please tick your appropriate *age* group box:

☐ ≤ 25 years ☐ 26 – 39 years ☐ 40 - 55 years ☐ 56 – 67 years

What is your gender?: ☐ male ☐ female

Chapter IV
The Influence of Constructivist E-Learning System on Student Learning Outcomes

Thanakorn Wangpipatwong
King Mongkut's University of Technology Thonburi, Thailand

Borworn Papasratorn
King Mongkut's University of Technology Thonburi, Thailand

ABSTRACT

In this article, the study of how a constructivist e-learning system affects students' learning outcomes was explored and a two-phase study was designed. The first study sought to create a constructivist e-learning environment (CEE) and discover how students expected their learning outcomes under CEE. CEE is composed of three constructs, which are exploration, collaboration, and construction. The statistical results showed the high level of student expectation on every construct. Consequently, constructivist e-learning system (CES) was developed. In the second study, CES was used in the actual classroom environment. The purpose was to compare the learning outcomes and knowledge development of students who studied the course using CES with those of students who learned it under a traditional learning environment. A T-test method was used to analyze the learning outcomes. The results showed that students who used CES had better learning outcomes and knowledge development than students who did not use CES.

INTRODUCTION

E-learning refers to an alternative method of teaching and learning using all electronic media, including the Internet, network, audio/video tape, and CD-ROM. For the past few years, the use of e-learning to conduct teaching and learning in educational institutes has rapidly increased along with the development of information technology. E-learning has an advantage of enabling students to learn from anywhere and at anytime. E-learning also provides a one-stop service for teachers and learners in order to create and deliver educational content quickly, effectively, and economically (Ong, Lai, & Wang, 2004).

In the past, e-learning researches have focused more on learning objects. The IEEE Learning Technology Standards Committee defines learning objects as "any entity, digital or non-digital, which can be used, re-used or referenced during technology-supported learning" (Shepherd, 2006). However, students may not improve learning outcomes by using only learning objects. Recently, the focus of e-learning has shifted from defining, sharing, and reusing learning objects to emphasizing learning activities based on the concept of learning design, which emerges as one of the most significant recent developments in e-learning (Yu, Zhang, & Chen, 2006). The key principle in learning design is to develop learning activities that are performed by different learners in the context of learning (Koper, 2006). The objective of learning design is also to assist students to effectively learn by creating and managing their learning processes (Pan & Hawryszkiewycz, 2004).

Regarding learning design, e-learning has blended with learning theory in order to try to improve learning outcomes. The learning methods, such as independent learning, active learning, self-directed learning, problem-based education, simulations, and work-based learning are based on constructivist learning theory (Reiser, 2001). The e-learning design in this study, therefore, is based on constructivist learning theory.

This study was divided into two phases. The first phase was to create constructivist e-learning environment (CEE) and to explore the expected student e-learning outcomes on CEE. The objective of this step was to investigate what learning outcomes students might expect by using a statistical method. Consequently, based on CEE, the constructivist e-learning system (CES) was developed. The second phase was designed as an experimental research that focused on the comparison of actual student e-learning outcomes. The CES was used in the classroom environment. The student learning outcomes between students in traditional classroom environment and constructivist e-learning environment were compared. The statistical method (t-test) was also conducted to test the significance in this study.

REVIEW OF LITERATURE

Constructivist Learning Theory

The constructivist learning theory has emerged as a prominent approach to teaching during the past decade. The research of Dewey, Piaget, Vygotsky, and Jonassen, among others, provides historical precedents for constructivist learning theory. Constructivist learning theory represents a paradigm shift from education based on behaviorist theory to education based on cognitive theory. In a constructivist learning environment, students have better learning outcomes than in traditional learning environment (Parker & Becker, 2003, Tynjala, 1999).

Among many definitions of constructivist learning theory, the most common characteristic is that they all focus on activities and environments rather than on learning objects. Knowledge is constructed by learners and not transmitted by an instructor. Dewey (1938) believes that knowledge emerges only from situations in which learners have to draw them out of meaningful experiences. Piaget (1960) indicates that learners are active and constructive in making sense of their environment. Piaget (1975) believes that learning should be attained through well-defined stages by active participation of a learner. Vygotsky (1978) focused more on learning activities. In addition, Jonassen (1994) suggested that the constructivist learning should emphasize less on the sequence of instruction and emphasize more on the design of the learning environment. He also pointed out that constructivist environments stress situated problem solving tasks. In conclusion, constructivist learning is an educational approach that effectively motivates learners by enabling a more active, explorative and interactive learning process. In other words, through the learning process, learners construct knowledge within a constructivist learning environment.

CONSTRUCTIVIST E-LEARNING ENVIRONMENT (CEE)

Constructivist learning is considered to be the ideal pedagogy for e-learning. First, constructivist learning focuses on a student's learning experience rather than an instructor lead teaching method. In an e-learning environment, an instructor's role is to help students develop their knowledge and give students a degree of choice such as what to study, where to study, and how to study. Students are placed at the center of the learning experience. Second, constructivist learning sees students as an active participant in their learning experience rather than a passive participant. In an e-learning environment, context is also an important part for student learning. E-learning forces students to explore information, make connections, and build knowledge. Finally, constructivist learning sees learning as a social experience.. E-learning easily enables communication among students without the barriers of time and place; collaboration is crucial.

In recent literature, the following constructivist e-learning researches can be found. Chuang and Tsai (2004) studied on the preferences toward the constructivist internet-based learning environments. Zhang, Zhou, Briggs, & Nunamaker (2005) studied the influence of interactive video in constructivist e-learning environment. Zualkeman (2006) designed the framework for developing authentic constructivist e-learning environments using game-based learning as a medium. Moreno, Gonzalez, Castilla, et al., (2006) applied constructivist e-learning to a computer architecture and engineering course using Moodle platform. Puntambekar (2006) developed constructivist, distributed learning environment (CoDE) and used an online graduate course in order to study the process of collaboration.

For constructivist e-learning applications, computer-supported collaborative learning (CSCL) is widely studied. Many systems have been developed based on CSCL and considered to be constructivist learning applications such as CSILE (computer supported intentional learning environments), which functions as a collaborative learning environment (Scardamalia & Bereiter, 1994). CoVis (collaborative visualization) is an integrated software environment that incorporates visualization tools for open-ended inquiry (Edelson, Pea & Gomez, 1996).

Puntambekar (2006) designed the constructivist e-learning environment called CoDE. CoDE uses two main cognitive tools to help students construct knowledge. The first tool is called Reflective Notebooks. Reflective Notebook was designed to help students analyze ideas and write a reflective essay into the system. The other tool is Discussion Tool, which is designed to support group working for students.

Figure 1. Constructivist e-learning environment (CEE)

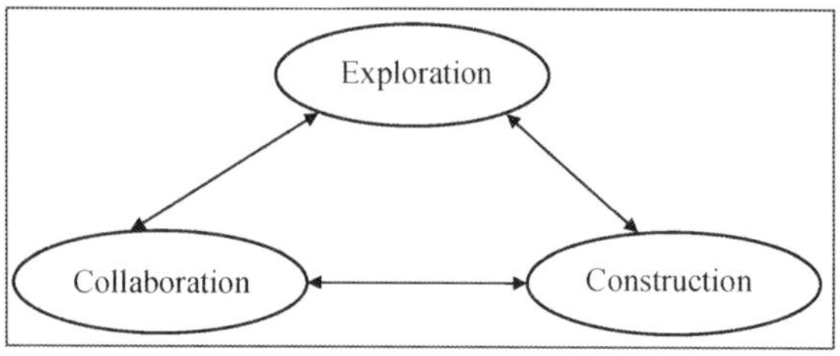

This research proposes an alternative design of learning environment called CEE. CEE consists of three constructs which are exploration, collaboration, and construction, as shown in Figure 1.

Exploration is one of the main approaches in the constructivist learning theory (Murphy, 1997, Stager, 2001). An exploration activity can be defined as searching information resources to comprehend the information and to acquire knowledge (Kashihara, Kinshuk, Oppermann, et al, 1998). The exploration activities include search, variation, experimentation, play, flexibility, discovery, or innovation (March, 1991). In a real world situation, students prefer to learn by exploration in the context of a real task. They need to perform, rather than taking time out to work through the documentation in a task-independent manner (Rieman, 1996). Exploration is also a favored approach to encourage students to seek knowledge independently and to manage the pursuit of their goals (Murphy, 1997). Knowledge that is generated by exploration activities is often new knowledge (Katila, 2002). To help in the exploration process, instructors can use hyperlinks to link other useful online resources to allow students to journey on a process of discovery. Search engine is also a knowledge retrieving tool that assists students to construct new knowledge (Liaw, 2005).

Collaboration is also considered to be a key feature of constructivist learning theory. Collaboration is a characteristic of a powerful learning environment which results in active construction of knowledge (Van Merrienboer & Pass, 2003). Through a process of collaboration, students have an active and constructive role in the learning environment (Dewiyanti, Brand-Gruwel, Jochems, et al., 2004). The interaction between individual and collaborative learning activities is part of knowledge construction (Puntambekar, 2006, Yli-Renko, Autio, & Sapienza, 2001) and contributes to higher learning performance in the learning environment (Wang & Newlin, 2000).

Writing is suitable for tasks where the aim is to foster understanding, change, and develop student thinking skills (Tynjala, 1999). Constructivist learning theory encourages student writing activities in order to reflect on student knowledge development. Therefore, the construction component, in this study, is designed for writing or recording students' reflection.

Reflective learning is another main characteristic of the constructivist learning theory (Jonassen, 1994). Reflection improves the knowledge creating potential of all students. When students have an opportunity to discuss and explain, they improve their learning (Pirolli & Recker, 1994). In addition, reflection helps students establish the linkage between theory, research, observations, and experiences (George, 2001).

STUDY 1: EXPECTED E-LEARNING OUTCOMES ON CEE

Hypotheses

The objective of this study is to understand the student's expectation on their learning outcomes under each component of constructivist e-learning environment. The hypotheses applied to the study 1 are:

H1: The average value of expected e-learning outcomes on collaboration is greater than 3

H2: The average value of expected e-learning outcomes on exploration is greater than 3

H3: The average value of expected e-learning outcomes on construction is greater than 3

Methodology

Participants

Participants were students of an introduction to computer course at Bangkok University. At the beginning of the semester, a random sample of 600 students out of the total population of 4200 received an e-mail that described the study and provided a link to where the questionnaire could be completed. Four hundred sixty three students (77.17%) responded to the e-mail. The size of valid responses conforms to finite population sampling formula (Yamane, 1973), along with a 95 percent confidence level and a 5 percent precision level. The demographic data of these participants are shown in Table 1.

Instruments

The data for this study were gathered by means of a questionnaire. The questionnaire included four major sections: (a) demographic information, (b) expected outcomes from collaboration process, (c) expected outcomes from exploration process, and (d) expected outcomes from construction process.

The participants were asked to determine the range of expected learning outcomes from each construct by indicating 1 as Very Low level; 2 as Low; 3 as Moderate; 4 as High; and 5 as Very High. Mean rating points were then distributed to the following scale classification levels such as Very Low (1-1.80); Low (1.81-2.60); Moderate (2.61-3.40); High (3.41 –4.20); and Very High (4.21 to 5.00). Findings from the questionnaires could

Table 1. Demographic data of participants in study 1

Demographic Characteristics	Frequency	Valid Percent (%)
Gender		
Male	169	36.5
Female	294	63.5
Age		
17	1	0.2
18	122	26.3
19	227	49.0
20	73	15.8
21	19	4.1
22	12	2.6
23	7	1.5
24	2	0.4
Faculty		
Accounting	5	1.1
Business Administration	175	37.8
Communication Arts	235	50.8
Economics	1	.2
Fine and Applied Arts	3	.6
Humanities	3	.6
Laws	35	7.6
Science	6	1.3
Number of Participants	**463**	**100**

Table 2. Cronbach's alpha reliability coefficients

Construct	Cronbach's Alpha	Number of Items
Collaboration	.883	6
Exploration	.781	2
Construction	.853	3

Table 3. Means of expected learning outcomes

	N	Mean	Std. Deviation
Collaboration	463	3.486	.752
Exploration	463	3.893	.886
Construction	463	3.647	.796

determine the current level of expected learning outcome from CEE.

Reliability Assessment

The internal consistency reliability of items in each construct was examined using Cronbach's alpha to confirm the adequacy of the measures for testing the hypotheses. The results in each construct are presented in table 2. The reliability of all constructs shows a high level of internal consistency above the recommended minimum level of .70.

RESULTS

Table 3 shows the average expectation of student learning outcomes for each construct. Referring to the hypothesis, the mean value of collaboration construct, exploration construct, and construction construct are 3.486, 3.893, and 3.647, respectively. The results show that the level of student expected learning outcomes is high for all constructs.

One sample t-test was used to test the hypotheses. Table 4 shows the results which indicate that the mean value of collaboration construct is significantly greater than 3 at significant level 0.05. The mean value of exploration and construction constructs are also significantly greater than 3 at significant level 0.001. Therefore, hypothesis 1, 2, and 3 are supported.

STUDY 2: THE COMPARISON OF E-LEARNING OUTCOMES

Research Model and Hypotheses

The results form study 1 show that the level of expected learning outcomes on each construct of CEE was high. This result could imply that students believed the CEE constructs could help them develop their knowledge. Therefore, the constructivist e-learning system (CES) was developed based on CEE as shown in Figure 2.

In this study, CES was used to create the constructivist e-learning environments. The

Table 4. Results of one sample t-test

	Test Value = 3.0		
	t	Sig. (2-tailed)	Mean Difference
Collaboration	2.459	.014*	.086
Exploration	11.979	.000***	.493
Construction	6.661	.000***	.247

$^{*}p<0.05$, $^{***}p<0.001$

Figure 2. Constructivist e-learning system (CES)

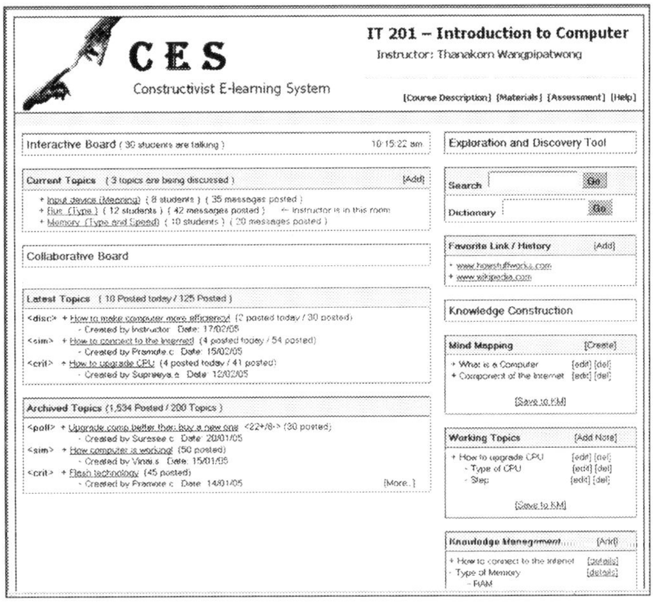

objective was to compare learning outcomes between students who used the CES and students who studied in the traditional learning environment. This study also examined the knowledge development between students who used the CES and students who studied in traditional learning environments. The subject used in the experiment is "Introduction to Computer," which is normally taught in the traditional classroom environment. This subject introduces basic concepts of computer and internet technologies. An instructor in the class was prepared to use CES as a tool, integrating with constructivist learning activities. The instructor was also prepared to understand the concept of constructivist learning theory and how to use the activity in the classroom. The same instructor taught the same content as in traditional classroom environment. The duration of the lecture in both classes was 13 weeks.

The hypotheses applied to Study 2 are:

H4: Students who use the CES in traditional face-to-face classroom environment will achieve better test scores than students who do not use it in traditional face-to-face classroom environment.

H5: Students who use the CES in traditional face-to-face classroom environment will achieve better knowledge development than students who do not use it in traditional face-to-face classroom environment.

Methodology

Participants

Two class sections were randomly selected for the research experiment—Section T and C. Section T was taught in traditional learning environment. Section C was taught in constructivist e-learning environment using CES as a tool. Both sections had the same amount of time for lectures. Students who enrolled in the course had no prior knowledge about the teaching method. There were 31 students in section T and 28 students in section C. The demographic data of students in both classes is shown in Table 5.

The average GPA of students from both groups are shown in Table 6. Although the average GPA values from each group are slightly different, the results from statistical analysis (t-test) displays no significance (p=.434). Therefore, GPA was considered not to be effective to the research experiment.

Instrument

The independent sample t-test analysis was used to compare the mean values of test results between two groups. There were five tests in this study—test 1, 2, 3, 4, and 5. Test 1 was designed as a pre-test and was administered at the beginning of the semester to measure student prior knowledge.

Table 5. Demographic data of participants in study 2

		Section	
		T	C
Year	1	31	28
Gender	Male	14	12
	Female	17	16
Age	17	0	1
	18	9	10
	19	16	15
	20	5	2
	21	1	0
Total		31	28

Table 6. Independent sample t-test of GPA

	Section T (Mean)	Section C (Mean)	t-value	Sig. (2-tailed)
GPA	2.38	2.50	-.788	.434

Table 7. Description and schedule for all tests

	Description	Schedule
Test 1 (Pre-test)	General information about computer and information technology	Before class started
Test 2	MS Word and MS Excel	One week before midterm exam
Test 3	Fundamentals of computer system, process concept, and storage devices	Midterm exam
Test 4 (Post-test)	General information about computer and information technology (more difficult than test 1)	Last week before final exam
Test 5	Introduction to computer network, the Internet, Web design and development	Final exam

Table 8. Independent sample t-test between section T and section C

	Section T (Mean)	Section C (Mean)	t-value	Sig. (2-tailed)
Test 1	55.484	58.571	-.796	.429
Test 2	69.301	83.334	-2.157	.036*
Test 3	66.258	70.214	-1.159	.251
Test 4	67.903	80.357	-3.522	.001*
Test 5	58.839	58.929	-0.31	.976

*$p<0.05$

Table 9. Paired sample t-test of test 1 (pre-test) and test 4 (post-test)

Pair T-test Between Test1 and Test4 of	Mean Paired Differences	Standard Deviation	t-value	Sig.(2-tailed)
Section T	-12.419	14.712	-4.700	.000***
Section C	-20.536	9.461	-11.485	.000***

***$p<0.001$

The other four tests were administered in order to measure the learning outcomes. Since the class had two sessions, test 2 was taken at the end of the first session before the midterm examination date. Test 3 was a midterm examination. Test 4 was designed as a post-test and was taken at the end of last session before final examination date. Finally, test 5 was a final examination. The test 3 and test 5 schedules were pre-announced to students before the test date. However, students did not know the test date of test 2 and test 4 before they took the test. The potential test scores ranged from 0 to 100. All tests were closed book and closed notes.

Test 1 and test 4 covered the general information about computer and information technology. The question types in both tests were similar, but questions in test 4 were more specific and difficult. The questions in test 2 were about how to use Microsoft Word and Microsoft Excel as they were part of the learning objective in the lecture content. Test 3 and 5 covered all the lecture contents of midterm and final, respectively. Table 7 summarizes all test descriptions and schedules.

Results

Table 8 displays a mean score for each test and section. Notice that all mean scores from section C are higher than the mean scores from section T, which means hypothesis 4 is supported. However, regarding t-test analysis, the mean score of section T from test 1 is not significantly different to the mean score of section C ($p=.429$). Because test 1 is a pre-test, it can be concluded that student backgrounds in computer and information technology of both sections are not different.

An independent-sample t-test was also used to determine if there were any significant mean differences between class sections from test 2 to test 5. As in Table 8, the results showed that test 2 and test 4 are significantly different ($p<.05$). For test 3 and test 5, despite the higher score on section C, the results show no significant difference between section T and C.

Considering the development of student knowledge from pre-test (test 1) to post-test (test 4), the results show significant improvement for both sections ($p<.001$), as shown in table 9. However, the mean difference of student improvement in section C is much higher. Therefore, hypothesis 5 is supported.

DISCUSSION AND CONCLUSION

This study proposed an alternative design of constructivist e-learning. The two-phase study was aimed at examining the extent of constructivist practices when applied to traditional classroom environment using constructivist e-learning system. According to the literature review, the constructivist e-learning environment (CEE) was designed. CEE is composed of three constructs, which are collaboration, exploration, and construction. Study 1 was conducted in order to evaluate what students expected from each CEE construct toward learning outcomes. The results from study 1 showed that students' expected learning outcomes from each CEE construct were significantly high. In addition, exploration construct had the highest mean scores. This implies that students believe most of their learning outcomes are from the exploration process. Possibly, the new knowledge gained from the exploration process is more obvious than the others. For example, when students needed information in the Internet, they searched and found it. They would feel immediately that they gained new knowledge. Collaboration construct, by the way, has the lowest effect on students' expected learning outcomes. This situation occurs because students may not receive the new knowledge every time they collaborate with others.

To strengthen the value of the research results, the constructivist e-learning system (CES) was developed and used in the actual classroom environment. The experimental research was designed in study 2. Two class sections were used in the experiment. According to the t-test analysis, the average GPA of students between sections were not significantly different. Therefore, GPA was not the factor that affected the study results. In other words, it could imply that the learning outcomes from each section occurred from the learning process.

The pre-test was also administered in both sections in order to evaluate student background knowledge in computer and information technology. The statistical findings showed that there were no significant differences of student background between each group. As a result, we could also eliminate student background on computer and information technology from the factor that affected the student learning outcomes.

The findings from study 2 showed that students who used the CES in traditional face-to-face classroom environment achieved better test scores than do students who did not use CES. However, it is interesting that only two tests (test 2 and test 4) had significant mean difference but the other two tests (test 3 and test 5) had not. Notice that the test date of test 3 and test 5 were announced before the test. It could be possible that students in both sections had a better preparation for test 3 and test 5 before the tests, which, therefore, yielded to closed results between section T and section C.

The significant difference of results between test 2 and test 4 would imply that students who learned in constructivist e-learning environment had a better knowledge development than students who learned in traditional learning environment. As stated before, student background knowledge on computer and information technology were not different, and student took the test by not knowing the test date. Therefore, students needed to use their knowledge developed from learning process to answer the tests.

The results from statistical analysis of pre-test and post-test also confirmed the previous findings. The results showed that the mean difference of students in section C was much higher than those in section T, although knowledge improvements of both sections were significant. Therefore, it could be concluded that students who studied in constructivist e-learning environment had a better knowledge development than students who studied in traditional learning environment.

Although the research results supported all hypotheses, many aspects should be discussed. First, we found that students were apparently unfamiliar with the constructivist learning environment. The expectation of knowledge acquisition from an instructor was high. However, students had more understanding on the new role as time went by. In fact, students seemed to be happier studying in the constructivist e-learning environment. Second, we found the same result as Puntambekar (2006) that students expected an instructor to take the lead in every discussion and waited for instructor feedback before they continued to other discussions. Therefore, the instructor should understand the facilitator role and encourage students to continue learning in constructivist e-learning environment.

Concerning to the use of the CES, we also found many interesting aspects. First, using collaboration tools in the CES helped students learn not only from their own group but also from other groups. Since the students were capable of seeing all of the ideas within the CES, they could build knowledge from a broader point of views. Second, students who were familiar with the computer and internet had more fun using and learning by the CES than students who were not. It seemed that they did not have to worry about how to use the CES. They just concentrated on how to study and learn by using the CES. Finally, students seemed to have more focus on studying when using the CES. With a traditional teaching, some students might use the web, or msn when listening to a teacher. These activities did not happen when students used the CES unless they already finished the assignment.

There are some limitations in this study that should be noted. First, the use of student subjects from Bangkok University may limit the generalization of the results. Second, the findings from using the CES in classroom environment were obtained from a single study. In fact, only the Introduction to Computer course is used in the experiment. Therefore, studying CES in other courses may be needed. Caution should be taken before generalizing the findings, as well. Finally, a teacher who intends to use CES should understand the constructivist learning concept in order to utilize the CES tools.

REFERENCES

Chuang, S., & Tsai, C. (2004). Preferences toward the constructivist internet-based learning environments among high school students in taiwan. *Computers in Human Behavior, 21*(2), 255-272.

Dewey, J. (1938). *Experience and education.* New York: Macmillan.

Dewiyanti, S., Brand-Gruwel, S., Jochems, W., & Broers, N. J. (2004). Students' experiences with collaborative learning in asynchronous computer-supported collaborative learning environments. *Computers in Human Behavior, 23*(1), 496-514.

Edelson, D., Pea, R., & Gomez, L. (1996). Constructivism in the collaboratory. In B. G. Wilson (Ed.), *Constructivist learning environments: Case studies in instructional design* (pp. 151-164). Englewood Cliffs, NJ: Educational Technology Publications.

George, S. E. (2001). Learning and the reflective journal in computer science. *Australian Computer Science Communications, 24*(1), 77-86.

Jonassen, D. H. (1994). Thinking technology: Toward a constructivist design model. *Educational Technology, 34*(3), 34-37.

Kashihara A., Kinshuk, Oppermann R., Rashev R., Simm H., & Toyoda J. (1998). Towards sharable intelligent assistance for learning-by-exploration. SIG Technical Notes in Japanese Society for Artificial Intelligence, SIG-IES9703, JSAI, (pp. 1-6)Shizuoka, Japan,.

Katila, R. (2002). New product search overtime: Past ideas in their prime? *Academy of Management Journal, 45*(5), 995-1010.

Koper, R. (2006). Current research in learning design. *Journal of Educational Technology & Society, International Forum of Educational Technology & Society, 9*(1), 13-22.

Liaw, S.-S. (2005). Developing a web assisted knowledge construction system based on the approach of constructivist knowledge analysis of tasks. *Computers in Human Behavior 21*(1), 29-44.

March, J. G. (1991). Exploration and exploitation in organizational learning. *Organization Science, 2*(1), 71-87.

Moreno, L., Gonzalez, C., Castilla, I., Gonzalez, E., & Sigut, J. (2006). Applying a constructivist and collaborative methodological approach in engineering education, computers & Education, in press.

Murphy, E. (1997). *Constructivist checklist.* retrieved August 18, 2005, from http://www.stemnet.nf.ca/~elmurphy/emurphy/cle4.html

Ong, C.-S., Lai, J.-Y., & Wang, Y.-S. (2004). Factors affecting engineers' acceptance of synchronous e-learning systems in high-tech companies. *Information & Management, 1*(6), 795-804.

Pan, W., & Hawryszkiewycz, I. (2004, December 5-8). A method of defining learning processes. In R. tkinson, C. McBeath, D. Jonas-Dwyer, & R. Phillips (Eds.), *Beyond the comfort zone: Proceedings of the 21st ASCILITE Conference* (pp. 734-742). Perth,. http://www.ascilite.org.au/conferences/perth04/procs/pan.html

Parker, J.R., & Becker, K. (2003). Measuring effectiveness of constructivist and behaviourist assigments. In CS102, Proceedings ITiCSE'03 ACM, (pp. 40-44).

Piaget, J. (1960). *The child's conception of the world.* Totowa, NJ: Littlefield, Adams.

Piaget, J. (1975). *The construction of reality in the child.* Ballantine Books.

Pirolli, P., & Recker, M. (1994). Learning strategies and transfer in the domain of programming. *Cognition and Instruction, 12*(3), 235-275.

Puntambekar, S. (2006). Analyzing collaborative interactions: Divergence, shared understanding and construction of knowledge. *Computers & Education, 47*(3) 332-351.

Reiser, R. (2001). A history of instructional design and technology. Part 2: A history of instructional design. *Educational Technology, Research and Development, 49*(2), 57-67.

Rieman, J. (1996). A field study of exploratory learning strategies. *ACM Transactions on computer-Human Interaction, 3*(3), 189-218.

Scardamalia, M., & Bereiter, C. (1994). Computer support for knowledge building communities. *Journal of the Learning Sciences, 3*(3), 265-283.

Shepherd, C. (2006). Objects of Interest. Retrieved from http://www.fastrak-consulting.co.uk/tactix/Features/objects/objects.htm.

Stager, G.S. (2001, July 7-25). *Constructionism as a high-tech intervention strategy for at-risk learners*. National Educational Computing Conference, Building on the Future 1. Chicago, IL

Tynjala, P. (1999). Towards expert knowledge? A comparison between a constructivist and traditional learning environment in the university. *International Journal of Educational Research 31*(5), 357-442.

Van Merrienboer, J. J. G., & Paas, F. (2003). Powerful learning and the many faces of instructional design: Towards a framework for the design of powerful learning environments. In E. De Corte, L. Verschaffel, N. Enstwistle, & J. J. G. Van merrienboer Eds.), *Powerful learning environments: Unravelling basic components and dimensions.* Oxford: Elsevier Science.

Vigotsky, L. S. (1978). Mind in society: The development of the higher psychological processes(Cambridge , MA: Harvard University Press).

Wang, A. Y., & Newlin, M. H. (2000). Characteristics of students who enroll and succeed in psychology web-based courses. *Journal of Educational Psychology, 92*,137-143.

Yamane, T. (1973). *Statistics: An introductory analysis*, (3rd Ed.). New York, NY:Harper & Row.

Yli-Renko, H., Autio, E., & Sapienza, H. J. (2001). Social capital, knowledge acquisition, and knowledge exploitation in young technology-based firms. *Strategic Management Journal, 22*(6-7), 587-613.

Yu, D., Zhang, W., & Chen, X. (2006). New ceneration of e-learning technologies. I*n proceedings of the First International Multi-Symposiums on Computer and Computational Sciences (IMSCCS'06).*

Zhang, D., Zhou, L., Briggs, R. O., & Nunamaker, J. F. Jr. (2005). Instructional video in e-learning: Assessing the impact of interactive video on learning effectiveness. *Information Management, 43* (1), 15-27.

Zualkernan, I. A. (2006). A framework and a methodology for developing authentic constructivist e-Learning environments. *Educational Technology & Society, 9* (2), 198-212.

This work was previously published in International Journal of Information and Communication Technology Education, Vol. 3, Issue 4, edited by L. Tomei, pp. 21-33, copyright 2007 by IGI Publishing (an imprint of IGI Global).

Section I.b

Practice

Chapter V
The Didactical Agency of Information Communication Technologies for Enhanced Education and Learning

Andreas Wiesner-Steiner
University of Applied Sciences Bremen, Germany

Heike Wiesner
Berlin School of Economics, Germany

Heidi Schelhowe
University of Bremen, Germany

Petra Luck
Liverpool Hope University, UK

ABSTRACT

This article presents substantial results from two projects that deal with teaching and learning with digital media in basic and higher education and offers a new perspective on the active role of technology in learning processes. The first case draws on the project "Roberta—girls conquer robotics," which was launched by the Fraunhofer Institute (AIS) with the aim to help promote girls' interest in sciences, mathematics and technology. It suggests a new pedagogical approach towards the use of robotics in education and discusses how didactics and technology (LegoMindstorms) interact and how the character of robotics itself plays an important role here, such as it already comes along as gendered material. The second case focuses on distance education teaching methods in childcare management. The space left for practitioners in Higher Education is either to embrace the new media or to watch its inevitable unfolding. We take a critical stance towards that perspective and suggest that the shape and learning effect of new media in higher education is contested and evolves in communities of practice. No technologies are neutral and it is more appropriate to speak of technological and societal features as interactively fostering e-learning processes through distributed actions (Rammert, 2002).

INTRODUCTION

Informed by a constructivist learning approach and the principles of gender mainstreaming, these two cases draw conclusions towards general educational concepts for digital media. If carefully used as a *didactical actor*, information communication technology not only suit learners' interest in technological messiness but enables them for a technologically mediated life instead of just feeling overwhelmed. Digital media can therefore serve as media for general education in the more comprehensive sense of developing personality, professional identity and agency.

The Didactical Agency of Robotics for Education

"Roberta—girls conquer robotics," a project funded by the German Federal Ministry of Education and Sciences (BMBF), was launched by the Fraunhofer Institute (AIS) with the aim to help promote girls' interest in sciences, mathematics and technology, and especially to encourage girls' curiosity for engineering and computer science (Müllerburg/Petersen/Theidig 2004)[1]. Scientifically escorted by the University of Bremen, Digitale Media in Education (DiMeB) and the Institute for Didactics of Natural Sciences (IDN). Roberta addressed 10-16 year old girls. The projects' basic assumption was that robot construction kits—offering possibilities to develop more self-confidence in one's skills—provide an attractive access to technology for girls. By offering substantial results from the qualitative evaluation of Roberta courses we suggest a new pedagogical approach towards the use of robotics in education.

The robot construction kits (Lego Mindstorm) consist of complementary mechanical, dynamic and electronic parts that allow the construction and programming of different types of robots. Basic models can be equipped with different engines and sensors (contact sensors and optical sensors). The programming can be done in two programming languages (RIS and NQC), the first offering easy to combine graphical blocks, the second requiring more teaching and explanations. The programmes are transmitted on to the RCX module, a programmable Legobrick with 3 input sockets for sensors and 3 for engines. In order to learn about informatics, the teaching of basic programming skills marks an important aim of the.Roberta courses.

While informatics is treated in Roberta as a constructivist science, the educational sciences provide the necessary orientation for both shaping and evaluating digital learning environments. Our evaluation thus focused on the following questions:

- How can the interest of girls and women in technology be triggered by the use of robotics?
- How is curiosity for technology generated?
- How should learning environments be designed in order to satisfy both girls and boys?
- Which didactical concept is appropriate in connection with robotics?
- Are robotics and didactics suitable to influence the self-concept of the students?

Results of the quantitative evaluation show that the course experience in longer Roberta courses are noticeably stronger influenced by the focus of the teacher (didactics, informatics, gender, technology) than in shorter ones (Rethfeld/Schecker 2005). The didactical focus stages as the most positive influence on the experiences of the participants—which is why the importance of the course-concept increases with the length of the courses. Although the self concept of informatics and occupational orientation are only sustainably affected in medium sized and longer courses, all Roberta courses help to develop a more positive attitude towards informatics with the participants—both concerning the self-estimation of their own competence and their occupational orientation (Hartmann/Schecker/Rethfeld 2005). The following qualitative exploration allows for a deeper and more detailed insight into these issues. Methodically, material- and video analysis, participative observation, as well as single and group interviews and expert interviews were combined (Wiesner 2004: p. 128 ff).[2]

De/construction of Gendered Materials—*"Now We Add Real Formula One Decorations to it"*

Not only does the importance of didactics increase with the length of a Roberta course, the materiality of robotics itself plays an important role. Right from the start the material speaks for itself, because the children handle lego bricks they already know. The programmable bricks, engines and sensors however provide an unknown means to make experiences, so that children of both sexes are usually confronted with something new, too. The Legomaterial in this sense is evocative i.e. it generates presumptions, experiences and actions by itself. Treated from a gender-sensitive perspective, it even appears to be "gendered material." A practical example:

The use of a car-like basic model often leads to car-like robots. Triggered by the impulse car-likeliness, boys—and often girls too—in no time construct vehicles. This phase of construction is often introduced by remarks such as: *"Now we add real formula one decorations to it."*

Were no (car-like) models are given, girls and boys often construct models with strong analogies to humans and animals. If the children are left to choose the models themselves, their constructions are observed to be less gender specific. Gender specific behaviour becomes more obvious if given materials, models and tasks already contain and enforce gender specific orientations. Although the (Lego-) material evokes gender specific behaviour[3], didactical interventions and gender conscious tasks allow to successfully deconstruct the gendered material. This point is made very clear by Deirdre Butler who is an experienced researcher from Ireland using robotics in education:

> *(..) a teacher. [...] noticed [...] the fact that the boys' ideas were dominating and they all centred around wheels. They all had to be vehicles that moved fast. Rather than separate the groups—[...] for the next project, [...] they simply should not made a wheeled robot. They could use wheels to make conveyor belts or create other moving parts [...] And that began to change things in her classroom, because they began to make other types of things.*

As a consequence, the task not to construct a wheeled vehicle can transform both internalised gender specific behaviour and the use of gendered materials. Combined with a gender sensitive view on team interaction and help, this can lead to new learning effects.

The Staging of Gender in Robotics—"Just Stick to the Construction Manual..."

Our observations could not support the assumption that girls tend to work more team oriented than boys. In small groups, both sexes are able to develop social skills and prefer to work in groups. Though in some boys' and girls' teams, alternations between team work and a hierarchical task division can be observed, to us, these differences had also strongly to do with the learning arrangement: the more intense a gender sensitive approach the more the boys and girls can work as teams. This aspect became evident in the practice of themes like moving the robot through a maze, a highly self-designable task where problem-solving strategies often got developed team oriented. "Gender-neutral" themes thus helped to prevent particular mixed teams from falling apart into two gendered groups.

Gender differences and gender specific behaviour brought into the learning environment as a precondition have been, in our view, actively transformed and thus co-produced by both the gendered material and the didactical design of Roberta courses. The consequential "staging of gender" (Wiesner 2002) thereby takes place, where material and discoursive worlds interact in specific ways. Due to that, technical materiality and didactical intervention constitute a switching relation, co-producing each other's effects in learning environments.

In that process, a gender-conscious didactical approach to both the technology and the students becomes essential. This is important to avoid girls being robbed of their fame on a crucial social point, when they present their robot in front of all participants. and to avoid boys from failing if they are driven by a "self-concept of the winner" (Buschmann 1994). Noticeably often, the boys disturbed the girls' presentations by letting their own robots drive into their presentations. This was amplified by a rather reserved behaviour of the girls that displayed gender stereotypes. (*"You go ahead starting, we can present our robot at the end"*). Nonetheless, the boys were also put under stronger pressure by the amount of teacher attention, as a row of unhappy presentations showed. In both cases, LegoMindstorms can work as an exclusive or inclusive technology.

As we know today, bringing technology into schools can be beneficial but very much depends on how the teachers mediate the new tools. No wonder, the necessity of a gender sensitive training concept for teachers[4] is an important conclusion from the results of the qualitative evaluation. In Roberta courses, such an approach particularly amplifies positive experiences with technology design by ensuring that girls and boys have equal access and starting points. What we think robotics in education offers new here is that technological interest, creativity and the discovery of new skills as well as the gaining of knowledge—perceived as a reflexive mixing of didactics and technology—can make our kids ready for an increasing sociality with technical objects and processes, which will accompany and transform their life constantly. It is through this very concrete relation and interaction between both the materiality and virtuality of building and programming robots and their didactical mediation, that, for us, the specific potential of robotics can get to the kids. The following sections will illustrate how robotics then can create potentials to initiate new forms of learning which deal with both the concrete and the abstract world.

Robotic Material and Constructivist Learning

Merging abstract programming and concrete construction worlds, robotics bear a character of challenge. Instead of following instructions, there is not just one way to a robot. Many possibilities are found to tie to one's own imagination. According to Seymour Papert, they ought to be things one can think with and that open specific and appropriate possibilities for the individual way of learning (Papert 1994, Ackermann 1996). Through concrete handling the Lego material supports the access to abstract concepts and vice versa, the transfer of abstract programming concepts into concrete motion. Meanwhile, it gives feedback on how successful a construction or programming process is. Such proceedings offer appropriate conditions to promote girls' and boys' technical curiosity. Girls often feel inferior with regard to technical constructions, and often fear the embarrassment. Robotics technology—combined with a gender-sensitive constructivist learning approach—instead allows for a less biased access.

Future Trends: Recommendations for Educational Concepts

In this respect, the experience with robotics to promote girls' interest in technology allows conclusions towards general educational concepts for digital media. Robotics as a *relational actor* not only suits boys' and girls' interest in the messiness of technologies but enables them for a technologically mediated life instead of just feeling overwhelmed. It is however also an appropriate medium for general education in the more comprehensive sense of developing personality and agency.

How does Robotics work as a Didactical Actor?

This adresses the question how information and communication technology effects and transforms identity (Turkle 1995). Resembling one of the major discourses in Science and Technology Studies, where technology is not understood as seperate from the behaviour and identity of human users, but as a productive (f)actor in hybrid sociotechnical settings (Latour 1998; Rammert 2002), where it works as an agent and a translator of human practice and experience.

The Lego-Mindstorms technolgy both in its virtual and material appearance offers exactely that—regarding the design and the use of its inscribed, activity-engaging potential for new learning strategies. During the Roberta courses students constantly receive system-feedbacks (either from the screen or from the robot) that structure their actions together with the teachers' didactical approach. Even if the technology does not act intentionally, different forms of the *attribution of agency* can be observed here, as Rammert et. al. (2002) put it. The system reports, or the robot does not work if incorrect programming commands were entered or technical malfunctions appear. The system suggests new action without pointing into the "right" direction. From the kids' view, the technology at hand "acts" on two levels: they experience an immediate physical, material processing that is executed through semiotic processes (programming). Only for this reason, students experience a process of the merging of abstraction and experimental interaction, of their own and of technological actions. System feedbacks are thus fundamental, simultaneously didactically constructed *and* mediatable agency-patterns, simply because they evoke and initiate reactions. They are also gender neutral, since they occur independent from whoever operates the computer and do not assess changes due to gender differences but on grounds of the given tasks. Many teachers agree that particularly the system feedback activated self-learning effects.

According to our case for gendered materials, gender sensitive didactics is however not realised automatically via system feedbacks. If you want to promote girls' interest in technology, LegoMindstorms only becomes a didactical actor through a sensitively designed learning environment. Learning effects, creativity and new actions thus not only evolve through successfully mastering the technology but by the *translation* of human agency through technological agency. In other words: learning emerges from the collaboration of technological and didactical worlds, the allocation of their different forms of agency and their active transformation by students. The potential of robotics as a didactical actor then lies in this possibility to frame the gradual development, allocation and attributation of both human and technological agency within learning processes (see Wiesner-Steiner/Wiesner/Schelhowe 2005).

While sociologists of technology often point to an increasing mangling of technology and sociality (see Latour 1998), our perspective on robotics as a didactical actor for learning processes is particularly driven by the idea to discover potentials that lie in a gradual development and allocation of *different* forms of agency:

Question: "I have noticed that the robots did not always drive through the maze independently with their light- and motion sensors. How come?"

Answer 1: "What I learned—the robots don't like you. You can't always master the technology." (boy)

Answer 2: "The robot is a technical product and then it is our fault, that we didn't understand this. (girl)

Answer 3: "I think it's got to do with both. Sometimes the robot didn't do something it was programmed to, but there were also program errors." (girl)

Answer 4: "We have not seen that the robot didn't do what we told him to at all. Sometimes he did something we thought to be wrong, but that was in fact somehow right." (girl)

Answer 5: "I think the most mistakes occured because we didn't really know the programming language and so the robot did what he was programmed to do but not what we wanted him to." (girl)

Particularly in NQC-related programming and presentation phases of Roberta courses, different forms of the allocation of agency and creatorship by the students and teachers (what does the technology, what should students do in order to make the robots do something ...) can be observed, that create a specific form of experimental interactivity with the Lego material: Acting emerges both from an impression of the relatedness of technical and human agency and a clear feeling of their distinctiveness. The allocation of human and/or technological agency by the students based on this ambivalence thus shows relatedness as well as it expresses boundaries (see Gieryn 1995). Learning experiences are categorised and processed within a highly differentiating cognitive framework. The allocation and distribution of different forms of agency not only display where insecurities regarding causalities and connections in the learning process exist but what students think is special with digital media in comparison to other learning experiences.

Based on their former experiences, boys and girls were most convinced of their own agency in the phase of construction, while in the programming phase they attribute the strongest form of technological agency. Nevertheless, the programming phase markes the time when they are most active in confronting the technology. This becomes evident if a group is observed throughout a long course. A robot programmed by a boy and a girl via NQC in same shares would do something slightly different, often together with what they programmed him to. Altogether responsible for over 60 tries in their programming phase (2 ½ days), that group wouldn't stayed on track without the attachment of their programming interest to the concrete object "robot," nor without any didactical intervention.

In opposition to the programming, in the presentation phase students are particularly interested not only to connect but to perform their own and technical activities. Amplified by the test-situation, this is why the allocations the students make in this phase are more goal- and control-oriented than in more experimental course phases. Here, the agency of the technology is perceived as failure or success in relation to the given task, while mastering tasks is seen as both a success of the team *and* the robot. But isn't this the place where research points to gender specific differences? Where boys rather allocate success to themselves, while girls tend to allocate failure to the circumstances? This clearness vanishes in the case of Roberta courses, because particularly in long courses it can be observed that boys like girls use *both* allocations. The importance of these allocations regarding the technology and themselves is thereby influenced likewise by the gender specific orientations and the didactical conception of the course.

In consequence, an important didactical task is to help the students find their own initiative. It is not only important to tackle the given task by the means of successful control of the technology, but also to provide insights into the connection of human means of action and learning processes with technological agency. A corresponding remark by a Roberta teacher:

"In the first two days the students already asked for more help. If this is interpreted, they only made limited use of their means of action." (teacher)

Knowledge of the allocation and gradual development of technical and human agency in learning processes that aim at both the shaping and use of technology can be very instructive for handling situations like this. Recommendations of activity for different learning phases can be given and the technology itself can be continually improved as a didactical actor. Hard- and software used in Roberta courses have a special, technically delegated potential to de- and to restructure established learning routines. Both assist or prolong not only learned behaviour and application routines but also enforce, through their inscribed agency, new and creative forms of appropriation, without the paths being outlined in detail. A gender sensitive approach can amplify this process (or even initiate it for girls), by ensuring that the students have equal opportunities to develop and to design.

Following this, examples for possible changes of the self concept are presented in the next section. Here we are less interested in whether the students develop sustainable profession-oriented interests in engineering and computer sciences but whether constructing and programming robots can emulate transferable knowledge that gives them a more active understanding of technology, which goes beyond surface-level familiarity. To be productive with highly interactive, multi-modal, adaptive and autonomous future applications, students need to apply their knowledge in a wide range of situations.

Possibilities and Limits of the Change of Self Concepts by Roberta Courses

Whether one's own activity in dealing with LegoMindstorms is perceived as strong or weak gives hints to the possibilities for changing the self concept. For this reason, we did not only ask students what they learned, but let them reflect on their learning processes. While the students referred to a special form of "experimental interaction" (Rammert 1999) with the robotics material, in courses of different lengths it's mainly the programming technique that interacts in specific ways with the social environment:

"I do think that this has something to do with learning. I wouldn't program just for fun at home." (boy)

"I think that it is important to connect this to something gamelike. Looking at this programming language, I would absolutely not want to have this written on the blackboard. I think it is important to try it playing." (girl)

"There weren't really disciplinary problems, because the students were challenged by the computers" (short course, teacher 1)

"After four days social problems are of greater importance than in short courses. In short course, the robot as a medium is such a challenge, that there is hardly a chance for social problems to occur in a team." (Long course, teacher 2)

Question: "Would you wish to learn more in this way?"

"I wouldn't only refer to it in Physics and Informatics, but to Natural Sciences in general. Chemistry is usually a subject that I really like,

but we don't have such good experiments. And this is important: that I don't just get a formula, but that I can also see: it really is like this. To get this confirmation." (girl)

"What I learned? To start small and proceed in small steps.." (boy)

Besides sequencing the tasks, two wishes are formulated here: the wish for more playful, experimental learning and the wish for transferable knowledge. The question of change of the self concept was finally adressed by the teachers themselves:

"We actually learned to enter a few commands in NQC. That is what can be tested. Can you program a loop? But we should also make clear that school education is more than the pure transfer of knowledge. That one can learn from mistakes. The students don't even realise that it is also an aim of the course to help them to help themselves. Not experiencing the learning at junctions and parting of ways is what is valuable." (teacher 1)

"This self-confidence that they gain in both technology and themselves cannot be taken away from them so easy again. This will keep them for a while and then in the higher grades, when a colleague takes this up in Natural Sciences in an appropriate way, it will stay like that. (...) those are very subtle mechanisms. That they bear that in mind and may consider it in future decisions in the higher grades, for their hobbies and in leisure time. That they believe in their capability to handle technical questions. In that respect some technology distance may have been taken away." (teacher 2)

That kind of knowledge about the allocation and gradual evolvement of technological and human agency in design-oriented learning processes is very instructive for pedagogical concepts dealing with digital media. Recommendations for action can be specified for different courses, settings and learning phases, and the technology as a didactical actor can be improved permanently. The material and virtual aspects of LegoMindstorms thus bear the potential to de- and reconstruct learning routines. They do not only assist or prolong routines of action but, through their inscribed agency, enforce new creative ways of learning that are not determined in advance. A gender sensitive didactical approach (implemented in a corresponding learning environment) can start and amplify this process by ensuring that the students have equal experimental opportunities. According to the Roberta aim of promoting girls interest in technology, the following recommendations for the creation of learning environments are given:

- Assist dynamic processes of team formation
- Promote open working environments
- Promote "gender neutral" project themes and work
- Reflect help and attendance in a gender conscious way
- Provide a flexible mix of open and structured learning
- Give opportunities for team work and promote team work
- Schedule for gender sensitive interventions during the project and presentation phases
- Consider the materiality, resistance and agency of technology in didactical approaches
- De/construct gendered material
- Observe allocation and distribution of technical and human agency gender sensitively

THE DIDACTICAL AGENCY OF E-LEARNING

The cultural and technical history of e-learning scenarios can be traced back to traditional forms of distance studies, CD-ROM learning programmes, audio-programmes or educational TV. But other than these forerunners, two closely related myths often shape policy towards ICT and education: the irresistible power of globalisation and the determining effect of technology. Both views present the success of e-learning throughout the education system as inevitable. The space left for practitioners in Higher Education is either to embrace the new media or to watch its inevitable unfolding. In this paper we take a critical stance towards that perspective and suggest that the shape and learning effect of new media in higher education is contested and evolves in communities of practice. No technologies are neutral and it is more appropriate to speak of economic, technological and societal features as interactively fostering the importance of e-learning through *distributed actions* (Rammert, 2002). From such a perspective, e-learning is perceived as a co-product of didactically and technically situated features (Wiesner-Steiner, Wiesner & Schelhowe, 2006) that foster and enable but don`t determine human learning through the use of digital technologies. Main characteristics are:

- Interactive and multimedial design of content
- Learning via digital networks
- Netbased communication

The EU-Leonardo-project "European Enhancement of Early Years Management Skills—EEEYMS" (http://www.eeeyms.org/) was intended to enhance employability of people employed in the Early Years Childcare management sector by providing access to a high level qualification in line with the emerging industry requirements. This was achieved by developing distance learning materials available via the World Wide Web and other forms of media including CD ROM`s, specific to the employment area which is also aligned to a degree pathway, and will be available within Europe. It was further achieved by the creation of a European network association for childcare to ensure sustainability after the project is complete. EEEYMS provides an accredited route for the attainment of a relevant degree level qualification for careers and managers within the childcare sector, and assist in attracting suitable people into this employment sector to meet the childcare demand over the next 10 years. With ODL materials, the project enhances employment opportunities and career status for a still predominantly female workforce. Research suggests that the increased status and professionalisation obtained through the availability of a high level qualification will make the industry more attractive to male employees. EEEYMS thus provided higher level qualification to people disadvantaged in the labour market and those who faced discrimination in accessing training due to disability, geographical location or family commitments. The use of ICT systems was thus thought to enhance knowledge and learning experience *and* the employability factors, as the knowledge will be directly transferable to the work environment.

The primary target group was that of childcare professionals actively working in the sector or entering this profession, where a niche in the market exists for a relevant specific degree award. EEEYMS thus wanted to attract more women into managerial positions, while encouraging more men to enter the profession by providing a credible award.

Because empirical evidence on the increase of e-learning-efficiency is both difficult and important, external evaluation of the EEEYMS e-learning modules via surveys has been an integral part of the entire project. The aim here was to include a more objective, independent feedback at every stage of the programme. According to the projects aims, the evaluation was conducted following the principle of gender mainstreaming (Wiesner, Kamphans, Schelhowe, Metz-Göckel, Zorn, Drag, Peter & Schottmüller, 2004) and considering intercultural inclusion-aspects (Zorn & Wiesner-Steiner, 2006).

Our second example is divided into three main sections. After introducing the use of VLE and a problem based learning approach, we discuss the effects of group work, the use of technology and the main learning experiences. As a result we come up with an overview of critical sociotechnical issues that show how distance learning technologies and materials can interact as didactical actors.

RATIONALE FOR THE USE OF A VLE AND A PROBLEM-BASED LEARNING APPROACH

In the development of e-learning for the early years sector through the EEEYMS partnership these key issues emerged: the importance of the use of a suitable VLE in delivering the learning programme, the use of problem based learning (PBL) to enhance student motivation through collaboration, the need of IT skills development and the role of context as it relates to student success.

The VLE in use is Granada's 'Learnwise'. This VLE has as one of its technical features collaborative 'Forums' in which participants take part in asynchronous discussion in small teams and work on specific management and education problems The partnership decided that these forums would provide a prime vehicle for student support through 'encouraging active learning', shifting from didactic to facilitative teaching' or 'building online communities' (Armitage, Brown & Jenkins, 2001)

The stated aim of the EEEYMS project is that early years practitioners will develop knowledge and understanding of the educational and management issues pertinent to their sector, and that they will also develop the requisite skills to critically analyse, evaluate and apply this knowledge. As professional knowledge requires functioning knowledge that can be put to work immediately, most module designers for EEEYMS choose to adopt a 'Problem Based Learning' approach.

Problem based learning simulates everyday learning and problem solving. Knowledge is acquired in a working context and is put back to use in that context. The learning and assessment on the programme will be aligned (Biggs, 1999) to learners everyday work experiences. Participants learn the skills for seeking out the required knowledge when the occasion arises during the process. They are motivated immediately by the interaction with a 'real' problem and are active early in the process.

Although on-line participants face time constraints as working practitioners and as parents with family responsibilities, the use of media-communicated communication has been used to build successful collaborative learning. As Salmon (2000) asserts, the Internet can change concepts of space and time:

Working and learning with others who happen to live in a particular locale may become less important than finding shared professional and personal interests in online environment. (p. 492)

The EEEYMS project aimed to provide learning opportunities at degree level, so that practitioners can develop the requisite skills to critically analyse, evaluate and apply knowledge. A large body of literature support the motivational aspects of collaboration on learning (Johnson & Johnson, 1989; Sharan & Shaulov, 1990). Wenger (1999) also offers a perspective on learning that emphasises social learning processes within *communities of practice* where individuals engage in the negotiation of meaning and the mutual construction of knowledge. The EEEYMS participants often refer to this 'community of practice' when expressing the relevance of the tasks to the everyday practice.

The issue of gender was also pertinent as with the exception of one male EEEYMS participant, all others were female. For example, a study by Kirkup & von Prumm (1990) comparing the experiences of women adult distance learners in Germany and the UK points to a pattern of preference for shared learning.

This type of social-technical interaction, learning and decision making is expected in the workplace today and this approach should ultimately therefore promote a desire for and ability to partake in 'life long learning'.

Meisalo, Lavonen & Juuti (2005) also emphasise the importance of Web based community formation for off-campus participants in their study of primary teachers taking a science education course. Dron (2005) in his paper on the construction of e-learning environments to cater for the needs of diverse learners utilises Michael Moore's theory of transactional analysis. For Moore (1980), distance is a pedagogical more than a physical phenomenon, and transactional distance measures the amount and nature of dialogue. Transactional distance is said to be low when there is a lot of dialogue between learners and teacher, but where transactional distance is high, teachers often provide a highly structured learning experience. The use of PBL appears to ensure that student autonomy flourishes and dialogue is high not only between student and teacher, but also student and student.

This importance of web based community and the need to maintain a low transactional distance through constant dialogue appears to be a critical outcome of the EEEYMS project. Donohue (2002) analyses the challenges of teaching the target group for EEEYMS online, as the Early Years sector is characterised by 'low tech/ high touch'. While many Early Years Managers and Practitioners might only have little involvement with high tech equipment such as computers in their work place settings, much of their practice is concerned with managing relationships with colleagues, children and families. Donohue (2002) suggests the use of learning approaches aiding the building of a community of practitioners such as collaborative knowledge construction and group work. The evaluation results discussed now show that it has successfully utilised learning approaches to mirror that 'high touch'.

PARTICIPANT EXPERIENCES WITH GROUP WORK AND E-LEARNING TECHNOLOGY

In our view, interlinking the scopes of didactics, evaluation and technology can help to increase the user's (long dated) commitment to e-learning modules. If electronic learning tools are perceived as "didactical actors" that not only bear their own action potential but influence and redirect participants belief systems and agency (Wiesner-Steiner, Wiesner & Schelhowe, 2006), new relations between learners and *didactical technology* come into focus. The results of the external evaluation thus mark important "passage points" for technical *and* didactical implementations of e- learning modules.

Methodically we used semi-standardized questionnaires that consist of a combination of closed yet multiple choice questions and open questions that leave room for participants to explain their more subjective learning experiences. Interpretation was done by means of the content analysis (Mayring, 2000; Gläser & Laudel, 2004).

GENERAL PARTICIPANT EXPERIENCES

Asking for general experiences, we could identify three points. Time management, intensity of tasks and work amount mark the most difficult aspects. Although quite challenging, the modules "were well designed" and "offered something for everybody"—learning outcomes were met throughout all the modules. Group work and tutors play a crucial and positive role in all modules.

GROUP WORK

Throughout our evaluation of EEEYMS, the overall importance of group work in a VLE became evident. Group work activities are not just a work form among others. For e-learning modules, they proved to be the work form par excellence! Group work was not only generally important, it does enhance group committments in virtual communities. Two of the most stunning *social* remarks about *electronic* group work thus stated: *"Without group work I don't think I would have managed the course."* (...) *"I feel as if I know my group members better without having the physical get-togethers than in previous studies."* Accordingly, students also pointed to the advantage of a shared workload, especially for part-time students or people who have to work full time and have a family life. They also mentioned that informal phone and e-mail contacts have been used in group work more often than informal chat rooms. Moreover, they found that e-learning group work offers different perspectives on an issue, allowing for a more holistic image and approach of the tasks. The relevance of group work can be summed up as follows:

- Electronic Group work needs blended learning; with face-to-face meetings at the beginning, "everybody had a face"
- Being part of the same group in different modules helps for getting used to different learning styles
- Changing groups can constrain learning processes

Both the combination of face-to-face with electronic group communication and the importance of group continuity mark important points for the appropriation of different learning styles and the development of social commitment. Didactically applied, they can improve the general learning experiences mentioned above.

Although most EEEYMS students agree that group work plays a very important role in e-learning modules, they also addressed some risks:

"The team leader's position is sometimes confusing because of individual's different aims—sometimes ignoring other needs
"You have to monitor yourself all the time in order to try to avoid communication misunderstandings"

Group members have to learn to work in a team. But as we know from everyday life and work practices, teams can be organized in more hierarchical or more symmetrical ways, depending on the members and social dynamics of a certain group as well as on the specific tasks and contexts. Within e-learning environments, students have to organize their group work mostly on their own. Moreover, they have to deal with social aspects like leadership and communication. Participants also found that group work is helpful not only for dealing with certain tasks but to navigate through both the technical and learning requirements of a module. Due to that, group work also functions as a method to downsize the drop-out quote in two directions—dropping out because of difficult (social) tasks or dropping out because of technical problems. But do the online students use the communication offers given by Learnwise? Studies in the area of e-learning and knowledge management systems conclude that communication offers are not used very often if they are not designed in a didactically carefully fashion. This is clearly perceived by EEEYMS participants as one of the most positive aspects of the e-learning modules. Thus group work—experienced as a new way of collaborative learning and in combination with team support by tutors—was perceived as *the* most positive aspect. Nonetheless, group work proved both important and challenging, creating mutual dependencies as well as commitments. It is important to note that both aspects can be perceived as very positive, depending on the student's experiences with team support and technical support.

Group work was also useful in cases when technology did not work as expected, trying to solve or bypass technical problems as a group. Because group work also works in informal, non-electronic ways, an important point stressed by the students is the unique form of electronic group work. Electronic group work is seen as a learning process in itself, requiring both commitment and easily accessible technology. In sum, the importance of electronic group work for learning processes was highlighted for all Hope modules, not the least because group work and electronic learning tools build learning communities and communities of practice (Lave & Wenger, 1991). We thus recommend that the concept of technically mediated group work should mark one of the central aspects for future e-learning modules and should be integrated into didactical approaches.

TECHNOLOGY

In accordance with the role of group work, we also evaluated the role of technology in e-learning environments:

- Electronic systems are great as long as they are running
- E-sources where sometimes difficult to find and time consuming
- Forums and chat rooms for group work were useful for communication among group members and contacts with tutors
- Online-sources were most useful in combination with supplementary materials (CD Rom), group work and tutorial help
- Tutors play an important role as they mediate between the requirements, technical possibilities and social dynamics of the e-learning modules
- Tutors were given excellent credits, have been responsible "even at simple questions," accessible most of the time and (often) replied promptly

For EEEYMS participants, websites and journals became high-rated whereas library resources seem to be "out." Nonetheless, in terms of provided materials the students often felt that hard-copy handouts gave more safety than other material. This has to do with technical problems that are perceived as very time-consuming, i.e. access-problems that occurred while trying to use e-journals, informal chat rooms with slow responding action or web-links with passwords (with the exception of tutor-directed web links). CD-ROMs and module handbooks thus became important when access to those materials failed. Moreover, they were also important in the modules' introductory phases. In addition, the role of the tutor became quite important in cases when technology failed, forcing tutors to organize a new or alternative learning environment. The facilitation and encouragement of electronic communication by tutors marks another important evaluation point. We thus recommend that for future e-learning modules, tutors should be especially trained in supporting online communities and group activities.

LEARNING

Not only did EEEYMS participants learn something new, but they became able to translate their new knowledge into their own professional contexts. In addition, the following features of their collective learning experiences point to the close interplay between didactical, social and technical e-learning issues:

- Discovering the unique style of e-learning
- Discovering group work as a learning experience
- Learning of IT skills and time management
- Discovering the advantages of problem based learning
- Discovering the possibility to study while working and being "old age"
- Discovering the possibility to choose one's own learning time (look at resources, listen to CD lectures etc.)
- Discovering many ways to act towards the same aims
- Discovering appropriate and comprehensive modules for day care managers"
- Discovering excellent experiences in communicating and learning with colleagues from different countries
- Discovering that children have the same basic demands and affairs in spite of cultural differences

FUTURE TRENDS

In sum, e-learning was clearly perceived as a new type of learning. Nonetheless, we might only use 20 % of the possibilities of e-learning. To some extent, that mirrors the development of television where at the beginning the actors did act like actors in a stage play and were not aware of the new technology and its influence on their performance. The same thing could be true for e-learning, when virtual group work and electronic learning tools are offered and mediated in a didactically carful fashion, creating new forms of life-long learning. As our informants mentioned, this process can be initiated and improved by

- Strengthening of group commitments
- Clarification of submission procedures
- Research guidance
- Obligatory contributions to the forums and chat rooms
- Language diversity
- Less academic language
- Small weekly activities with one large assignment at the end of a module
- Reflections on different professional backgrounds and qualification levels

CONCLUSION

A few conclusions can be drawn from our two case studies: If electronic learning tools are perceived as "didactical actors" that not only bear their own action potential but influence, reshape and redirect participants' actions, belief systems and agency, new relations between learners and didactical technology come into focus.

The techniques of programming and constructing robots with LegoMindstorms thus not only offer the potential to enhance the room for activity. Purposefully used as a "didactical actor," the interaction with the "social machines" of the Roberta technology even offers possibilities for the change of self concepts *before* (female) students make educational choices and withdraw from computer science, engineering or math courses.

Similar conclusions can be drawn from our second case, where interlinking the scopes of didactics, evaluation and technology helped to increase the user's (long dated) commitment to e-learning modules in higher education.

In this respect, Table 1 summarized important issues of distance learning materials, issues that are *at the same time* technical, didactical and social. Throughout these issues, participants clearly articulated how working in groups and tutorial (i.e. didactical) support 'kept them going' with the technology at hand.

From our perspective, these cases clearly imply that a both social and technology sensitive didactical approach (implemented in a cor-

Table 1. A summary of critical sociotechnical issues

A summary of critical sociotechnical issues for future distance learning materials: • Time management, intensity of tasks and social context (work amount, family) create learning problems • Integration of working duties and job related perspectives into the content of teaching and into the structure of e-learning modules is important (e-learning adds the load to daily work!) • Interaction between didactical and technological issues within sociotechnical support: tutors and group work play a crucial role in e-learning-modules as they mediate between the requirements, technical possibilities/problems and social dynamics of the e-learning modules • Electronic Group Work needs to be discovered as a learning experience of its own; • Group work bears risks and opportunities, is at the same time socially challenging and creates new learning experiences; • Intercultural Aspects are linked to technical aspects: Time differences, quality of technology at hand and language barriers (academic language) can create communication problems between participants from different countries • Online-sources are most useful in combination with supplementary materials (CDROM) and group work

responding learning environment) can play a key role for learning with information communication technologies, as they combine and mediate the requirements, technical options and social dynamics of the technologies at hand. One of the prime requirements for success in that fields thus is to take the agency of technology itself seriously and to put it into relation to our own agency.

REFERENCES

Armitage, S., Brown, T. and Jenkins, M. (2001). Management and Implementation of Virtual Learning Environments: a UCISA funded Survey, *UCISA*

Biggs, J. (1999). Teaching for Quality Learning at University, *SRHE*, Open University.

Buschmann, M. (1994). Jungen und Koedukation. Zur Polarisierung der Geschlechterrollen. *Die deutsche Schule*, 86. Jg. 11.2, 192-213.

Donohue, C. (2002). It's a small world: Taking your first steps into online teaching and learning, *Childcare Information Exchange, 9*, 20-25.

Dron, J (2005) Control Termites and E-Learning, *IADIS International Conference Web Based Communities*, 23-25.

Gieryn, T.F. (1995). Boundaries of Science, Jasanoff et. al. (Ed.), *Handbook of Science and Technology Studies*, 393-444.

Gläser, J. and Laudel, G. (2004). Experteninterviews und qualitative Inhaltsanalyse als Instrumente rekonstruierender Untersuchungen, *Verlag für Sozialwissenschaften.*

Hartmann, S., Schecker, and H., Rethfeld, J. (2005). Mädchen und Roboter—Ein Weg zur Physik? Anja Pitton, (Ed.) *Tagungsband der Jahrestagung in Heidelberg 2004 der Gesellschaft für Didaktik der Chemie und Physik—Relevanz fachdidaktischer Forschungsergebnisse für die Lehrerbildung*. Münster: Lit Verlag.

Hartman, S. and Schecker, H. (2005). Mädchen im Umgang mit Informatik, Technik und Naturwissenschaften—Externe Evaluationsergebnisse zu dem Projekt Roberta, *Zeitschrift für Didaktik der Naturwissenschaften*, ZfDN

Joerges, B. (1996). Technik Körper der Gesellschaft. Arbeiten zur Techniksoziologie, Suhrkamp, Frankfurt am Main.

Johnson, D. W. and Johnson, R. T. (1989). Cooperation and Competition: Theory and research, *Interaction Book Company.*

Kirkup, G and von Prummen, C. (1990). Support and Connectiveness: The needs of Women Distance Education Particpants, *Journal of distance Education*, Retrieved from http://cade.athabascau.ca.vol5.2/7_kirkup_and_von_prummer.html

Latour, Bruno (1998). Über technische Vermittlung. Philosophie, Soziologie, Genealogie. Rammert, Werner (Ed.) *Technik und Sozialtheorie*. Campus-Verlag, 29-83.

Lave, J. and Wenger, E. (1991). Situated Learning: Legitimate Peripheral Participation, *Cambridge University Press.*

Mayring, P. (2000). Qualitative Inhaltsanalyse. Grundlagen und Techniken, *Deutscher Studien Verlag.*

Meisalo, V., Lavonen, J. and Juuti, K. (2005). A Case Study on a group of unqualified Primary Teachers taking a science education course in a web based environment, *IADIS International Conference Web Based Communities.*

Moore, M.G. (1980). Independent study, Redefining the Discipline of Adult Education, *Redefining the Discipline of Adult Education*, Boyd and Apps (eds), San Francisco, 16-27.

Müllerburg, M., Petersen, U., and Theidig, G. (2004). Mit Robotern spielend lernen. VDI (Ed.) *ROBOTIK 2004*. VDI Berichte Nr. 1841, 393-400.

Rammert, W. (2002). Technik als verteilte Aktion. Wie technisches Wirken als Agentur in hybriden Aktionszusammenhängen gedeutet werden kann. *Technical University Technology Studies Working Papers*, TUTS-WP-3.

Rammert, W. and Schulz-Schaeffer, I. (2002). Technik und Handeln. Wenn soziales Handeln sich auf menschliches Verhalten und technische Abläufe verteilt. Rammert et.al. (Ed.) *Können Maschinen handeln? Soziologische Beiträge zum Verhältnis von Mensch und Technik.* Campus Verlag, 11-65.

Rammert, W. (2002). Technik als verteilte Aktion. Wie technisches Wirken als Agentur in hybriden Aktionszusammenhängen gedeutet werden kann. *Technical University Technology Studies Working Papers*, TUTS-WP-3-2002

Rammert, W. (1999). Weder festes Faktum noch kontingentes Konstrukt: Natur als Produkt experimenteller Interaktivität. *Soziale Welt* 50 (3), 281-296.

Rethfeld, J. and Schecker, H. (2005). Roberta—Mädchen erobern Roboter. Ergebnisse der quantitativen Begleitforschung. *3. Qualitätsmanagement-Workshop*, 20-21 Juni, Fraunhofer-Institut AIS, Birlinghoven.

Salmon, G (2000). Computer Mediated Conferencing for Management Learning at the Open University, *Management Learning*, 31 (4), 491-502.

Sharan, S. and Shaulov, A. (1990). Cooperative learning, motivation to learn and academic achievement, *Co-operative Learning: Theory and research*, S. Sharan (ed.), 173-202.

Turkle, S. (1995): Life on the Screen: Identity in the age of the Internet. *London: Phoenix*

Wenger, E. (1999). Communities of Practice: Learning, Meaning, and Identity, *Cambridge University Press.*

Wiesner, H. (2002). Die Inszenierung der Geschlechter in den Naturwissenschaften. Wissenschafts- und Geschlechterforschung im Dialog. *Campus-Verlag.*

Wiesner, H., Kamphans, M., Schelhowe, H., Metz-Göckel, S., Zorn, I., Drag, A., Peter, U. and Schottmüller, H. (2004). Leitfaden zur Umsetzung des Gender Mainstreaming in den "Neuen Medien in der Bildung—Förderbereich Hochschule," *Bremen—Dortmund.*

Wiesner, H. (2004). Handlungsträgerschaft von Robotern: Robotik zur Förderung von Chancengleichheit im schulischen Bildungsbereich. *His-*

torical Social Research, Zentrum für Historische Sozialforschung, 29, 120-154.

Wiesner-Steiner, A., Wiesner, H., Schelhowe, H. (2006). Technik als didaktischer Akteur: Robotik zur Förderung von Technikinteresse, *Hochschulinnovation. Gender-Initiativen in der Technik. Reihe: Gender Studies in den Angewandten Wissenschaften. Gender Studies & Applied Sciences*, Gransee, C. (ed.), LIT-Verlag Hamburg, 89-115.

ENDNOTES

1 Three different types of courses have been offered: short courses (2 to 5 hours), intermediate courses (5 to 15 hours) and long courses (15 +). Up to 2005, a total of 153 courses have been conducted (1.880 students, of which 1.605 were girls; for more information about Roberta visit http://alex.ais.fraunhofer.de/zeno/web?action=content&journal=16413&rootid=15465

2 The database contains qualitative interviews with a total of 11 tutors, 6 group discussions with students, 2 expert interviews and minutes, photo- and video analysis from 6 courses. All types of courses (short, medium, long) were analyzed.

3 The offered combinations of wheels and engines often leads to the exclusion of other functional lego bricks.

4 Gender sensitve didactics consist of:

- performance-related praise (particularly girls)
- gender conscious reflection on the given attention and help
- gender sensitive intervention during the project phases
- "gender neutral" tasks (e.g. circus-indstead of soccer scenarios)
- open learning scenarios
- de-/construction of the "gendered material"
- use of designable technology

Chapter VI
Comparative Analyses of Online and Traditional Undergraduate Business Law Classes: How Effective is E-Pedagogy?

Daniel J. Shelley
Robert Morris University, USA

Louis B. Swartz
Robert Morris University, USA

Michele T. Cole
Robert Morris University, USA

ABSTRACT

E-learning and e-pedagogy continues to grow in importance in the delivery of higher education, due in part to the cost of higher education, a changing student profile, scarcity of traditional classroom space, and the recognition that distance learning has created a genuinely new paradigm of instruction. To respond to the changing student demographics, working adults, students in the military and residents of rural communities as well as of other countries, more and more universities are including online (internet-based) course offerings to their core offerings. As they do, the question arises whether online instruction is, or can be, as effective as classroom instruction. Investigating the question has been the focus of several studies. Our studies compared students enrolled in both online and traditional classroom versions of one business law course where all elements were the same except for the instruction format. The first study found no significant difference between the two formats with regard to student satisfaction and student learning, supporting earlier comparisons of online and traditional instruction modes. However, the second study did find statistically significant differences between the online and the traditional course formats with regard to student satisfaction with the instructor, and student satisfaction with the course structure.

INTRODUCTION

Robert Morris University (RMU) in Pittsburgh, Pennsylvania, has continued to develop and offer an increasing number of online course offerings to meet the needs of its traditional student base, working adults, as well as those of a growing number of international and off-campus students. Since its first online offerings in 1999, RMU has added 246 new online and partially online courses. In academic year 2006-07, there were 145 totally online courses university–wide. Of these, fourteen were offered in the School of Business. In that year, there were an additional 136 courses partially online, forty-three of which were in the School of Business. As the University expands its offerings and more and more instructors and students become involved in online education, ensuring instructional quality and learning effectiveness assumes a central role in course planning.

RMU is a private university with an enrollment of approximately 5000 students. Founded in 1921, the university has experienced rapid growth in the last two decades. It supports six schools with the School of Business being the largest. A large number of undergraduate and graduate course offerings in this school have had online course development as a focus for several years. A number of the courses are available to the students in both the traditional and the online formats. For the past three years, *Legal Environment of Business* (BLAW 1050) has been a popular course in both formats.

Overview of Legal Environment of Business (BLAW 1050)

The course is designed to enable students to develop an understanding of the American legal system and to attain a working knowledge of ethics, contract law and consumer protection to a degree sufficient to be useful in business and consumer transactions. The course also helps students to better comprehend the rules of conduct they can reasonably expect others to follow, as well as the conduct others may expect from them in various business situations. In this course, students acquire an awareness of their legal rights and responsibilities and gain the ability to apply legal principles to help solve business and consumer problems.

Online vs. Traditional Instructional Issues

In any discussion of online and traditional course delivery and development, some obvious and fundamental differences will be acknowledged by instructors. In general, the traditional course is taught in a structured classroom, the students are physically there, all instruction is in real time and the instructor is present for the class meetings. In the online format, the class is taught in a cybernetic environment, instruction does not have to be in real time, the students are not present in one place, and the instructor monitors most of the activity from a distance.

In defining distance education, Desmond Keegan (1996) identified six significant elements of online learning. These were: the separation of the teacher from the student; placement with an educational organization; use of technology to convey content and unite instructor with the learner; two-way communication that facilitates student-initiated conversation; potential for face to face meetings for social as well as instructional purposes; and participation in an "industrialized form of education"(Keegan,1996, p. 44).

The fundamental differences between online and traditional instruction pose some major challenges and concerns for course instructors and educational institutions. Chief among these is student learning and perhaps to a lesser degree, student satisfaction as it affects learning in an online environment. Online teaching, or e-pedagogy, forces the instructor to assume a new teaching role and necessitates a reappraisal of the traditional teacher-student relationship.

Online teaching requires the instructor to rethink and reorganize the existing teaching paradigm. Online instruction also forces the institution to evaluate its strategies for ensuring quality education. Wang (2003) notes that e-learning is one of most significant developments in the IS industry. We argue that e-learning, and with it, e-pedagogy, are becoming two of the most significant developments in higher education as well.

In most cases, conveying the basic content to the students in the online format is easily accomplished. A greater challenge is getting the instructional quality of the online course to match, or exceed, the instructional level of the traditional classroom course. It is not sufficient for the online instructor to have an understanding of the technological skills and course development tools alone. He or she must have a strong sense of course design and an understanding of good pedagogy as well. Good pedagogy is generally accepted by educators to involve: 1) a high level of learner activity, 2) a high level of student interaction, 3) a format for motivation and, 4) a well-structured knowledge base.

As online instruction gains acceptance, researchers have begun to test the proposition that online instruction can indeed incorporate the principles of good pedagogy and effective course design. Schulman and Sims (1999) studied students enrolled in five separate courses, each offered in both the online and traditional format. Both sections of each course were taught by the same instructor. In their sample, they found that students learned as well online as they did in the traditional classroom environment. Schulman and Sims compared course assessments and final outcomes in both instructional scenarios.

In *The No Significant Difference Phenomenon,* Thomas Russell (1999) reviewed 355 research reports, papers and summaries on the subject of the online versus traditional learning. He found no significant difference in grades, satisfaction or effectiveness when "E-learning" was compared to traditional teaching. R. C. Ryan's study at the University of Oklahoma compared the online and traditional versions of the course entitled, *Construction Equipment and Methods (CNS 4913).* The final grades for the two groups were not significantly different and survey results indicated that students perceived no difference in the quality of the instruction (Ryan, 2000). Other studies have found little or no difference between online and classroom learning when such issues as race, gender, technological and academic backgrounds, and socioeconomic status were taken into account (Navarro & Shoemaker, 2000).

However, Rivera and Rice (2002) reported that while several studies (including Russell's 1999 work) have demonstrated that online and traditional courses were found to be comparable with regard to the cognitive factors (learning, performance and achievement), the same could not be demonstrated consistently with regard to student and instructor perceptions and satisfaction with online learning.

Our studies (2007, 2008) relied on satisfaction surveys and grade comparisons to assess whether online instruction was as satisfactory as traditional instruction and if student learning were the same or better with online versus traditional instruction in one specific area, Business Law.

Online Business and Law Courses

Discussing the challenges to the instructor and developer of online law-related courses, Kathy Marcel noted that the best online courses were instructor-facilitated, student-centered and highly interactive (Marcel, 2002). The design of an online law course, as with the design of any online course, is critical. The instructor's role is one of designing a learning experience and guiding the students through the process. Marcel found that in fact, many law instructors tend to work very well with the facilitative aspect of good online course development. Marcel argued that because of the nature of their profession, law professors teaching online courses tended to expect students to be engaged and not merely passive learners. The suitability of teaching law courses online was even more evident, she found, with regard to upper-level law courses, because these courses themselves often rely on case studies, projects and Socratic dialogue.

Suanpang, Petocz and Kalceff (2004) addressed the comparison of student attitudes when taking a Business Statistics course in the online and traditional formats. Working with 230 students (with an N=112 in the online format and an N=118 in the traditional format) both quantitative and qualitative data were analyzed. The study concluded that "...students taught online develop strongly positive attitudes towards learning statistics, which influence their learning and make understanding statistics easier for them than for students taught in the traditional mode" (Suanpang et al., 2004, p. 17).

E. Cassel (2003), after having taught law online for over six years, concluded that online learning matched or exceeded traditional environments in several respects. In her experience with online learning, the level of student-professor and student-student interaction through asynchronous (Threaded Discussion) and synchronous (Chat/E-mail) was higher than in the traditional classroom setting. Additionally, the various audio and video options enhanced the learning environment for students. Cassel also points out a consideration often overlooked as an advantage of the online format; that is, that with online learning, classroom and classmate distractions, interruptions and basic annoyances are not present, thus allowing the learner to focus more completely on the subject matter and activities. Shelley concurs. His pilot study of Robert Morris University's move to place its entire core undergraduate history courses online demonstrated student satisfaction with the online format, the course content, sequencing, as well as with the textbook (Shelley, 2005).

Both Cassel (2003) and Marcel (2002) describe the advantages of online instruction for effective legal instruction. Although focusing on the use of voice–recognition software to enhance online law courses, K. H. Miller (2004) also found that legal education, thoughtfully designed, could be delivered effectively online. Some would argue, as Kristine Ellis does in *A Model Class* (2000) that designing a law course requires going back to the basics. That would mean constructing an online law program that would teach students how to formulate and deliver a legal argument and to analyze and systematize case decisions.

Why These Studies?

Bernard, Abrami, Lou, Borokhovski, Wade, Wozney, Wallet, Fiset, and Huang (2004) note in their analysis of studies comparing distance and classroom instruction that the value of such studies lay in their usefulness in determining the impact on desired outcomes, lending credibility to the innovation (online learning in this case) and providing focus for further developments.

The available evidence seems to indicate that, if carefully designed, an online course would offer at least a comparable, if not better, learning environment for students than the same course presented in the traditional format. However, little has been published on the online delivery of undergraduate business law courses. In a post-Enron environment, incorporating the principles underlying Sarbanes-Oxley into undergraduate law courses intensifies the need for effective instruction in business law. But is teaching business law online as effective as teaching business law in the classroom?

Weaver-Kaulis and Crutsinger (2006) cite considerations of accreditation, budget and accountability as stimulants in the increased attention on documentation of student learning beyond the traditional grading system and the impetus for faculty driven assessment programs.

In their study of student performance, Frantz and Wilson (2004) note that the increased scrutiny of legislators and accrediting bodies, particularly in business schools, has intensified the need for research into determinants of success. Specifically, they remark on the lack of research on legal studies courses in business schools – "a surprising void given the importance of legal studies to business education" (p.225). These studies begin to address that void by examining the effectiveness of one core business law course taught both online and in the classroom.

Determining how well students are learning is critical in any educational setting. It is of particular significance to RMU's School of Business, which is in the midst of its AACSB (Association to Advance Collegiate Schools of Business) accreditation process. Measurement of student learning is central to the review of current course offerings and to the development of new ones. Student satisfaction with the learning environment not only contributes to student retention, but it also serves as a measure of faculty performance and pedagogical effectiveness.

RESEARCH QUESTIONS

The studies looked at four research questions:

- **Research Question 1:** Does student satisfaction with the course overall differ significantly between the online format and the traditional class format?
- **Research Question 2:** Does student satisfaction with the instructor differ significantly between the online format and the traditional class format?
- **Research Question 3:** Does student satisfaction with the course structure differ significantly between the online format and the traditional class format?
- **Research Question 4:** Does student learning differ significantly between the online format and the traditional class format?

METHODOLOGY

Structure and Sample

The course examined was *Legal Environment of Business (BLAW 1050)* which is required for every business major at Robert Morris University. The course is offered in both the online and the traditional classroom formats. In both studies, the same professor taught each section of BLAW 1050 surveyed, using the same textbook, required readings, activities, projects, exams, and assessment for both groups.

In the first study, comparative data was drawn from four online sections of the course (two in 2004, one in 2005 and one in 2006) and two traditional sections in the spring of 2005. Fifty-eight of the sixty-four enrolled students completed the online sections of BLAW 1050 (N=58) or 90.6%. Forty-six of the forty-nine enrolled students in the traditional sections completed the course (N=46) or 93.8%. The total number of students receiving grades for BLAW 1050 during the study period was one hundred and four (N=104) or 94.5%.

In the follow-up study, comparative data was drawn from two online sections of the course (fall, 2006 and spring, 2007) and from one traditional classroom section in the spring of 2007. Forty students from the online courses responded to a web survey which duplicated the paper surveys given to the students in class (N=40). Twenty-seven of the students from the traditional class section participated in the study (N=27).

Thirty–nine of the online students completed the course and received a grade (N=39). Thirty-five of the students in the classroom section completed the course and received a grade (N=35).Thirty-nine of the forty-four enrolled students completed the online sections of BLAW 1050 (N=39) or 88.6%. Thirty-five of the thirty-nine enrolled students in the traditional section completed the course (N=35) or 89.7% of those enrolled. The total number of students receiving grades for BLAW 1050 during the study period was seventy-four (N=74) or 89 % of those who enrolled.

Of the sixty-four students enrolled in the online sections of BLAW 1050 in the first study, six withdrew, for a retention rate of 90.6%. The retention rate for the traditional sections was higher, at 93.8%; of the forty-nine who enrolled, three withdrew. Of the forty-four students enrolled in the online sections of BLAW 1050 in the second study, five withdrew, for a retention rate of 88.6%. The retention rate for the traditional sections in the second study was slightly higher as well, at 89.7%. Of the thirty-nine who enrolled, four withdrew.

Course Design

The online sections of BLAW 1050 were developed using the eCollege™ format. All students taking an online course at Robert Morris University are required to complete the Online Learning Training Module prior to being registered for the class. All online sections of the course were developed and maintained by the instructor involved in this study. The online format employed available instructional tools, including digital drop boxes, document share areas, synchronous and asynchronous dialog, e-mail and online assessment. The textbook readings were enhanced and supplemented with lecture notes and illustrations of key points. The classroom sections of BLAW 1050 used the same syllabus as the online course and had the same assignments and assessments. The same topics used in the threaded discussions in the online format were used in real time in the traditional classroom format.

Instrumentation

In the first study, a twenty-four question satisfaction survey with a five-point Likert Scale was distributed in each class (attached as Appendix (A)). The survey was administered by the instructor after grading was completed. Participation was voluntary. Thirty-three of the fifty-eight online participants responded, for 56.9% return rate. Thirteen of the forty-six students in the traditional courses completed their surveys for a return rate of 28.2%.Total number of students participating in the survey was forty-six (N-=46).

For the students in the traditional classroom section, the same process using the same survey was followed in the second study. For the students in the online courses, the same survey was uploaded as a web-based instrument (websurveyor). This was done to facilitate student participation and accuracy of data conversion for analysis. Forty of the forty-four students enrolled in the online course responded using websurveyor, for a response rate of 90.9%. Twenty-seven of the thirty-five students who completed the classroom course participated in the survey for a response rate of 77.1%.

Questions one through thirteen applied to students in both the online and classroom courses and were answered by both groups in both studies. A comment section was provided on the survey itself for qualitative input. In both studies, question one asked if the student felt he/she had learned the subject material. Questions two and ten focused on the performance of the course instructor. Questions three and four which focused on the quality of the selected textbook, were not used for the analysis. Questions five through nine and eleven through thirteen dealt with issues involved directly with the course structure.

Participant responses from the online and classroom sections were aggregated and compared for both studies. Responses to question one formed the basis for comparison for Research Question 1. Responses to questions two and ten formed the basis for comparison for Research Question 2. Responses to questions five through nine and eleven through thirteen formed the basis for comparison for Research Question 3. Questions fourteen through twenty-five were designed specifically for the online format and thus, were not used for the analysis. Final grades from the online and traditional classroom courses formed the basis for comparison for Research Question 4 in both studies.

In both studies, the structure of the survey allowed for both quantitative and qualitative data to be analyzed. Each study used SPSS for data analysis. In both, independent-samples t-tests were run for each research question. In the second study, ANOVAs were run to determine if there was a significant difference between the results of the two studies.

Within the context of both studies, "satisfaction" is defined as having met expectations as demonstrated by the student responses. "Learning" is defined as having acquired knowledge of the subject matter as evidenced by the course grades. The study controlled for what Benbunan-Fich, Hiltz and Harasim (2005) refer to as moderating factors that influence the outcomes when measuring learning. These are technology, course, instructor characteristics and student characteristics.

RESULTS

Research Question 1: Does student satisfaction with the course overall differ significantly between the online format and the traditional class format? See Table 1.

Research Question 2: Does student satisfaction with the instructor differ significantly between the online format and the traditional class format? See Table 2.

Research Question 3: Does student satisfaction with the course structure differ significantly between the online format and the traditional class format? See Table 3,

Research Question 4: Does student learning differ significantly between the online format and the traditional class format? See Table 4.

One-way ANOVAs were run to determine if there was a significant difference between online and the traditional classroom instructional formats when the study group responses were combined for research questions 1-3 (Tables 5 and 6). One-way ANOVAs also were run to determine if there was a significant difference between the two study

Table 1. Student satisfaction with the course overall

2004-2006	t-test for Equality of Means	
	t N=46	Sig. (2-tailed) N=46
VAR0002 Equal Variances Assumed	- .885	.381

Aggregated mean score for the online sections 4.4242
Aggregated mean score for the traditional sections 4.6154

2006-2007	t-test for Equality of Means	
	t N=67	Sig. (2-tailed) N=67
VAR0002 Equal Variances Assumed	- 1.146	.256

Aggregated mean score for the online sections 3.9500
Aggregated mean score for the traditional section 4.1481

Table 2. Student satisfaction with the instructor

2004-2006	t-test for Equality of Means	
	t N=46	Sig. (2-tailed) N=46
VAR0002 Equal Variances Assumed	-.460	.647

Aggregated mean score for the online sections *4.5385*
Aggregated mean score for the traditional sections *4.6154*

2006-2007	t-test for Equality of Means	
	t N=225	Sig. (2-tailed) N=225
VAR0002 Equal Variances Assumed	-4.673	.000

Aggregated mean score for the online sections *4.1310*
Aggregated mean score for the traditional section *4.6375*

Table 3. Student satisfaction with the course structure

2004-2006	t-test for Equality of Means	
	t N=46	Sig. (2-tailed) N=46
VAR0002 Equal Variances Assumed	.053	.957

Aggregated mean score for the online sections *3.8920*
Aggregated mean score for the traditional sections *3.8846*

2006-2007	t-test for Equality of Means	
	t N=536	Sig. (2-tailed) N=536
VAR0002 Equal Variances Not Assumed	-3.424	.001

Aggregated mean score for the online sections *3.4125*
Aggregated mean score for the traditional section *3.7500*

Table 4. Student learning

2004-2006	t-test for Equality of Means	
	t N=104	Sig. (2-tailed) N=104
VAR0002 Equal Variances Assumed	1.299	.197

Aggregated mean score for the online sections *2.9871*
Aggregated mean score for the traditional sections *2.7609*

	t-test for Equality of Means	
	t N=74	Sig. (2-tailed) N=74
VAR0002 Equal Variances Not Assumed	.912	.365

Aggregated mean score for the online sections *2.6859*
Aggregated mean score for the traditional section *2.4500*

Table 5. Student satisfaction with the course -04-07 studies

ANOVA

q1

	Sum of Squares	df	Mean Square	F	Sig.
Between Groups	.475	1	.475	.935	.336
Within Groups	56.427	111	.508		
Total	56.903	112			

Mean score for students (N=73) in the online courses 4.16
Mean score for students (N=40) in the classroom course 4.30

Table 6. Student satisfaction with the instructor – 04-07 studies and student satisfaction with the course structure – 04-07 studies

ANOVA

		Sum of Squares	df	Mean Square	F	Sig.
mean2.10	Between Groups	6.473	1	6.473	15.750	.000
	Within Groups	45.624	111	.411		
	Total	52.097	112			
mean5to9.11to13	Between Groups	.699	1	.699	2.880	.092
	Within Groups	26.942	111	.243		
	Total	27.641	112			

RQ2: Mean score for students (N=73) in the online courses 4.1370
RQ2: Mean score for students (N=40) in the classroom course 4.6375
RQ3: Mean score for students (N=73) in the online courses 3.6293
RQ3: Mean score for students (N=40) in the classroom course 3.7938

groups' results for research questions 1-3 (Tables 7 and 8). There were seventy-three students responding who were in the online courses (N= 73) and forty who were in the traditional classroom instruction courses (N=40). Study Group I was composed of forty-six students (N=46); Study Group II was composed of sixty-seven students (N=67).

There was no significant difference between student responses to Research Question 1, student satisfaction with the course overall (.336) or to Research Question 3, student satisfaction with the course structure (.092). However, there was a significant difference between student responses in the online and classroom courses to Research Question 2, student satisfaction with the instructor (.000).

Comparing results from each study, there was a significant difference between Study Group I and Study Group II with regard to each of the three of the research questions measuring student satisfaction. In each case, the results from the first study were more positive. Significance levels were .001, .001, and .000 for research questions 1, 2 and 3 respectively. Survey comments for the second group were mixed (see Appendix B).

Learning, as measured by grades, was higher for online students overall and higher in Study Group I overall. The mean score for the students in the online courses was 2.8365. For students in the classroom courses, the mean score was 2.60545. The mean score for Study Group I was 2.874. For Study Group II, the mean score was 2.56795.

One-way ANOVA Results Comparing Combined Online and Classroom Responses are shown in Tables 5 and 6.

Table 7. Student satisfaction with course overall- 04-07 studies

ANOVA

q1

	Sum of Squares	df	Mean Square	F	Sig.
Between Groups	5.484	1	5.484	11.839	.001
Within Groups	51.419	111	.463		
Total	56.903	112			

Mean score for Group I (N=46) 4.48
Mean score for Group II (N=67) 4.03

Table 8. Student satisfaction with the instructor – 04-07 studies and student satisfaction with the course structure – 04-07 studies

ANOVA

		Sum of Squares	df	Mean Square	F	Sig.
mean2.10	Between Groups	4.890	1	4.890	11.498	.001
	Within Groups	47.207	111	.425		
	Total	52.097	112			
mean5to9.11to13	Between Groups	3.180	1	3.180	14.429	.000
	Within Groups	24.461	111	.220		
	Total	27.641	112			

RQ2: Mean score for Group I (N=46) 4.5652
RQ3: Mean score for Group II (N=67) 4.1418
RQ2: Mean score for Group I (N=46) 3.8899
RQ3: Mean score for Group II (N=67) 3.5485

One-way ANOVA Results Comparing Study Group I to Study Group II Responses are shown in Tables 7 and 8.

DISCUSSION

Several studies of effectiveness of online learning appear in the literature. Fjermestad, Hiltz and Zhang (2005) reviewed published empirical studies which compared the effectiveness of course delivery, the authors conclude that the evidence is overwhelming. Online delivery is at least as effective as traditional classroom delivery (p.39).

Arbaugh and Hiltz (2005) discuss the difficulty in reaching definitive conclusions when measuring learning because of variations in measurement tools and methodologies. Their study found either that there was no significant difference between learning as measured between online and traditional courses or that there were significantly higher results from the online course instruction. With regard to Research Question 4, learning, our results support Arbaugh and Hiltz's findings.

Results from other studies are mixed as well. Bernard et al. (2004) concluded that the differences between the two modes of instruction were not significant. Their study was a meta-analysis of the empirical literature comparing distance (online) and classroom instruction in which they analyzed 232 studies measuring student achievement, attitude and retention. They found the effect sizes to be basically zero on all three measures and found wide variability due in part to the disparity in the degree of rigor in the studies analyzed. Some applications of distance education were better than classroom instruction; some were worse than classroom instruction.

Fjermestad, et al (2005) presents the results of thirty empirical studies comparing online and traditional course delivery. Those that looked at student satisfaction and student learning had findings similar to our studies with regard to research questions one and four. Of the twelve studies on student satisfaction, 41.6% were positive for online, while 25% were negative. In a third of the studies, student satisfaction as measured resulted in no difference between the two modes. With regard to objective measures of learning, 61.7% resulted in a finding of "no difference." 34% positive for online learning mode and four percent negative for online learning. The sample size was forty-seven (pp 45-46).

Our original study (2007) found no statistically significant difference between the online and traditional instructional/learning formats with regard to any of the four research questions on student satisfaction and student learning. These results were consistent with earlier studies (Schulman and Sims, 1999, Navarro & Shoemaker, 2000, Suanpang, Petocz and Kalceff, 2004, Bernard et al., 2004) and supported the proposition that a course provided online would offer a comparable, if not better, learning environment for students than the same course presented in the traditional format.

While the results from our first study clearly fall into the "no significant difference" category and support the majority of the earlier studies, the results from the second study present more mixed results. For example, although there was no significant difference between the online and traditional formats in student satisfaction with the course overall or in student learning, there was a significant difference found in student satisfaction with the instructor and with the course structure. In both cases, the mean scores for the traditional classroom course students were higher than for the online course students. Yet, student learning, as measured by final course grades, was higher for the online course students. Since none of the elements from the first study were changed in the second, i.e., the same instructor, course materials, course structure, and exams, the results are puzzling.

In the first study, student satisfaction with the course structure was slightly higher in the online format as opposed to the traditional instructional format. Student learning in the online courses were slightly higher than for those in the traditional classes. Those results reinforce Russell's "no significant difference phenomenon." In the second study, student satisfaction with the course overall, the instructor, and with the course structure was higher for students in the classroom course than it was in the first study.

The first study's survey results also supported findings in the earlier work by Schulman & Sims and by Ryan with regard to research questions on student satisfaction with the course, the instructor and the course design of BLAW 1050, *Legal Environment of Business.*

In the earlier study, student input under "Comments/Suggestions" from both groups was comparable with two exceptions, that students in the online courses also referenced the online features (positively) and that students in the traditional class setting commented on the outside assignments and exams. Seventy-two percent of the online students who participated in the study also added comments compared with 69% from the students in the traditional classroom setting. This feature was used less in the second study (two comments from those in the traditional instruction course, fifteen from those in the online course). Comments ranged from enthusiasm about the experience to some complaints about the text and the delivery platform.

Study limitations in the first study, sample sizes and the difference in participation rates, were ameliorated in the second study. In the first study, 59.6% of the students in the online courses participated while only 28% of the students in the traditional courses participated. Participation was higher in the second study, in part due to extra credit given for participation, the ease of use of a web-based survey instrument, and the ability to complete the survey during class time. Forty of the forty-four enrolled students in the online course responded to the survey for a 90.9% response rate. At least one of these withdrew prior to the end of class however. Participation was higher in the classroom course as well. Twenty–seven of the thirty-five students who completed the course completed surveys for a response rate of 77.1%.

CONCLUSION

While our studies broadly support the conclusions drawn by others with regard to the comparative effectiveness of online learning, a more nuanced study of online learning compared with classroom learning of business law as taught in *Legal Environment of Business* is needed to explore the differences found in the second study that were not present in the first. Further investigation of online instruction versus traditional classroom instruction on the whole needs to be done before any definitive conclusions can be drawn as to whether online should replace or simply supplement classroom learning. Both the Bernard et al. analysis (2004) and the Phipps and Merisotis study for the Institute on Higher Education Policy in 1999 argue that more rigorous studies need to be designed for researchers to be able to answer the question, "Is there a difference?".

As Rudestam and Schoenholtz-Read (2002) suggest, access to, and use of the internet for knowledge transfer present challenges and opportunities for creating new paradigms for learning. Access to, and use of the internet for knowledge transfer also presents challenges and opportunities for teaching.

REFERENCES

Arbaugh, J.B. & Hiltz, S. (2005). Improving quantitative research on aln effectiveness. In S. Hiltz & R. Goldman (Eds.), *Learning together online: Research on asynchronous learning networks* (pp 81-102). London: Lawrence Erlbaum Associates.

Benbunan-Fich, R., Hiltz, S., and Harasim, L. (2005). The online interaction learning model: An integrated theoretical framework for learning networks. In S.Hiltz & R.Goldman (Eds.), *Learning together online: Research on asynchronous learning networks* (pp. 19-37). London: Lawrence Erlbaum Associates.

Bernard, R.M., Abrami, Lou, Borokhovski, Wade, Wozney, Wallet, Fiset, & Huang (2004). How does distance education compare to classroom instruction? A meta-analysis of the empirical literature. *Review of Educational Research, 74*(3), 379-439.

Cassel, E. (January 2003). Teaching and learning law online. *Modern Practice: FindLaw's Practice and Technology Magazine.* Retrieved October 27, 2004 from http://practice.findlaw.com/archives/teaching_0103.html.

Ellis, Kristine. (December 2000). A model class. *Training,* 37(12).

Fjermestad, Hiltz, S. & Zhang, Y. (2005). Effectiveness for students: Comparisons of "in-seat" and aln courses. In S.Hiltz & R.Goldman (Eds.), *Learning together online: Research on asynchronous learning networks* (pp. 39-79). London: Lawrence Erlbaum Associates.

Frantz, P.L. & Wilson, A.H. (2004). Student performance in the legal environment course: determinants and comparisons. *The Journal of Legal Studies Education, 21*(2) 225.

Keegan, D. (1996). *Foundations of distance education.* (3rd ed). London: Routledge.

Marcel, K. (2002). Can law be taught effectively online? *JURIST,* December, 2002. Retrieved May 5, 2005 from http://jurist.law.pitt.edu/lessons/lesdeco2.php.

Miller, K.H.(2004).The law catches up with distance learning. *THE Journal, 31*(7), 31-34.

Navarro, P & Shoemaker, J. (2000). Policy issues in the teaching of economics in cyberspace: research design, course design, and research results. *Contemporary Economic Policy, 18*(3), 359-366.

Phipps, R. & Merisotis, J. (1999). *What's the difference? A review of contemporary research on the effectiveness of distance learning in higher education.* Washington, D.C.: The Institute for Higher Education Policy. *18*(3), 359-366.

Rivera, J.C. & Rice, M.L. (2002). A comparison of student outcomes and satisfaction between traditional & web based course offerings. *Online Journal of Distance Learning Administration, V*(III).

Rudestam, K.E. & Schoenholtz-Read, J. (2002). The coming of age of adult online education. In K.E. Rudestam & J. Schoenholtz-Read (Eds.), *Handbook of online learning: Innovations in higher education and corporate training* (pp. 3-28). Thousand Oaks, California: Sage Publications.

Russell, T. (1999). *The no significant difference phenomenon.* Office of Instructional Telecommunications, North Carolina State University Chapel Hill, N.C.

Ryan, R.C. (2000). Student assessment comparison of lecture and online construction equipment and methods classes. *THE Journal, 6*(27).

Shelley, D. J. (2005). Developing, implementing and assessing a university level online U.S. history course. *Proc. 8th IASTED International Conf. on Computing and Advanced Technology in Education,* Oranjestad, Aruba, 2005, pp. 312-314.

Shelley, D.J., Swartz, L. B. & Cole, M.T. (2007). A comparative analysis of online and traditional undergraduate business law classes. *International Journal of Information and Communication Technology Education, 3*(1), 10-21.

Shelley, D.J., Swartz, L. B. & Cole, M.T. (2008). Learning business law online vs. onland: A mixed methods analysis. *International Journal of Information and Communication Technology Education, 4*(2), 54-66.

Schulman, A.H. & Sims, R.L. (1999). Learning in an online format versus an in-class format: An experimental study. *THE Journal, 26*(11), 54-56.

Suanpang, P., Petocz, P. & Kalceff, W. (2004). Student attitudes to learning business statistics: comparison of online and traditional methods. *Educational Technology & Society, 7*(3), 9-20.

Wang, Y. (2003). Assessment of learner satisfaction with asynchronous electronic learning systems. *Information & Management, 41,* 75-86.

Weaver-Kaulis, A. & Crutsinger, C. (2006). Assessment of student learning outcomes in FCS programs. *Journal of Family and Consumer Sciences, 98*(1), 74-81.

APPENDIX A

Survey
Legal Environment Of Business
Louis B. Swartz, JD
Assistant Professor Of Legal Studies

Please answer each question based on the following scale:

5=Strongly Agree 4=Agree 3=Moderately Agree 2=Agree Slightly 1=Do not Agree

1. I feel I learned a great deal about the Legal Environment of Business 5 4 3 2 1

2. I feel that the instructor was well prepared for this course 5 4 3 2 1

3. I feel that the course followed the text book 5 4 3 2 1

4. I feel that the text was a good choice for the course 5 4 3 2 1

5. I feel that the overall layout of the course was easy to follow 5 4 3 2 1

6. I feel that the weekly assignments were fair and reasonable 5 4 3 2 1

7. I feel that there should be more outside assignments for this course 5 4 3 2 1

8. I feel that the quizzes that were given in the course were fair 5 4 3 2 1

9. I feel that the course examinations created anxiety 5 4 3 2 1

10. I feel that the instructor was accessible and easy to contact 5 4 3 2 1

11. I feel it was easy to respond and participate in discussions 5 4 3 2 1

12. I feel that I was able to concentrate and pace myself throughout the course 5 4 3 2 1

13. I feel that the course format allowed for easy interaction with my classmates 5 4 3 2 1

The following questions should be answered by **online students ONLY:**

14. I feel that the online quizzes and exams with the ability to review correct answers helped me understand the material 5 4 3 2 1

15. I feel that the threaded discussions added to the course quality 5 4 3 2 1

16. I feel that the curriculum in the course was well-organized and followed a logical progression 5 4 3 2 1

17. I feel that the Announcements and emails set forth clear instructions and expectations 5 4 3 2 1

18. I feel that the instructor made it clear what work was required and what work was optional 5 4 3 2 1

19. I feel that quizzes are a beneficial part of an online course 5 4 3 2 1

20. I feel that the mini-lectures and text provide the appropriate information to achieve the goals set forth on the instructor's syllabus 5 4 3 2 1

21. I feel that the course "due dates" made it easy for me to plan my schedule 5 4 3 2 1

22. I feel that the Doc Sharing was useful and helpful to me 5 4 3 2 1

23. I liked that the email responses from my instructor were private 5 4 3 2 1

24. I feel it is easier for me to learn in an online course than in an on land course 5 4 3 2 1

25. Besides this course, how many other online courses have you taken? _______

Comments/Suggestions

Other Comments

APPENDIX B

BLAW 1050 – LEGAL ENVIRONMENT FOR BUSINESS

Survey Responses to "Comments/Suggestions"

Study Group I

Twenty-four of the thirty-three students who took the course online and who responded to the survey also added comments:

- Really liked the course; would have liked more threaded assignments and readings; text made for students; did not like "cute" names for cases
- Due date feature and private e-mail instructor responses a big plus; whether online is better depends on the subject matter
- Online worked well; learned more in threads and reading than in taking the exams
- Loved this class
- Really enjoyed the class
- Schedule for exams and quizzes too rigid for busy lives
- Enjoyed class and instructor; one of the best at RMU; learned a lot
- One of the better classes; open discussions fostered by instructor incorporated current events into the course
- Too restrictive on access dates for assignments and exams
- Overall liked the course; problem with "proofreading" and phrasing of questions
- Pace good; glad did not use "chat" features; most organized taken; liked due date check list to plan ahead
- Enjoyed the course; learned a lot through the threaded discussions and weekly quizzes; online – fantastic, will continue to sign up
- Enjoyed course tremendously; first completely online course- hope others go as smoothly
- Enjoyed first online course very much; workload a bit heavy, but that is to be expected in a fully online course
- Loved this course; threaded discussions most beneficial; appreciated that assignments graded promptly; based on this online experience, wish could have taken entire degree online
- Informative instructor; course planner helped a lot – better than any syllabus; more interaction with this professor than with any other at Robert Morris; would recommend; tough class, need to pace yourself
- Need more time for essays for poor typists
- Enjoyed course; learned a lot, although grade did not reflect it
- Liked the way course set up. Easier to learn and to say what I wanted. "..am now even considering becoming a lawyer.."
- Great instructor, accessible and willing to answer questions; first online course, felt very comfortable; threaded discussions forced creativity; course harder than originally thought "…wish I would have taken it in a classroom just so I could get more interaction and asked questions as they arose, but for my first online class, I think it went pretty well."

- Course a success "...wish I had the option of taking another online class..."
- Learned a lot; "...will remember a lot of the content due to the online class environment" ; liked due dates, well organized; liked individual responses to threads; greatly enjoyed the class ; online allowed for flexibility needed
- Great class; loved having it online
- Enjoyed doing the work in the online setting; able to concentrate more and work at own pace; test taking better without other students around.

Nine of the thirteen students who took the course in the traditional classroom setting who responded to the survey also added comments:

- Really enjoyed the class; not easy, lot of information; only suggestion would be to add some visual aids
- There should be more outside assignments for those who are not great test takers; that would allow for more points while grasping the material
- Appreciated the abridged book; text and instructor informative; homework manageable; outside assignments not necessary, am a good test-taker
- Within time limits, course taught very well; basic understanding of the law; easy to follow using the book; professor's knowledge of real life situations made it easier to follow
- Really enjoyed the class; exams were difficult since the questions were long and needed to be reread; should be more assignments to compensate for the exam grades
- Course not that hard; lot of reading and studying; one of the best classes so far – enjoyed it thoroughly
- Really liked the class; learned a great deal and was challenged; material interesting, examples helped; still remember a great deal of what was taught
- More out of class assignments to add to the experience and ability to retain the information and would eliminate the anxiety of cramming for an exam
- Had to study a lot, but remember pretty much.

Study Group II

Two of the twenty-seven students who took the course onland who responded to the survey also added comments:

- Text has a lot of errors. Suggest a new one
- Good course, learned a lot

Fifteen of the forty students who took the course online and who responded to the survey also added comments:

- More online classes should be available
- Overall I thought the course was well done. I however did not like how the TD's were graded. I felt that many of the remarks made on my discussions were unfair and at sometimes very petty. I think that if a student is asked a question about their TD's they should be able to respond and defend their answer.
- Sometimes if I had a question I don't feel like I got a response from the teacher and the exams were always extremely hard.
- I think that this online class was very well and I enjoyed it very much. Thanks again
- This was the first online course I took and I feel that I've taken the most from this one because it forced me to read every chapter. Where as if I was in a classroom I would never read and just listen to the lectures. As for the threaded discussion, I liked them too because it forced everyone to interact where as in a classroom you only have one maybe two students give a response.
- I felt that this class was designed much better than my last online course. It was structured well and was much more interactive.
- I enjoyed the class, but was disappointed with the application layout. I have taken many online courses and this is by far the weakest layout. I believe this to be true due to the fact that important dates, announcements etc. were accessible after three links. I feel that important items should pop-out at one's face and not be hidden. In addition, I believe there is too many folders and links... maybe blackboard, webcet, or embanet would be a better choice?
- There were assignments due on Spring break and Finals week. I think it should be changed so that weekly assignments are not due on those 2 weeks and a quiz and a test should not be scheduled within 2 days of each other.

Chapter VII
Student Perceptions of Data Flow Diagrams vs. Use Cases

Ido Millet
Penn State University Erie, USA

Robert Nelson
Penn State University Erie, USA

ABSTRACT

Data Flow Diagrams and Use Cases are two popular methodologies in teaching as well as in practice. For the last 4 years, we have been using both methodologies in our Systems Analysis course. Questionnaire results indicate that students find the Use Cases methodology slightly easier to understand. However, students believe that Data Flow Diagrams are significantly better at communicating with users and programmers.

INTRODUCTION

The Data Flow Diagram (DFD) technique had been introduced in the late seventies (DeMarco, 1978; Gane & Sarson, 1979) and has become a popular process modeling tool for information systems. Research has shown that DFDs are also one of the most common tools taught in Systems Analysis and Design courses (McLeod, 1996).

While some believe that object-oriented design methodologies provide an "easier modeling process" and "improved communication" among developers as well as between developers and users (Johnson, Hardgrave, & Doke, 1999), empirical studies seem to disagree. Empirical research by Vessey and Conger (1994) shows that DFDs are easier to learn and to use, at least by novice users. An empirical study by Freeman (2003) indicated that a short review of the methodology tends to improve the accuracy and process satisfaction for novice users. Agarwal et al. showed that DFDs produce higher-quality solutions in process oriented tasks and are not inferior to object-oriented methodologies even in object-oriented tasks (Agarwal & Atish, 1996).

In our Systems Analysis course we have been using a simplified version of DFDs, as proposed by Millet (1999), whereby a single data store symbol represents a whole database rather than a single table. This modification makes DFDs easier to create, understand, and maintain. It also reduces the overlap with the Entity-Relationship Diagram technique. The CASE tool we have been using for DFDs is Sybase ProcessAnalyst. In the Fall 2003 semester, we added the Use Case methodology and Rational CASE tools to the course. Rational Rose was chosen because it was the primary UML CASE tool offered by Rational Corporation, the company (later purchased by IBM) whose name is most closely associated with UML (Grossman, Aronson, & McCarthy, 2005). Since starting to teach Systems Analysis with both methodologies, we've been using a questionnaire to evaluate student responses of these two competing methodologies.

We published initial results of this study in the *International Journal of Information & Communication Technology Education* (Millet & Nelson, 2007). This chapter provides an updated analysis based on a larger data set (4 years, 8 semesters, 15 course sections, and 309 observations).

To our knowledge, this is the first empirical investigation of how novice users perceive the Data Flow Diagram methodology compared to the Use Case methodology. Since both methodologies aim to model the services provided by a system, and since many instructors face the question of which of these methodologies they should use, such a comparison is both meaningful and warranted.

Unlike the conclusions reached by Vessey and Conger (1994), our results indicate that students perceive Use Cases as equally easy to use and slightly easier to understand than DFDs. However, students believe that DFDs are better for communicating with users and programmers. Another key result is that, if instructors elect to teach both methodologies, it does not matter which methodology is introduced first.

We start this chapter by describing design of our empirical research and questionnaire. We then discuss the quantitative results and provide qualitative context through examples of student comments. After providing design suggestions for course assignments, we summarize the implications of this study for the coverage and sequencing of the DFD and Use Case methodologies in the IT curriculum.

RESEARCH DESIGN

From Fall 2003 through Fall 2007, fifteen sections of our Systems Analysis course were introduced to structured analysis techniques as well as object-oriented methodologies. The same instructor taught all fifteen sections.

We assigned each section to either a "DFD First" or a "Use Case First" treatment group. This was done in order to balance and investigate the sequence effect of introducing one methodology before the other. For example, in the Spring 2006 semester, we assigned one section with 26 students to the "DFD First" treatment group and the other section with 19 students to the "Use Case First" treatment group.

As shown in Table 1, the "DFD First" group was introduced to data flow diagram concepts during Lecture #1. In the next class (Lab #1), this group was given a lab session and an assignment on Data Flow Diagrams using Sybase ProcessAnalyst as the CASE tool. During Lecture #2, this group was introduced to Use Case concepts and, again, this was followed by Lab #2 where these students were given a lab session and an assignment on Use Cases using Rational Rose as the CASE tool.

Table 1. Different course sections were exposed to the methodologies in different sequences

Group	DFD First (N=139)	Use Case First (N=112)
Lecture #1	DFD	Use Cases
Lab #1	Process Analyst	Rational Rose
Lecture #2	Use Cases	DFD
Lab #2	Rational Rose	Process Analyst
	Questionnaire	

The individual assignment we used for both methodologies was a Work Order System case adapted from Shelly, Cashman, & Rosenblatt (2006). This is a small case with three main processes and three external entities/actors. The students had two days to complete each assignment.

During the 5th class meeting, students were asked to complete a questionnaire (See Appendix A) comparing the two methodologies.

As shown in Table 1, the same approach but in reverse sequence was taken with the "Use Case First" group. This group was exposed to Use Cases before they were exposed to Data Flow Diagrams. Questionnaire results from all fifteen sections yielded a total of 309 observations. Questionnaire

As shown in Appendix A, the questionnaire consisted of student classification by semester standing and major as well as five questions about the two methodologies. The classes were composed of mostly business MIS students, both majors and minors. Most students were juniors or seniors because the systems analysis course is recommended in the 6th or 7th semester of the MIS curriculum. To solicit student reactions to the two methodologies, the questionnaire used a Likert-type scale ranging from '1' to '7' with '1' designated as "Strongly Disagree" to '7' indicating "Strongly Agree". The five questions about each of the two methodologies were:

1. This methodology is easy to *understand.*
2. This methodology/software is easy to *use.*
3. This methodology really helps *Users* to communicate and confirm requirements to the Systems Analyst.
4. This methodology really helps *Systems Analysts* to extract and validate requirements from the *user.*
5. This methodology really helps *Systems Analysts* to communicate requirements to the *programmer.*

As shown in Appendix A, these questions were followed by prompts inviting clarifying comments from the students.

RESULTS

It is important to differentiate between those results that reflect how easy it is for the students to learn and use the methodology (Questions 1 and 2) and those results that reflect what students think of the methodology (Questions 3, 4, and 5). There is little reason to doubt the validity of responses to the first type of question. However, since novice users are not in a good position to compare the value of these methodologies, we must exercise care in using the results from the second group of questions. We used questions 3, 4, and 5 merely to assess whether students were able to recognize the benefits from these methodologies rather than to render judgment on the true value of these methodologies.

Appendix B provides the results from 309 questionnaires from eight semesters and fifteen course sections of our Systems Analysis course. The information is presented in a way that facilitates comparisons within group (DFD First or Use Case first) as well as overall.

The students rated Use Cases as slightly easier to understand when compared to DFDs (5.49 compared to 5.31). While the difference is not large, it is statistically significant ($t = 2.24$; $p = .03$). However, the two methodologies rated equally well on the question of Ease of Use ($p = .17$).

The students rated DFDs as significantly better at helping system analysts extract and validate requirements from the user (5.41 compared to 5.16, $t = 3.44$, $p = .00$). However, the students rated the two methodologies equally well on the question of helping users communicate and confirm requirements to the Systems Analyst ($p = .76$).

The most extreme differences in ratings across the two methodologies were in relation to their effect on helping systems analysts communicate requirements to programmers (5.60 compared to 4.85, $t = 9.38$, $p = .00$). The students clearly perceive Data Flow Diagrams as a better mechanism in that respect. It seems that students react positively to how DFDs allow them to model required system services in a hierarchy of functionality that terminates with relatively small chunks of written descriptions ("primitive specifications"). In addition, as reflected by their comments, students seem to perceive the explicit depiction of data flows as an advantage.

Another interesting finding is that exposing students to one methodology before the other does not lead to significant changes in student perceptions of these methodologies. This suggests that Systems Analysis courses can cover these two methodologies in any sequence.

Student Comments

As part of the questionnaire, we asked students to provide written comments about each question. The following list provides sample comments we received for each question. We believe these comments reflect and explain the quantitative results described earlier.

Question #1 – Easy to Understand

- Use cases make more sense to me, but DFDs really explain how things are separated better.
- The DFD methodology is fairly easy to understand and they allow a system to be planned out and understood with significantly less explanation.
- DFDs can be confusing the more data flows, processes and sources/sinks there are.
- Use Cases use less terminology and concepts are easier to understand; however, because of this, it seems vaguer compared to DFDs.

Question #2 – Ease of Use

- The software for both methodologies is easy to use and self explanatory. They are straight forward and only a few buttons are needed to navigate
- Both are easy to use. I like using DFDs because of the different levels; and you can see more clearly what it should look like.
- Getting the system to look right in DFD was harder and keeping processes with the decomposition was hard.

Question #3 – Help Users Communicate Requirements to the Analyst

- I think the DFDs explain the process much better and explain what is expected of all the components of the system.

- I think Use Cases are easier for users to understand and thus they will be able to communicate to the analyst with ease. Users may find DFDs too complicated.

Question #4 – Help Analyst Extract & Validate Requirements from Users

- DFDs might help get the requirements a little better because they can be decomposed.
- DFDs can breakdown user requirements more specifically and represent the data flow more precisely.

Question #5 – Helps Analyst Communicate Requirements to the Programmer

- DFDs allow programmers to begin at a higher level, and the decomposition helps them make classes for each process at a time, and test them for inputs/outputs.
- The DFD's work well if they are properly created with a satisfactory amount of detail. The Use Cases work well, but the structure of the diagram can be distracting.
- DFDs show a tier-based approach that programmers will like. They are more descriptive and show more of what the system is and what it needs to do.

Design Suggestions for Assignment

Based on our experience we can offer several suggestions for designing good assignments for both methodologies. First and foremost, as suggested by Newby and Nguyen (2007), the same case should be used in the assignment for each methodology. This not only reduces orientation time but also allows the students to better understand, discuss, and compare the two methodologies. For the same reasons, the instructor use similar examples when introducing both methodologies.

Each assignment should contain a partial solution to help students begin using the tool. However, it is important to intentionally seed these partial solutions with format and logic errors and warn the students that they should not follow these partial solutions blindly. By warning the students that grading will emphasize quality rather than length of specifications, a similar focus on analysis and critical thinking rather than on the mechanics of each methodology is achieved.

Assignments should be designed to require 1-2 hours of work and be assigned to individual students rather than teams. This ensures each student is actively engaged in acquiring the methodological skills.

As graded assignments are returned to the students, a review of typical mistakes helps students avoid committing the same mistakes during later project phases. At the same time, it is useful to also show correct solutions as benchmarks.

After students complete assignments for both methodologies, a class discussion comparing the two methodologies allows students to think critically about the methodologies and revisit the notion of how models help system analysts.

CONCLUSION

Our students found Use Cases easier to understand but Data Flow Diagrams more effective at helping systems analysts communicate with users and developers. They did not see the two methodologies as significantly different in ease of use or in their ability to help users communicate with systems analysts.

Rather than replacing DFDs with Use Cases, there may be a place for both methodologies in the IT curriculum. This, obviously, raises the question of how to structure our courses to accommodate both methodologies (Gabbert, 2000). According to comparisons across the two treatment groups (DFD first versus Use Case first) it does not matter which methodology is introduced first.

REFERENCES

Agarwal, R., & Atish, S. P. (1996). Cognitive fit in requirements modeling: a study of object and process methodologies. *Journal of Management Information Systems, 13*(2), 137-162.

DeMarco, T. (1978). *Structured Analysis and System Specification*. Englewood Cliffs, NJ: Prentice-Hall.

Freeman, L. A. (2003). A refresher in data flow diagramming: an effective aid for analysts. *Communications of the ACM, 46*(9), 147-151.

Gabbert, P. (2000). Systems analysis: the challenge of integrating two competing technologies. *Proceedings of the fourteenth annual consortium on Small Colleges Southeastern conference*, 193-200.

Gane, C., & Sarson, T. (1979). *Structured Systems Analysis: Tools and Techniques*. Englewood Cliffs, NJ: Prentice-Hall.

Grossman, M., Aronson, J. E., & McCarthy, R. V. (2005). Does UML make the grade? Insights from the software development community. *Information and Software Technology, 47*, 383-397.

Johnson, R. A., Hardgrave, B. C., & Doke, E. R. (1999). An industry analysis of developer beliefs about object-oriented systems development. *ACM SIGMIS Database, 30*(1), 47-64.

McLeod, R. J. (1996). Comparing undergraduate courses in systems analysis and design. *Communications of the ACM, 39*(5), 113-121.

Millet, I. (1999). A Proposal to Simplify data Flow Diagrams. *IBM Systems Journal, 38*(1), 118-121

Millet, I., & Nelson, R. (2007). Data Flow Diagrams versus Use Cases - Student Perceptions. International Journal of Information & Communication Technology Education, 3(1), 70-78.

Newby, M., & Nguyen, T. (2007). Using the Same Problem with Different Techniques in Programming Assignments. Journal of Information Systems Education, 18(3), 279-282.

Shelly, G. B., Cashman, T. J., & Rosenblatt, H. J. (2006). *Systems Analysis and Design* (6th ed.). Boston: Course Technology.

Vessey, I., & Conger, S. (1994). Requirements Specification: Learning Object, Process, and Data Methodologies. *Communications of the ACM, 37*(5), 102-113.

APPENDIX A: MISBD 430 – METHODOLOGY QUESTIONNAIRE

Semester Standing: Sophomore Junior Senior

My Major: Accounting Finance ECNS/BECON Management
Marketing MIS Undecided Other

	Use Cases							DFDs						
	Strongly Disagree			Neutral			Strongly Agree	Strongly Disagree			Neutral			Strongly Agree
	1	2	3	4	5	6	7	1	2	3	4	5	6	7
This methodology is **easy to <u>understand</u>.**														
Comments (what aspects in what methodology are difficult or not intuitive?														
This methodology/software **is easy to <u>use</u>**.														
Comments (what aspects are difficult or not intuitive?)														

Student Perceptions of Data Flow Diagrams vs. Use Cases

	Strongly Disagree 1	2	3	Neutral 4	5	6	Strongly Agree 7	Strongly Disagree 1	2	3	Neutral 4	5	6	Strongly Agree 7
This methodology really **helps <u>Users</u>** to communicate and confirm requirements to the Systems Analyst.														
Comments: why do you think one methodology would be better or easier for the user to communicate requirements to the Systems Analyst and then verify that the Systems Analyst understood the requirements?														
This methodology really **helps <u>Systems Analysts</u>** to extract and validate requirements from the **<u>user</u>**.														
Comments: why do you think one methodology would be better or easier for the Systems Analyst in understanding user requirements and verifying with the user that the requirements are correct?														
This methodology really **helps <u>Systems Analysts</u>** to communicate requirements to the **<u>programmer</u>**														
Comments: why do you think one methodology would be better or easier for the Systems Analyst in communicating specifications to the programmer?														

APPENDIX B: QUESTIONNAIRE RESULTS

Treatment	DFD First (n = 166)		Use Case First (n = 143)		ALL (n = 309)	
Question	DFD	U.C.	DFD	U.C.	DFD	U.C.
1. Easy to Understand						
Mean	**5.35**	**5.53**	**5.27**	**5.44**	**5.31**	**5.49**
(Variance)	(1.3)	(1.3)	(1.1)	(1.1)	(1.2)	(1.2)
Paired t-test α (two-tail)	**0.11**		**0.12**		**0.03****	
t Stat	-1.60		-1.57		-2.24	
2. Easy to Use						
Mean	**5.60**	**5.72**	**5.73**	**5.78**	**5.66**	**5.75**
(Variance)	(1.4)	(1.3)	(0.9)	(1.0)	(1.2)	(1.2)
Paired t-test α (two-tail)	**0.20**		**0.56**		**0.17**	
t Stat	-1.28		-0.58		-1.38	
3. Helps Users Communicate & Confirm Requirements with Systems Analyst						
Mean	**5.42**	**5.43**	**5.27**	**5.20**	**5.35**	**5.32**
(Variance)	(1.4)	(1.4)	(1.1)	(1.0)	(1.2)	(1.2)
Paired t-test α (two-tail)	**0.96**		**0.61**		**0.76**	
t Stat	-0.05		0.51		0.30	
4. Helps Systems Analysts Extract & Validate Requirements from Users						
Mean	**5.47**	**5.22**	**5.35**	**5.09**	**5.41**	**5.16**
(Variance)	(1.2)	(1.4)	(1.2)	(1.2)	(1.2)	(1.3)
Paired t-test α (two-tail)	**0.01****		**0.02****		**0.00*****	
t Stat	2.57		2.28		3.44	
5. Helps Systems Analysts Communicate Requirements to the Programmer						
Mean	**5.55**	**4.93**	**5.65**	**4.76**	**5.60**	**4.85**
(Variance)	(1.6)	(1.8)	(1.2)	(1.4)	(1.4)	(1.6)
Paired t-test α (two-tail)	**0.00*****		**0.00*****		**0.00*****	
t Stat	6.04		7.26		9.38	

* $p < .10$ ** $p < .05$ *** $p < .01$

Chapter VIII
Promoting Undergraduate Education with Agent Based Laboratory

Hong Lin
University of Houston-Downtown, USA

ABSTRACT

Agent-oriented design has become one of the most active areas in the field of software engineering. The agent concept provides a focal point for accountability and responsibility for coping with the complexity of software systems both during design and execution (Yu, 2001). It is deemed that software engineering challenges in developing large scale distributed systems can be overcome by an agent-based approach (Paquette, 2001). In this approach, a distributed system can be modeled as a set of autonomous, cooperating agents that communicate intelligently with one another, automate or semi-automate functional operations, and interact with human users at the right time with the right information.

INTRODUCTION

A distributed learning system typically involves many dynamically interacting educational components, each with its own goals and needs for resources while engaged in complex coordination. It is very difficult to develop a system that could meet all the requirements for every level of educational hierarchy since no single designer of such a complex system can have full knowledge and control of the system. In addition, these systems have to be scalable and accommodate networking, computing and software facilities that support many thousands of simultaneous users concurrently working and communicating with one another (Vouk et al, 1999).

We have studied the implementation of Collaborative Agent System Architecture (CASA) (Flores et al, 2001) with Chemical Reaction Model (CRM) (Banatre & Le Metayer, 1990 & 1993).

CASA is a model that can catch the interactive and dynamic nature of e-learning systems. Our research results are published in (Lin, 2004 & Lin & Yang, 2006). Following our existing work on the design methodology of multi-agent systems, we exploit this methodology in a project that aims at a grid system for laboratory use in undergraduate education. The new method will provide a solution to current problems in design of comprehensive environments to support lab activities in teaching courses on parallel/distributed systems and networks. The unified model in chemistry-inspired languages will enable formal specification of an evolving system and provide a framework for top-down design of the entire system.

BACKGROUND

With the fast innovation of computer and communication technologies, computer curriculum is being adapted to accommodate teaching modules that enhance teaching effectiveness by utilizing frontier technologies. For example, the Department of Computer Science, University of Houston-Downtown (UHD), is building Information Technology (IT) option, which consists of courses in modern computer technologies defined by the current industrial desires, in its Computer Science degree program, to respond the increasing need for effective convey of the knowledge of current technology to students to equip them for a career in the modern fast-changing computer industry. One of the most important parts of this project is designing labs that can be performed through the Internets. Our first step is implementing lab packages for our parallel computing and computer networking courses in a grid that encompass lab facilities centered at a Beowulf cluster. We will then extend our lab environment to include other CS and Mathematical courses.

The challenge we are facing, however, is that we need to build an infrastructure that will accommodate multiple courses in different disciplines. The problems we are solving include: (1) an interface that is extensible to incorporate more lab modules and customizable to different course structures; and (2) an computational backbone that provides services for various lab activities, such as testing a parallel program, production of network phenomena, performance analysis. Performing these activities requires coordination among multiple nodes. Also, the architecture of the system requires extensibility and scalability to accommodate multiple course modules. To address the first problem, we follow the practice we had when we built the lab package for our CSI course. Outstanding features of this package include a lab explorer that allows students to browse through lab activities and the ability to invoke programs through the interface. We adopt the same structure in the lab package we designed for our parallel computing and networking courses. To address the second problem, we need to build an array of servers that run on a computational grid. A grid is a system of networked computing and storage sources (see Grid.org) that allows the sharing of information and computational powers. The grid is also a platform on which experiments of distributed data processing and computation can be exercised. Services are provided by different nodes of the grid system. The design of the grid must meet certain criteria so that the incorporation of any unit fits into our long term blueprint. For example, as aforementioned, the underlying infrastructure must support incremental and dynamic addition of lab exercises into the lab package. This is to support our ongoing construction of closed labs for our courses in parallel computing, computer networking, and other courses (Lin & Nguyen, 2005). On the other hand, however, the complexity of the system makes the design of its infrastructure difficult. Our existing research results suggest that the agent model is a powerful tool to solve problems in a distributed system. Therefore, we use agent technology to build the architecture of the grid system to manage the coordination and communication among the nodes and handle the load balancing issues. We envision that our practice will provide a solution to the problem of immersing current technologies into educational efforts which have been continuously made at UHD through the development of a comprehensive lab environment.

THE PROJECT

Goals and Objectives

The barrier in front of us is the integration of various networking technologies into one client/server model to provide a uniform lab environment for different lab activities. Given the targeted use of this solution, we need to define and implement the infrastructure that balances functionality and reliability. Based on our existing research experience, we desire a formal system to define the architecture of the grid system so that the development of the services and lab modules will no longer be pursued on case-by-case basis. The formal system must provide a language for the architecture specification, and a derivative method for system refinement. Architectural design should focus on system topology, interactions among system units, and dynamic features of the system, without involving proprietary platform information such as the operating systems on individual nodes, programming languages for program units, and vendor specific machine features. With the formal definition of the architecture on hand, interfaces among system units will be formally specified and design and deployment of each functional unit, such as a lab module, will not affect other units or cause any revision on the overall system.

Unfortunately, traditional formal methods in computer sciences are usually oriented to typical statically defined problems and not suitable for large-scale dynamic systems. Although there are attempts for developing formal methods in parallel and concurrent programming, no formal methods have ever been systematically used on evolving areas such as grid computing. We need a new model that can address the dynamic nature of a complex system without any presumption on the computation model.

As described above, agent system provides an architectural model for distributed networking system. As an active research area, the study in agent technology strives to apply intelligent information processing technologies to complex software systems. Features of an agent system have been summarized in the literatures, for example, according to Griss and Pour (2001), an agent shows a combination of a number of the following characteristics: autonomy, adaptability, knowledge, mobility, collaboration, and persistence. These features exist in different types of agent systems such as collaborative agents, interface agents, reactive agents, mobile agents, information agents, heterogeneous agents, and economic agents (Weiss, 2003). Because of the Gamma language's higher-order operations and its closedness to specifications (no artificial sequentiality), these features can be described directly without being adapted to fit into proprietary frameworks. Since this paper focuses on the architectural design of the grid system, we omit some technical details about CRM. Interested readers can refer to our publications for explanations of our methods. In (Lin, 2004), a sequence of case studies shows that features of various agent systems can be grasped by the Gamma language succinctly. In (Lin & Yang, 2006), we give a comprehensive example of specifying a course material maintenance system using the Gamma language. In addition, part of our work in constructing the cluster is presented at the 16th IASTED International Conference on Modeling and Simulation (MS 2005) (Lin & Nguyen, 2005).

The Design

The project includes a sequence of major steps: grid construction, lab design, client/server model definition, definition of the interface of functional units, agent-based architecture construction, a module language for program refinement, and architecture specification in the Chemical Reaction Model. Our plan can be described as a pyramid-shaped model illustrated in Figure 1.

The system will be designed using a bottom-up strategy (the Design Theme). We construct the grid and design lab modules using existing toolkits, such as Globus Toolkit 3, Java, and Apache Server. The services provided by the system are implemented in client/server architecture. A Java based user interface delivers the services on the web. Servers run on the clusters. Multiple servers interact with one another in the agent based infrastructure. A formal definition of the interfaces of functional units of the system forms the basis for multi-agent system design. Each agent is then designed in the Module Language we have proposed for specifying multi-agent systems (Lin & Yang, 2006). The overall system is specified in the Chemical Reaction Model. In Figure 1, we can see the multi-agent system is the conceptual model for implementing grid services, and the interfaces of functional units define the interaction among functional units and are the central part of the agent system. The interface also separates the architectural design from the design of individual functional units.

Adding/deleting services or features in the grid can be done in a top-down strategy (the Application Theme). If a service of a new type is to be added into the system, for example, it is added into the architectural specification. Through an automatic transformation procedure (see Lin & Yang, 2006), the specification is re-written into a multi-agent system in the module language. The actual program that codes the services is then incorporated into the system through the standard interface. Therefore, updating services or lab exercises in the system will not cause any change in other parts of the system and correctness and reliability of the system can be ensured to the maximum extent.

A Show Case

The following is a list of labs we are using in our parallel computing and networking courses. These labs are carefully designed based on the goals of

Figure 1. The pyramid model of the project

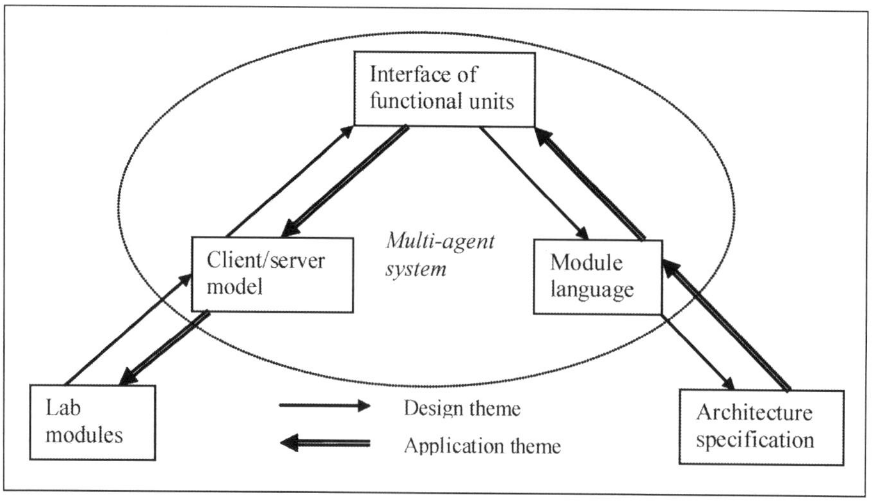

the course set forth in its syllabus and pursuit in our teaching experience. Lab topics are either typical topics of the area or problems we tackle within the course projects. Our lab design emphasizes the operability and vividness as well as the manifestation of the basic concepts and typical technologies. We also address the role played by the cluster when we design the labs.

- *Topology*: Circuiting messages in a ring
- *Collective communications*: Matrix transpose
- *Group management*: Matrix multiplication with Fox's algorithm
- *Scientific computation*: Solving linear systems with Jacobi's algorithm
- *Combinatorial search*: Traveling salesman problem
- *Parallel I/O*: Vector processing - Summation
- *Performance analysis*: Visualization with Upshot—Trapezoidal rule problem
- *Parallel library*: Solving linear system with ScaLapack
- *Scalability analysis:* Bitonic sorting
- *LAN configuration*: The use of NICs and hubs
- *Network analysis*: Monitoring a chat room
- *Address resolution*: Experiment with ARP burst
- *IP masquerading*: Clustered web servers
- *WAN configuration*: The use of routers
- *Performance tuning*: Deal with congestion
- *Service configuration:* The configuration of a networked file system:

Here we show one example lab we have designed. This lab allows students to use standard metrics to analyze the performance of a parallel program. The students predict the performance of the parallel program they choose, load the program onto the cluster, compile and run the program, and then compare the predicted results to the experimental results. As illustrated in Figure 2, one lab session is organized in a series of tasks and each task a series of activities. In this lab, students study some standard measurement criteria, viz. speedup and efficiency, for performance analysis of parallel algorithms in Task Activity 1 and 2, and predict the speedup and efficiency of the chosen program given the size of the problem input and the number of nodes in Activity 3. Task 2 requires the students to load the chosen program onto the cluster and then compile the code. The students can click on the C++ Compiler button in the bottom of the page to compile the code once the loading is finished. Task 2 Activity 1 walks the students through the program loading process. Activity 2 asks the students to compile the code. The code is then checked in Activity 3 by a program to ensure its correctness. Erroneous code causes the students to be asked to correct the code till it is errorless. In Task 3, the students can analyze the experimental performance of the program by using MPICH JumpShot profiling software and compare the experimental results to the theoretical predicts, which have been done in Task 1. In Activity 1, the students are required to insert profiling commands into the program and obtain a profile of the program by running it. In Activity 2, the students start up the JumpShot program from the program menu to obtain a Gantt chart of the program. The students then calculate the actual performance data by using the logged timing data and compare the experimental results to the predicted. This is done in Activity 3. Figure 3 shows some snapshots of the lab activities. Figure 3(a) shows the window that takes the student's response to performance prediction questions. Figure 3(b) shows the moment when the student opens a program through a dialog window and monitor the execution of the program through a popup window. Figure 3(c) shows a text window in which the student adds profiling statements into the program.

Figure 2. The main window of the Lab platform

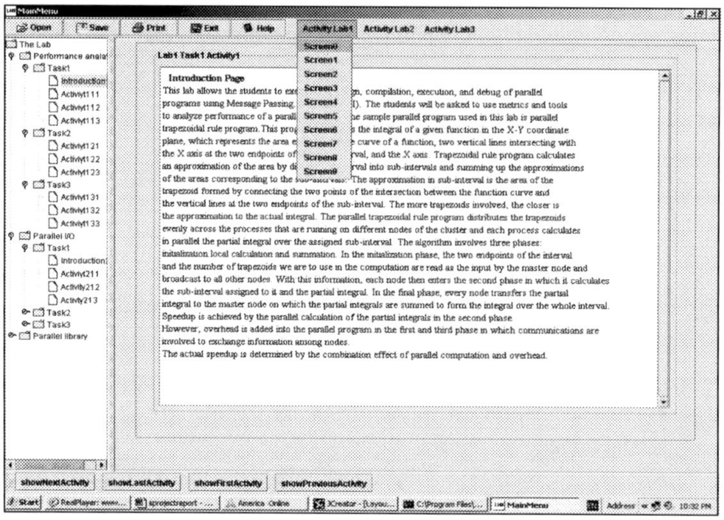

Figure 3. Snapshots of Lab—Performance analysis

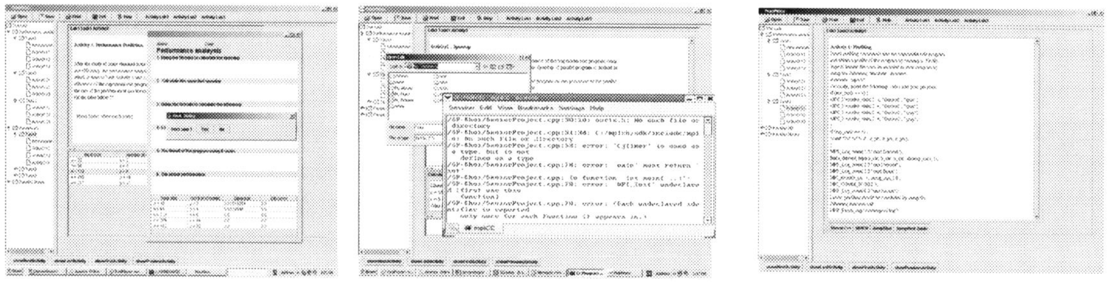

a. Performance b. Compilation and execution c. Profiling

STUDENT PROJECTS

Research at UHD is tightly coupled with its educational programs. Student involvement is an indispensable part of our research. For years, UHD's Scholar's Academy (SA) has been pairing up faculty and students and hosting organized research. Outstanding students are invited to present their work at the annual Student Research Conference (SRC). The Department of Computer and Mathematical Sciences has also widely recruited students in building the Labs and developing lab software. Senior student projects have been carried out throughout the design of the laboratory. Also, volunteering student research assistants constantly work in the UHD Grid Computing Lab to configure the clusters and implement research modules. With the limited resources of an undergraduate institution such as UHD, it is very important to involve students in research programs, not only to create activities for students to obtain hands-on experiences, but to couple research and education seamlessly. In the following, we present three student projects that are directly related to the project of building an integrated lab environment.

A Pioneer Work: Cluster Configuration and Testing

Parallel Computing course is an important part of Computer Science curriculum. We teach students to use Message Passing Interface (MPI) to design and test parallel programs. Since the Parallel Computing course is a writing course, students are also given a writing project which requests the students to write a report about applications of parallel programming. We are building a lab environment which can give the students hand-on experience in solving real-world large scale applications so that the students can get an image of the real performance of the parallel programs. To this end, a Beowulf cluster is constructed, configured, and tested using 15 similar computers and 1 newer, faster server using the MPI. The operating platform is Fedora 2 (Linux Red-Hat).

The Beowulf cluster is a rather simple architecture that many could recognize. There are two different configurations for the Beowulf: Class I and Class II. The Class I Beowulf is built entirely of commodity hardware and software. This type of cluster is usually less expensive than the Class II clusters that use specialized hardware to achieve higher performance. Intuitively, this project is geared towards a Class I Beowulf cluster. As shown in Figure 1, it consists of 15 nodes (computers) and 1 server. The server operates as the *master* node, and the 15 nodes serve as computational *slaves*. They are connected via a high-speed Ethernet and switch. All day-to-day operations and coding are done on the server.

This project focuses on the MPICH implementation on a Class I Beowulf cluster running on Fedora 2. Since MPICH is mirrored across the cluster, users can send MPI commands such as *mpirun* on the cluster. Packets of data are sent to a desired number of nodes on the cluster that are in turn sent back to the server to accomplish a given task.

The whole purpose of this Class I cluster is to achieve equivalent or greater processing compared to specialized computers and/or scalable parallel machines. Due to limited resources, a performance comparison to a Class II cluster or a scalable parallel computer (SPC) could not be made. Benchmarks on this cluster were made using MPI programs on a time-based analysis. Whatever program ran on the cluster, timings from start to finish of calculations were recorded.

Two MPI programs were used to benchmark MPI. The sorting algorithms: Mergesort (aka Binary sort) and Quicksort. The Mergesort is recursive sorting algorithm that begins with comparing the item being sought with the middle item in the data list. If the item is larger than that from the list, only the upper half of the list needs to be sorted; if smaller, then only the lower half needs to be sorted. The time complexity for this sorting algorithm is: $O(n\log_2 n)$—aka the Big *O* notation. Mergesort does $\log_2 n$ splits while doing *n* work at each split/layer. This way of dividing work and gathering the results presents a quite natural way implement a parallel version. The work is divided to in two - to 2 processes. Each of these processors divides their work again, until either no data can be split again or no processors are available anymore. Each process individually sorts the "n/p" elements. When all the processes are finished sorting individual arrays, the "p"

Figure 4. Class I Beowulf Cluster

partitions are merged parallel into a single array to form the final, sorted list.

A parallel version of Mergesort was successfully created and tested on the Beowulf clusters. A time-based analysis was conducted to measure the cluster's performance. As mentioned above, the cluster achieves optimum performance on 2n nodes; therefore, data was collected for Mergesort running on 1, 2, 4, 8, and 16 nodes. The list of elements contained 10^4, 10^5, 10^6, and 10^7 elements which were randomly generated by the program. Results were shown in Figure 5.

The partitioning of the elements into two groups by Mergesort illustrates the time complexity $O(n\log_2 n)$. Since this is a parallel implementation of this algorithm, the time complexity is improved because the partition is no longer two partitions in a single layer. The random generated list is divided into "n/p" partitions where work is done at each layer. For example, the time for Mergesort to sort 10^7 elements with one processor was 31.75 seconds, and its time with 16 processors is 12.14 seconds. The performance is calculated by taking the ratio of the difference between the times.

$$Performance(np) = \frac{Time(1p) - Time(np)}{Time(np)}$$

$$Performance(16p) = \frac{31.75 - 12.14}{12.14} = 161.53\%$$

Hence, the time for Mergesort to sort 10^7 elements resulted in a 161.53% performance increase! Furthermore, the same performance calculation for 10^6 elements resulted in only a 118.85% performance increase. This is due to the time complexity of Mergesort - $O(n\log_2 n)$. As n significantly gets large, the efficiency and performance increases

Figure 5. Performance of merge sort

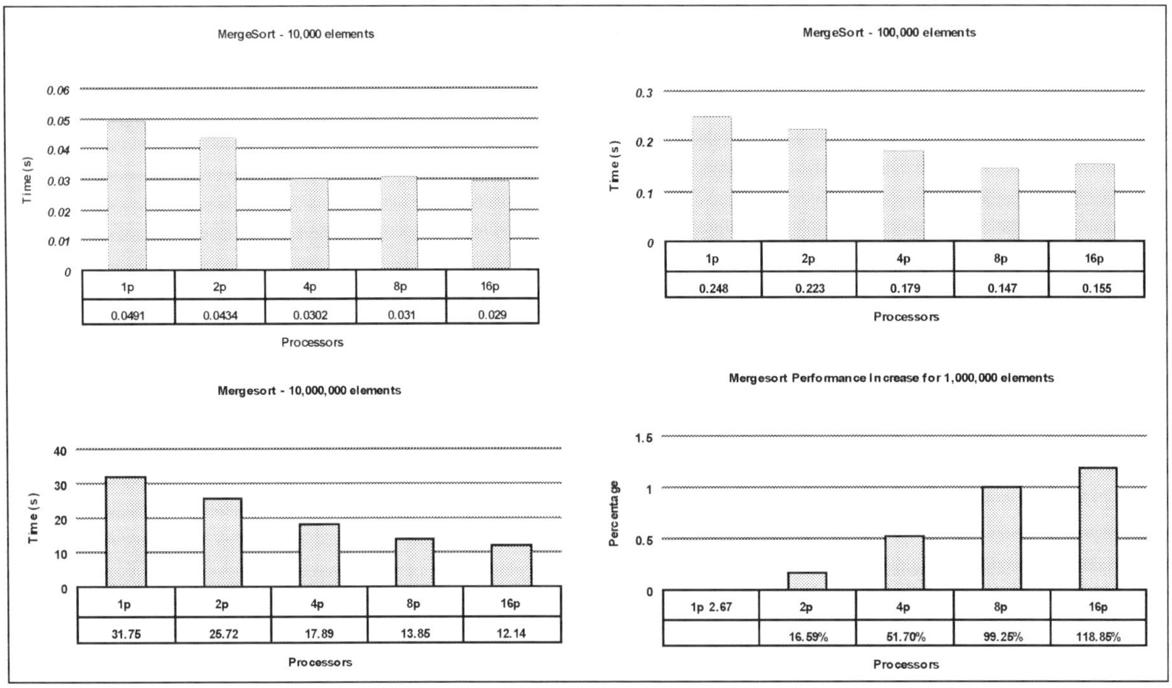

Figure 6. Performance data for quicksort

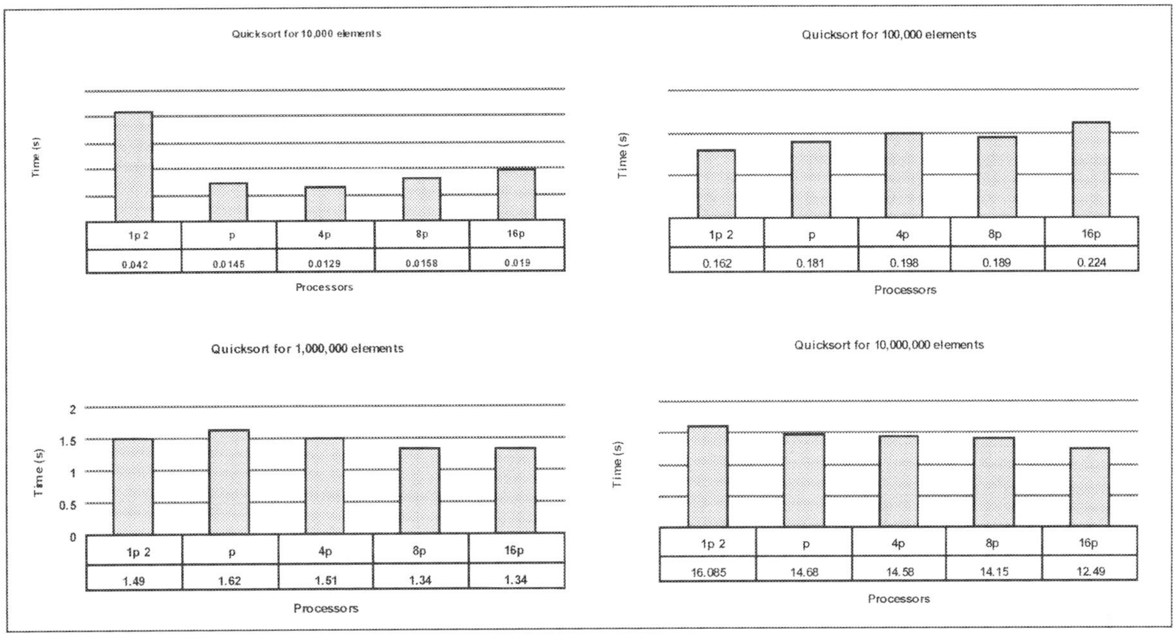

due to the parallel implementation. Below is a graphical representation of the performance increase for 106 and 107 elements.

Another sorting algorithm was implemented to test the performance of the cluster. Quicksort has the same time complexity, $O(n\log_2 n)$ as Mergesort, and its algorithm is also quite similar. However, Quicksort is an in-place, "divide-and-conquer", massively recursive sort. The list is partitioned by choosing a pivot—an element in the list. One part of the list is all elements less than the pivot, and one part is greater than or equal to the list. These partitions are sorted recursively using the same algorithm until there is only one element in the list. At this point the sub-lists are recombined to yield the final sorted list. Since each step of the algorithm executes two recursive calls, one can be tasked out to another processor. When the next recursive call is made in the first sub-list, it can be tasked out again. Data was collected for Quicksort running on 1, 2, 4, 8, and 16 nodes. The list of elements contained 10^4, 10^5, 10^6, and 10^7 elements which were randomly generated by the program. Results are shown in Figure 6.

Here, the performance is not as dramatic as Mergesort. Quicksort relies on picking an optimum pivot element the partitions the list evenly. The optimum pivot number for a list number 1 through 100 would be 51. Picking a bad pivot point, say the smallest value of the list, would result in an algorithm with time complexity $O(n^2)$—like that of Bubblesort. The performance calculations were carried out in the same manner as before.

There is a significant increase in processing power when dealing with large amounts of data. Network traffic also played a major role in the performance of the cluster. As mentioned, this is a Class I cluster where resources are limited and outdated. With the cluster complete and tested, the notion of use of the cluster comes to play.

Design of Lab Interface

The project goal is to design a layout interface for performing lab activities on a cluster of PCs. The main program is stored on the root node of the cluster. Students can upload a program onto the cluster, run it, and monitor the result. The lab allows students to use existing parallel computers, high-performance workstations, and vector computers to experiment using Linux operating system and java interface program. Simple parallel architecture ideas and basic analytical models of parallelism will be presented. The students will be able to run sample C++ program and see and analyze the result.

For this project, the first step was to carefully and manually design the lab layout and sketch the main menu layout. The second step was to add the lab's tasks and lab's activities to the main menu. The following step was to add activities such as print, close, open, save for the labs. The next step was to start thinking about how to automate the process. Java introduced the layout for the GUI (Graphical User Interfaces). It allows fields to automatically grow and shrink depending on how much screen is available. In this project some of the features of great quantity objects are combined with features from Java Layouts to create labs layouts.

The targeted versatility of the use of the lan package is ensured by the following criteria in our development plan:

- **Scalability:** We can add new nodes into the grid or delete nodes from the grid.
- **Extensibility:** The design of the lab environment makes it possible to incorporate other software packages to enhance the functionality.
- **Customizability:** Object-oriented design of the architecture and standardized interfaces of objects make the lab composition easy.
- **Accessibility:** The lab software creates an interface for the users to control node activities through the web browser.
- **Robustness:** The client-side component of the lab software ensures the correctness of the program before loading it onto the grid for execution.

The architecture of the lab package is an extension of the framework of the labs designed for our CSI course. The overall design for this GUI was that it would be simple, reliable, and portable. Although currently the software package developed is merely a prototype, it allows for further extension in accordance with our architectural design depicted as above. The GUI was developed with Java. Java is well known for its stability and portability. The GUI was developed with the Java 2 SDK with netBeans by Sun Microsystems. The main points for the flow of the GUI are:

- Card/Tab-Layout style Window
    - Menu bar with options
    - exit, help
- Tabs will contain:
    - Introduction (background information on clusters/MPI)
    - User window for loading, compiling/building, and running their MPI programs
    - Demonstration Programs
- Sorting Algorithms & sample distribution programs

Layout manager is an abstract class which handles constraints and simplifies implementation of new layouts. It's used as the super class for most of the other layouts. It provides a configurable horizontal and vertical margin around all components. In addition, it has an option to allow the layout to include invisible components in its layout policy. The main menu is designed in a way that it contains all other layouts screens. The main menu frame obtains the tree, the panel, and the upper and lower toolbars. Each toolbar has some action activity provide by the menus or buttons. The panel obtains the lab tasks and activities. The screens cards are controlled by the tree, the panel or the toolbars items (Figure 7).

The GUI was coded utilizing the JSwing packages and forms. All layouts were created as NULL layouts, for it offered more flexibility in placing items such as text windows and action buttons. The GUI was built upon a *Frame* form. Frames are typically used as stand-alone top-level windows as the main user interface to the application. Therefore, it was the optimal choice for the GUI.

The *Introduction* window/tab has general information on the cluster, MPI, and the GUI. A simple JTextpane was added to the base frame that contained the text. Figure 8 is the actual screen shot of the *Introduction* window.

For a user to load, compile/build, and run MPI programs on the cluster, a separate window, called *Your MPI*, was created. Here users can open their MPI program source code, compile & build it, and execute it. This window was added as a JPanel. JPanels are used to place other objects such as buttons and text areas on. On this panel, *Open, Build,* and *Run it!* buttons were added. Actions were assigned to each button to execute the said tasks. Also, a JTextArea was also added to display important messages and instructions. Mouse event listeners were added as reminders for the action buttons. Figure 9(a) shows the window displaying instructions. Figure 9(b), (c), and (d) show actual screenshots of a "Open File", "Build", and "Run it!" action, respectively, in the *Your MPI* window:

The "build" action button builds the opened file. Hence, the "Run it!" button executes the program. Also added was the *Demo* window for users to run sample programs. Within this window are a set of more tabs—one for each sample program. The sorting programs, Mergesort and Quicksort are located here to sample. Also included are the Cpi and a basic I/O program. This GUI offers the bare minimum features for running MPI programs on the cluster. Further development can be made on the GUI in areas of graphics and content.

Figure 9. Interface layout design

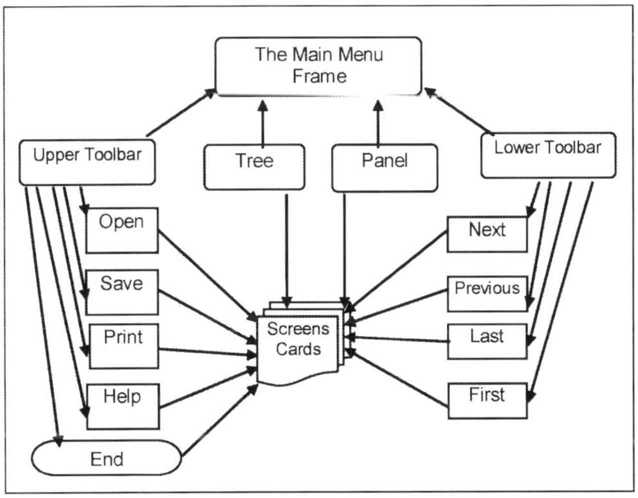

Figure 8. The introduction window

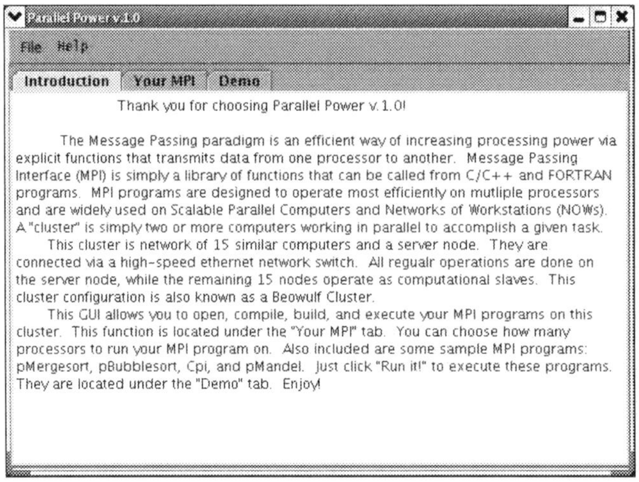

Figure 9. Interface to run a program

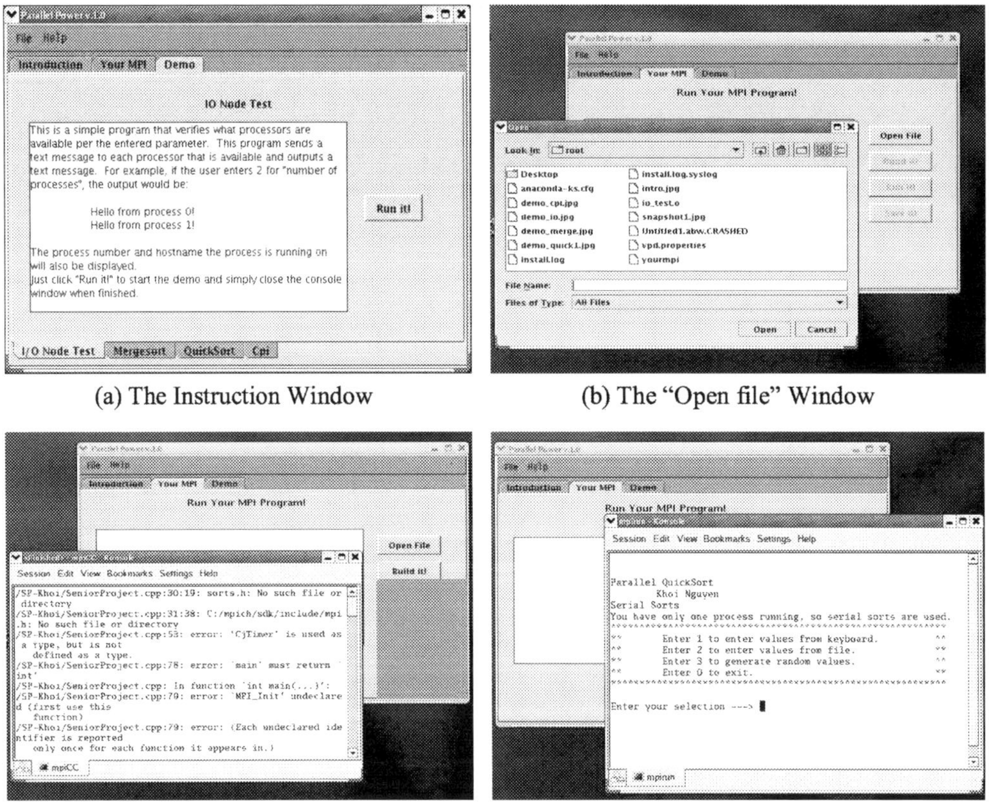

(a) The Instruction Window (b) The "Open file" Window

(c) The "Build" Window (d) The "Run it" Window

Stepping into Applications: Multi-Agent System Simulation

This project approaches Agent technology from the perspective of an e-learning environment. With the prevalence of high speed home networks and increased usage of the Internet, the ability to conduct classes across the Internet is now more feasible. Also, taking courses via the Internet gives people all ready a part of the workforce the ability to complete a college education or continue their education without leaving their job or making sacrifices that affect their family. As universities implement programs for conducting online courses, the need for a distributed system, in order to deliver customized course content, is apparent. Using Agents as a foundation for their distributed learning environment satisfies that need.

For this project, an Agent system is implemented and simulated using the Message Passing Interface (MPI). This document details Agent technology, the design of the e-learning agent system, the design and architecture of the agents themselves, and explanation of MPI and how MPI facilitates the simulation, and use cases for the agent system. The implementation of the e-learning system for this project demonstrates an agent system, using different types of agents to accomplish the goals of the system.

The goal of an Agent system is to address the inadequacies of modern distributed data systems, and this project focuses on distributed learning environments. In order to provide a comprehensive e-learning environment, the system needs to account for:

- Changes in the curriculum—adding or removing course content
- Solving problems that occur—reporting broken web links, and fixing them
- Customizing content presentation—displaying course content based on the individual user's needs
- Matching student models—delivering course content based on the specific student's requirements

Using current technology, the implementation of a distributed system, that satisfies these requirements, results in a complex, difficult to administer, and cost prohibitive solution. The complexity of the system is derived from individual components competing for resources and engaging in complex communications in order to deliver data and satisfy their individual goals.

The last piece to discuss about an agent system relates to the agents themselves. For this project, there are two types of agents to discuss; interface agents and information agents.

Interface Agents

Interface agents are designed to interact directly with the user or other agents. In the e-learning system developed for this project, they are referred to as control agents. They provide a mechanism for accepting commands from a user, processing the command, and delivering the resulting content back to the user. There is an interface agent associated directly with each information agent, and the commands from the user are sent to the information agents to gather the requested data.

Information Agents

Information agents facilitate the mechanisms for data manipulation. They receive command requests from the interface agents and execute the commands. The information agents primarily handle requests relating to adding, updating, and deleting data based on the commands sent by the user to the information agent's corresponding interface agent.

This project implements an agent based e-learning system. Its goal is to simulate the interactions among several agents in an agent based system to facilitate the delivery and manipulation of course content. The e-learning system developed for this project exists in a hierarchical structure. All requests from users are received into a top level agent. The top level agent then process the specific command received and sends the request to a control agent that corresponds to the information agent required to process the command. Once the command is executed, the result of the command is sent back to the top level control agent who then delivers the result to the user. Each information agent has a corresponding control agent that filters the command requested and passes the command on to the information agent. This creates a pairing between a control agent and information agent. The relationship between control agents and information agents is one to many. A single control agent processes requests for all of the information agents it connects to. Figure 10 depicts a high level diagram of the hierarchical structure of the agent system.

Figure 10 also shows the distribution of work based on the user that will be interacting with a specific agent. The Instructor, Student, and Registrar designation help describe the type of processing that will be done by that agent. Also depicted in Figure 10, are 5 types of agents that make up this system. They include the Master Agent Control Client, Agent Control Client, Maintenance Agent, Notification and Recommendation Agent, and Student Information Agent. Communications among the agents occur through the control clients. The system is designed such that, no agent can communicate directly with any agent on the same level in the tree hierarchy. For example, in order for the Maintenance Agent to request information from the Student Information agent, it must pass the request to the Agent Control Client, which then passes the request to the Master Control Client. The request is then sent down to the Student Information Agent. This structure insures an Agent's autonomy by not creating inter-dependencies among the Agents.

Master Control Agent

The master control agent serves multiple purposes. First and foremost, it is the entry point into the

Figure 10. Hierarchical structure of the agent system

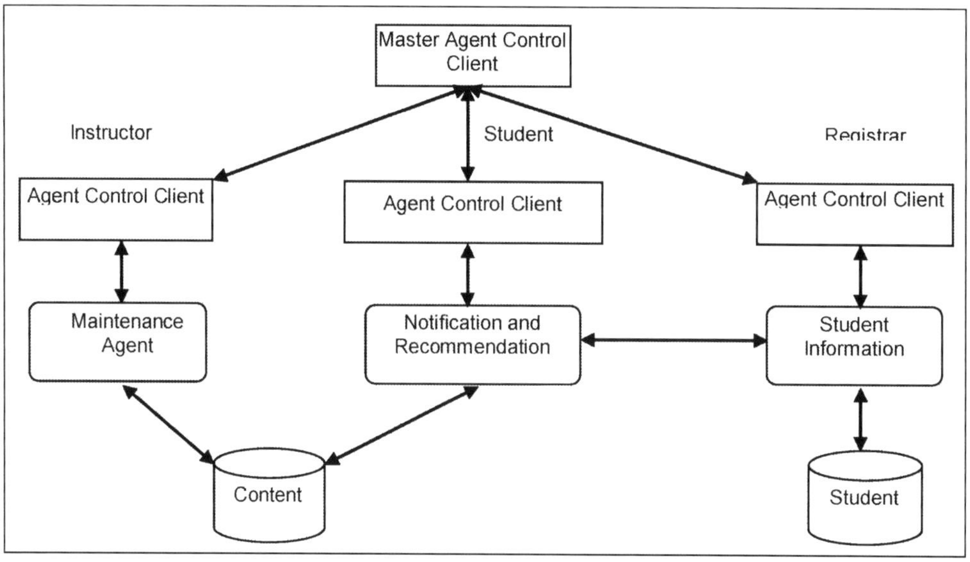

system. All user requests are filtered through this Agent and passed along to the appropriate Control Agent to process the request. There are two methods that a request can be sent to the system. The first, to facilitate online courses, is through a network socket used to receive requests from a web server into the system, and the second method is to pass commands into the system via the command line in the form of a text file. Second, the Master Control Agent is the primary MPI IO process; this is discussed in greater detail in the *Implementing MPI in an Agent System* section. Third, the Master Control Agent receives the result of a command and passes it back to the user. Lastly, the Master Control Agent controls the creation and deletion of the Control Agents.

Control Agent

The control agent manages requests to a specific information agent. With respect to the implementation for this project, the Control Agent receives a command from the Master Control Agent. Once the command is received, the Control Agent verifies that the command can be executed by one of its information agents, and then passes the command to the information agent. After the command has been executed, the control agent receives the result from the Information Agent and passes the result to the Master Control Agent. Also, the Control Agent manages the creation and deletion of its information agents.

Maintenance Agent

For this implementation, the maintenance agent corresponds directly with functions executed by an instructor. It processes requests directly relating to the management of course content and class interactivity. In addition, the Maintenance Agent handles the messages sent by students to an instructor and provides the ability for an instructor to send a message to a class or individual student. The Maintenance Agent also has the ability to retrieve a student listing from the Student Information Agent for a specific class.

Notification and Recommendation Agent

The Notification and Recommendation Agent relates to the activities of a student. It delivers the course content to the student as well as any messages sent to the class or from another student. In addition, the Notification and Recommendation allows a student to send messages to the instructor or another student.

Student Information Agent

The student information agent corresponds to the operations of the university registrar. This Agent manages all of the student and class data. The Student Information Agent allows students to be added and removed from a class, retrieve the list of students, retrieve a student's information, and the addition and deletion of a class.. This Agent has the highest level responsibility simply because it manages the student information and any changes to a student. In the case of adding a student, the Student Information sends the request to create a Notification and Recommendation agent for the individual student. This operation allows the Notification and Recommendation Agent to begin processing requests on behalf of the student and customizing class content for delivery.

The Master Control Agent and Control Agents both have the responsibility of managing the creation and deletion of agents. This behavior insures the reliability of the system by not allowing a command request to go unprocessed. This behavior is also handled automatically, meaning that specialized code does not be added to additional agents to control agent creation and deletion.

Message Passing Interface (MPI) is the technology used in this project to simulate the environment that an Agent system would be operated in. Traditionally, MPI is the foundation for distributed parallel communications between running processes on the same machine or on different machines. MPI is primarily used to build applications for problem solving on datasets that would require large amounts of time to calculate using a single a process. However, the versatility of MPI has enabled this technology to be used for this project.

The role of MPI for this project is to facilitate the communication mechanism between the agents. A request sent to the system is packaged into an MPI data structure and sent to the correct information agent using the *MPI_Send* and *MPI_Recv* commands. This communication behavior is common for all of the agents.

The way MPI handles IO also corresponds to the structure of the Agent System developed for this project. MPI allows for a primary IO source that divides the workload and initiate communication with the other process. In the case of this project, the Master Control Client corresponds to this IO source for MPI, thus all job requests originate from the Master Control Client.

The one limitation presented to this project from MPI is the inability for MPI to spawn or create processes. As described earlier, the control agents control the creation and deletion of agents. With the inability to create a new Agent process, a flag was added to each Agent to indicate whether or not an agent existed. In this case that the agent was created, a command was sent to create the agent, thus emulating the procedure required to create a new Agent. Due to the inability to create a new Agent process, the Control Agents have now way maintaining a one to many relationship with the corresponding information agents. Instead the MPI limitation keeps the ratio as a one to one relationship.

The E-Learning Agent system was developed exclusively using the C++ programming language. In order to maintain compatibility with MPI, all of the agents are derived from a single program executable. Each Agent's job function is determined by the process ID provided to each process by the MPI system. The Master Control Agent is assigned process 0. Agents use: if(process ID % 2 == 1) to determine the control agents, and the information Agents for each Control Agent use the result of process ID / 2 to determine which information agent they will be. Figure 11 illustrates the startup sequence for each agent and the determination of the type of agent.

In order to provide a mechanism for a web server to communicate with the Agent system, a CGI program was required to gather the data

Figure 11. Flowchart of agent startup

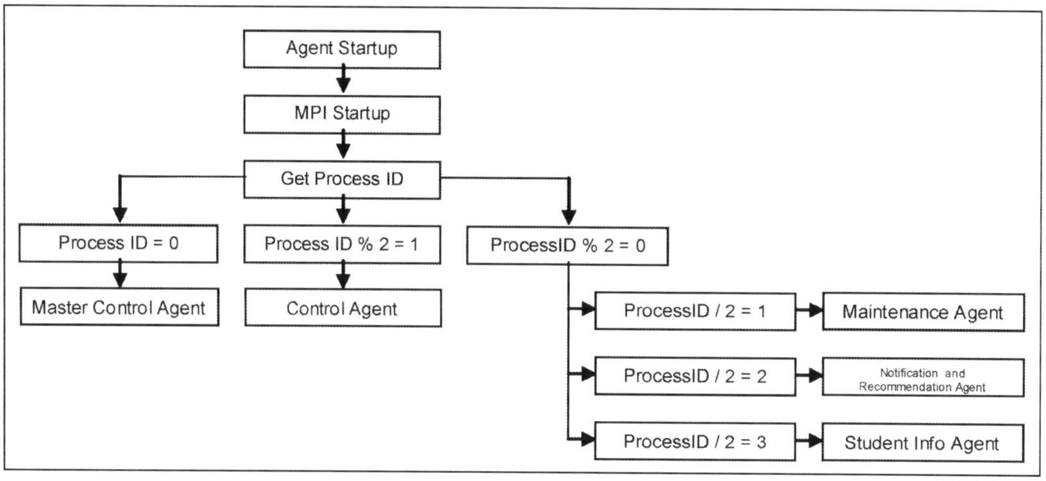

submitted by the user and send the data to the agent system. WEBAGENT.EXE is the program developed to satisfy this requirement.

The Master Control Agent provides a socket interface that can be used to communicate with the agent system. Described below is the process that WEBAGENT.EXE uses to send a request to the Agent system.

- The data from the web forms are copied from the QUERY_STRING environment variable
- The QUERY_STRING values are translated into an INPUTJOB structure
- The INPUTJOB structure is translated into a string value
- The string value containing the INPUTJOB structure is sent to the Agent system
- WEBAGENT.EXE receives a string from the Agent system containing the result of the command operation
- The string received from the Agent system is translated into a Result Code and Result Message pair
- From the Result code and Result message pair, the web page is built and sent back to the web server

WEBAGENT.EXE provides the mechanism for connecting the web server to the Agent system.

Figure 12. Student posters

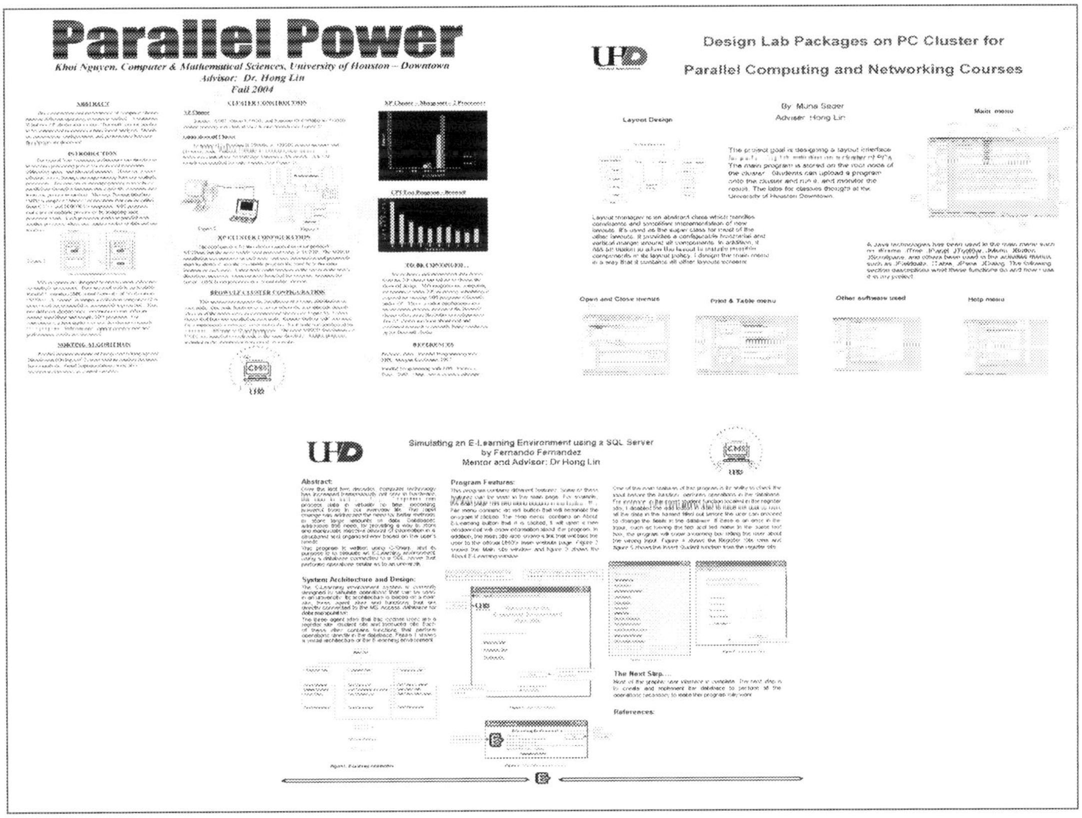

All the student projects have been presented in Scholar's Academy Student Research Conferences. Figure 12 shows a collection of posters of the above projects.

CONCLUSION

We present a method for designing a computational grid that supports online lab exercises, as part of our Information Technology track of curriculum design. A lab package is designed to support the learning process in courses of parallel computing and networking. The grid is centered at a Beowulf cluster, which provides a computational backbone of the grid, and services are deployed in distributed nodes of the computing networks and organized by a multi-agent system. To address high level architectural design issues, such as scalability, extensibility, and modularity, we use Chemical Reaction Model to formally specify the architecture and we facilitate a transformational method for implementing the system to the module interface level. We have developed the lab with an interface that accommodates different lab activities in different courses and demonstrated the design by show cases. Students have been involved in the implementation of the laboratory in forms of senior student projects and SA sponsored research projects. This makes our research coupled with education tightly.

ACKNOWLEDGMENT

This research is partially supported by NSF grant "Acquisition of a Computational Cluster Grid for Research and Education in Science and Mathematics" (#0619312). Some of the student research projects are supported by U.S. Army Research Office Award #W911NF-04-1-0024 through Scholars Academy of UHD.

REFERENCES

Banâtre, J.-P., & Le Metayer, D. (1990). The Gamma model and its discipline of programming. *Science of Computer Programming, 15*, 55-77.

Banâtre, J.-P., & Le Metayer, D. (1993). Programming by multiset transformation. *CACM, 36*(1), 98-111.

Flores, R.A., Kremer, R.C., & Norrie, D.H. (2001). An architecture for modeling internet-based collaborative agent systems. in Wagner, T., & Rana, O.F. (Eds.), Infrastructure for Agents, Multi-Agent Systems, and Scalable Multi-Agent Systems (pp. 56-63), LNCS1887, Springer-Verlag, London, UK.

Griss, M., & Pour, G. (2001). Accelerating development with agent components. *Computer, IEEE, 34*(5), 37-43.

Lin, H. (2004). A language for specifying agent systems in E-Learning environments. in: Lin, F.O. (ed.), Designing Distributed Learning Environments with Intelligent Software Agents (pp. 242-272). IGI Global, Hershey, PA, USA.

Lin, H., & Nguyen, K. (2005). Classroom simulation of massive parallel computing. Proc. The 16th IASTED International Conference on Modeling and Simulation (MS 2005), Cancun, Mexico, May 18-20 (pp. 45-50). ACTA Press, Calgary, AB, Canada.

Lin, H., & Yang, C. (2006). Specifying Distributed Multi-Agent Systems in Chemical Reaction Metaphor. *The International Journal of Artificial Intelligence, Neural Networks, and Complex Problem Solving Technologies, Springer-Verlag, 24*(2), pp.155-168.

Paquette, G. (2001). Designing Virtual Learning Centers. In H. Adelsberger, B. Collis, J. Pawlowski (Eds). Handbook on Information Technologies for

Education & Training (pp. 249-272), Springer-Verlag, Berlin et al.

Vouk, M.A., Bitzer, D.L., & Klevans, R.L. (1999). Workflow and end-user quality of service issues in web-based education. *IEEE Trans. on Knowledge and Data Engineering, 11*(4), 673-687.

Weiss, M. (2003). A gentle introduction to agents and their applications. Online presentation at http://www.magma.ca/~mrw/agents/, 2004.

Yu, E. (2001). Agent-oriented modelling: software versus the world. Agent-Oriented Software Engineering AOSE-2001 Workshop Proceedings (pp. 206-225), Montreal, Canada, May 2001. LNCS 2222. Springer-Verlag, Berlin, Germany.

Chapter IX
Supporting Arguments for Including the Teaching of Team Competency Principles in Higher Education

Tony Jewels
Queensland University of Technology, Australia

Rozz Albon
Curtin University of Technology, Malaysia

ABSTRACT

For optimum workplace effectiveness in knowledge-intensive industries in which principles of knowledge management need to be applied, it is necessary to take into account not only the competencies of individuals themselves but also the competencies of the teams in which they must operate. Although the incorporation of various types of group work into pedagogies is already widespread within institutes of higher education, many examples fail to embrace a rationale for, or the potential benefits of, multiple contributor environments. We present in this chapter arguments for including the teaching of team competency principles in higher education, supported by an original multi-dimensional team competency teaching model, a taxonomy for assessing team competency levels and an example of the implementation of these principles.

INTRODUCTION

Though the importance placed on knowledge is increasingly being recognized, applications of knowledge management principles are still inconsistent, the topic and even its definitions are still being widely interpreted (Von Krogh, Ichijo & Nonaka, 2000). The complexity of problems in our

Figure 1. Multi-dimensional team competency teaching model

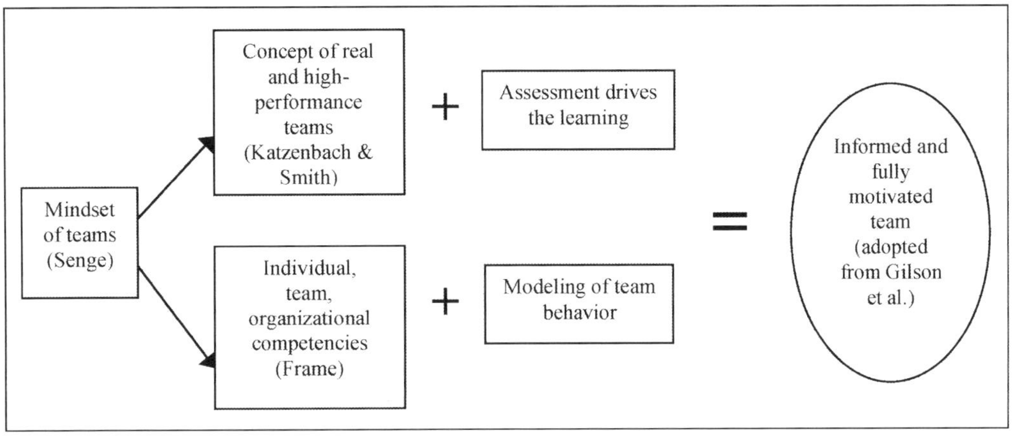

knowledge society requires that problem-solving activities be shared across disciplinary, cognitive, geographic and cultural boundaries (Leonard-Barton, 1995), with Jewels and Underwood (2004, p. 1) synthesizing these and providing a definition of knowledge management as *the collection and processing of disparate knowledge in order to affect mutual performance.*

It is expected that when most graduates enter the professional workplace, their ability to work as a team member will contribute to the team's immediate levels of productivity. Assumptions could once be made that graduates would enter a university or the workforce with an adequate degree of 'teamness' or team competencies acquired through a childhood of formal and informal team activities (such as sport). Over many years, team competencies were practiced and developed by the individuals themselves: they did not require teaching intervention of any kind. However, the advent of computers and the Internet has impacted on social activities of children, along with the already felt impact of television. It appears less time is now spent in team sport activities, and when considered cumulatively over a period of many years, such children are now entering universities less skilled in team competencies. Coupled with the increased need for team skills in the information age, as outlined further in this paper, we believe that it is important to attend to the development of team skills in training and university curricula.

Though various types of group work have already been incorporated into higher education pedagogies, many examples fail to embrace the potential benefits of multiple contributor outputs in knowledge-intensive environments. While perhaps being ideal candidates to capitalize on the benefits of knowledge-sharing behaviors, higher education, has generally not realized its potential. There has, according to Senge (1992), never been a greater need for mastering team learning in organizations.

Team learning is vital because teams, not individuals, are the fundamental learning unit in modern organizations (p. 10).

Until we have some theory of what happens when teams learn (as opposed to individuals in teams learning) ... Until there are reliable methods for building teams that can learn together, its occurrence will remain a product of happenstance (p. 238).

Synthesizing works from multiple authors (Senge, 1992; Katzenbach & Smith, 1993; Frame, 1999; Gilson, Pratt, Roberts & Weymes, 2000), we propose an original multi-dimensional teaching model that provides a foundation for discussion of the rationale for teaching team competencies (Figure 1).

Further, we propose a taxonomy for assessing these team competencies at different levels of team maturity.

Knowledge Within Teams

Forty years ago, Drucker (1964) defined knowledge workers as those people with a high degree of formal education who apply knowledge to work, rather than manual skill or brawn. There is now an increasing awareness that the knowledge that had always been residing tacitly with workers can be made explicit by capturing and codifying it for the purposes of re-use, transfer and the creation of new knowledge (Nonaka, 1991).

However, Taylor's scientific management principles (1967) support the notion that it was only management who understood both the processes that workers undertake and the links between all the various processes in the production chain. The workers themselves needed to be instructed on how to perform these tasks more efficiently and were not encouraged to develop more efficient ways of performing their tasks (Grant, 1997). In this type of cultural environment, management hardly felt a need to capture and codify the knowledge of the workers, as it was believed that the workers themselves contributed little to the knowledge processes embedded in their work, exemplified by the quotation:

I can say without the slightest hesitation that the science of handling pig-iron is so great that the man who is fit to handle pig-iron and is sufficiently phlegmatic and stupid enough to choose this for his occupation is rarely able to comprehend the science of handling pig-iron (Dubofsky, 1975, quoting Taylor in Grant, 1997).

Figure 2. Schema of pedagogical options (Barnett, 2004)

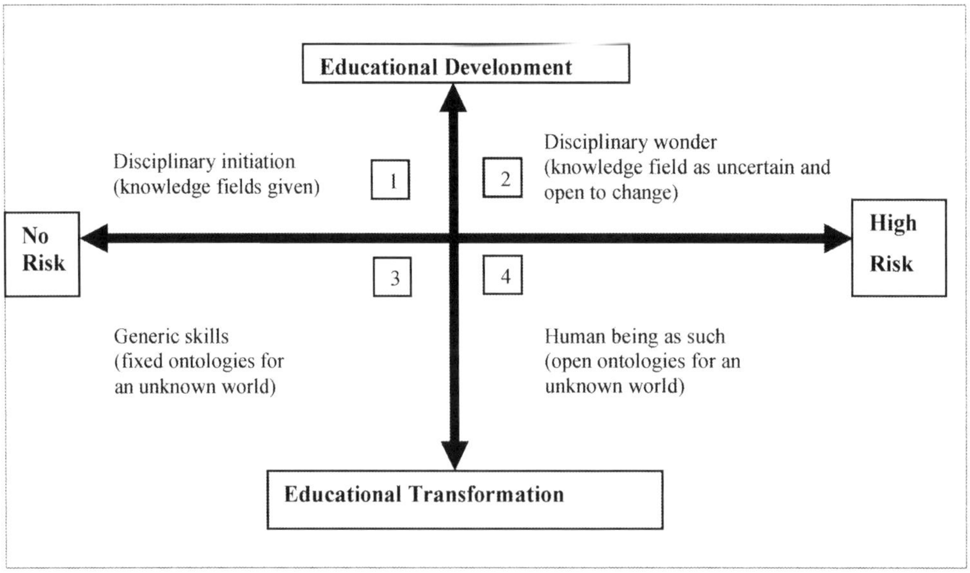

There is now, however, according to Grant, an implicit acceptance by management that workers are able to provide worthwhile knowledge regarding their activities. Though much of the current literature discusses the role and importance of the type of work that knowledge workers perform, there is still relatively little literature that contradicts the fundamental scientific management approaches of Taylor that places little value on the knowledge contributions of workers. These approaches may be outdated, having been developed for an industrial era, yet still being incorporated in pedagogies within our present knowledge society. Team competencies underpin the effectiveness of knowledge workers, thereby creating an imperative by higher education institutions to incorporate them into teaching practices. Newer and more innovative approaches are required to enable graduates to be effective knowledge workers and producers.

Pedagogy

It would appear that collaborative learning as a group approach as distinct from cooperative learning continues to monopolize the intention of teaching students to learn to work with others, a goal synonymous with team learning. The emergence of newer online learning approaches, such as 'inter-group collaboration,' still emphasizes knowledge access as distinct from knowledge sharing (Palloff & Pratt, 1999) dependent on the co production of knowledge, which itself is dependent on particular contexts or environments in which learning is socially situated (Brown, Collins, & Duguid, 1989); that is, learning cannot be separated from the situations in which it is to be used. Abstractions, or in many cases, theories, if not grounded in multiple contexts, will not transfer well, with Brown et al. (1989) emphasizing

it is not learning the abstraction, but learning the appropriate circumstances in which to ground the abstraction that is difficult (p. 19).

In addition to the arguments presented for the need to develop team competencies, Barnett (2004) has approached the same needs of society and the future workforce from a curriculum and pedagogical perspective.

Neither knowledge nor skills, even high-level knowledge and advanced technical skills, are sufficient to enable one to prosper in the contemporary world. Other forms of human being are required (p. 253).

These other forms, Barnett believes, are associated with individuals themselves, such as the confidence seen in students' willingness to speak to enable them to go forth into a challenging world. Confident and successful students know their knowledge and skills may be contested and yet they know, too, that they

... have to launch themselves forth into a world that will furnish responses that cannot be entirely anticipated (Barnett, 2004, p. 253).

Educators must prepare students to survive in this kind of future world and advanced knowledge. Curriculum designers must create pedagogical practices that ensure students can launch themselves into this unknown future. Barnett recognizes this in stating *"one's being has a will to go on"* (p. 254) and suggests that this self-energizing and self-propelling human dimension is included in curriculum design. To this end, Barnett has developed a schema of pedagogical options and we are proposing that within his options we include team learning and the development of team competencies. Barnett proposed a schema based on four quadrants developed through the crossing of two axes. The vertical axis spans educational development at one end and educational transformation at the other. The horizontal axis spans no risk to high risk. The crossing of the axes creates four quadrants, as illustrated in Figure 2.

Barnett reasons that the pedagogy of the first three quadrants can only carry society so far; even creative and imaginative chemists have their limitations. A greater transformation is required if we intend to prepare students for an unknown future.

Barnett highlights a paradox of this pedagogy:

that claims to be able to bring students out of their academic domains into forms of human being more adequate for a changing world than a more purely academic curriculum could offer (no matter how creative), but it does so by attempting to specify clearly the skills that are to be developed among the students. In short, we are confronted in this idea of education with the nonsense belief that we can generate human being for uncertainty through a new kind of certainty in the curriculum (p. 256).

The core of success for the future appears to reside in this element of the human being. We have assumed that many types of teams feature in the pedagogy, as it is in teams that the human element is grown and nourished, but the reference to individual marks would also feature.

Curriculum in quadrant 4 offers the possibility of what is required to equip students for the future being underpinned by transformation and high risk. Achieving in this quadrant requires a matching pedagogy, one in which outcomes are not tight and specific. Barnett believes that in the heart of curricula in this quadrant there will be an exposure to dilemmas and uncertainties emerging from complexities within a discipline, but requiring the engagement of the human being itself. This quadrant calls on the functioning of the highest performing teams to develop the human qualities Barnett speaks about.

This conceptualization of curriculum development by Barnett aligns itself to our position presented in this article. Dilemmas and uncertainties can best be addressed by crossing disciplinary boundaries, ensuring the world is seen from different perspectives. Taking this further, students must be able to work in high-performance teams to arrive at the resolutions of dilemmas in the most productive and efficient way. Educators have a responsibility to not only include the development of team skills as process in learning, but develop an assessable and authentic approach to team learning.

If we accept the logic and framework proposed by Barnett for workers of the future to be experienced in high-performance teams together with the arguments previously cited, it is even more imperative that team learning be developed and incorporated into university assessment approaches and pedagogy. We have offered the taxonomy as a model with which to develop team competencies as well as a model for the development of learning experiences and assessment.

Teaching team competencies should extend beyond, for example, merely requesting groups of students to produce a report in which individuals can adopt a jigsaw approach (Biggs 2003), where each individual places his or her piece in the final task or puzzle.

The traditional and popular belief is that it is the individuals within organizations, and not the organizations themselves, that learn (Simon, 1976; Weick, 1978). Yet there is now a proliferation of the use of 'teams and communities' in the literature according to Ferrán-Urdaneta (1999), with Senge (1992) describing the types of teams we are discussing:

...where new and expansive patterns of thinking are nurtured, where collective aspiration is set free, and where people are continually learning how to learn together (p. 3).

Management approaches developed for an industrial era are still applied in a new environment, widely referred to as an 'information or knowledge era.' Described by Toffler and Toffler (1995) as 'third wave,' and by Drucker (1993) as 'post-capitalist society,' this era demands new and innovative teaching practices that truly reflect multi-contributor environments in professional practice.

It would appear that in order to promote team competencies educators not only need to incorporate the core tenets of sharing knowledge but need also to understand the fundamental differences between teams and groups. Group work undoubtedly has a place in learning as one strategy that develops particular skills, such as communication and providing avenues to practice small and discrete skills. Contrasting with traditional group work, learning using team principles is a significantly different approach to knowledge sharing that harnesses the synergy of collective knowledge.

Traditionally, higher education predominantly assesses only at an individual level, yet for optimum effectiveness it is necessary to also take into account the competencies of the teams in which those individuals operate. To advance the teaching of team competencies and its inherent shared knowledge, a conceptual framework is required; one that will embrace the synergy and energy created when individuals aspire to excellence and are intrinsically motivated to accept challenge in dealing with conflict, in order to arrive at new knowledge.

Comparing Group Work with Team Work

Team development includes both enhancing the ability of individuals to contribute to the team as well as enhancing the ability of the team to function as a team (as distinct from a group of individuals). In effect, the higher-level or most productive teams are looking for solutions that no individual could identify, but which a team could. In his use of the term 'egoless team,' Weinberg (1971) refers to a team in which individuals are able to subordinate their desires to that of the team. This concept of subordination encapsulates the ethos that must underpin the teaching of team competencies in the higher education contexts, aptly termed team-centered learning, as distinguished from teacher-directed or student-centered learning.

Table 1. Five types of teams and their characteristics (Katzenbach & Smith, 1993)

Group/team type	*Characteristics*
Working groups	Members interact primarily to share information, best practices or perspectives to help each individual perform within his or her own area of expertise
Pseudo teams	No interest in shaping a common purpose and interactions detract from each individual's performance without delivering any joint benefit.
Potential teams	Require more clarity about purpose, goals or work products and lack discipline in approach. Still to establish collective accountability.
Real teams	Small number of people with complementary skills equally committed to a common purpose, goal and working approach for which they hold themselves mutually accountable.
High-performance teams	Meet all the criteria for real teams and have members deeply committed to one another's personal growth and success. That commitment usually transcends the team, which significantly outperforms all other like teams and reasonable expectations, given its membership.

In their book *The Wisdom of Teams,* Katzenbach and Smith (1993) describe five levels of group/team and their key characteristics (Table 1). Much of the multiple contributor work currently being conducted within institutes of higher education is not, in reality, team work at all, but reflects the characteristics of the poorest of the five levels of group/team performance described by Katzenbach and Smith (1993), where effective knowledge management practices are unlikely to occur.

In describing effectiveness in project management environments, Frame (1999) discusses levels of competency relating to the individual, team and organization. These levels are incorporated into our proposed teaching model.

It is acknowledged that assessment impinges on the learning outcomes for students. Assessment is integral to the learning cycle as students continue to process information over the duration of the task. However, the depth and kind of learning as an outcome from engaging in any assessment task relates to the nature of the task. If assessment is founded on students completing memory tasks only, then learning is shallow. If assessment is structured around responding to or building case studies, problems and solutions, scenarios and simulations, then learning is deeper, and more meaningful. The latter kinds of assessment foster additional skills and attitudes not normally acquired in memory-only tasks. It would appear, therefore, that a deep approach is essential to the assessment of team competencies. It would follow that assessment drives the learning; that is, if team competencies are valued, there is little alternative but to structure assessment to provide students the opportunities to feel and respond to functioning as a team. Opportunities must be made available to students to acquire and practice team competencies, and these in turn must be assessed according to team rather than individual output.

Teaching team competencies takes time.

It is not possible to wave a magic wand and create a high-performing, self-managed team overnight. A self-managed team needs a culture of life-long individual and team learning (McCann 2005).

McCann identified four components to team learning - questioning, valuing diversity, communicating and learning as an iterative process - that compliment Katzenbach and Smith's (1993) collective accountability, each of which (with practice) develops over time.

Students learn to be a team by functioning as a team. Designing assessments to include trust and commitment to the welfare of the group is an essential step in teaching team competencies. When there is trust, there will be cohesion, and a cohesive team in turn enables students to function productively and effectively. There are likely to be disagreements and debate, but the outcomes will be healthy debates in which members hold the good of the team as their prime concern. The 'egoless team' will evolve.

To accommodate the process and time it takes for the development of team competencies, we suggest that the acquisition of team competencies be included in all years of study in a degree program, beginning with the first year of study. Emerging from the necessary struggles and adaptations of working in teams over an extended period of time should be a culture of team functioning. Implementing assessment to reflect this culture will ensure that teams are built and maintained.

Developing a Team Mindset

It is paradoxical that when referring to team competencies we must also acknowledge the individual competencies of each team member. Although the literature does not clearly identify personal qualities that might contribute towards 'teamness,' Belbin (1981) has proposed an ideal mix of individuals that might contribute to an effective team. While still supporting Belbin's 'bottom-up' approach, we are more concerned with developing a strategic 'top-down' approach for identifying and providing the individual qualities necessary for contributing to team competency.

Much of the literature still discusses how leaders of teams can motivate team members into behaviors that will effectively contribute to more effective organizational outcomes (Jay, 1995; Wellins, Byham & Wilson, 1991), whereas Frame (1999) provides an alternative perspective of team competency, listing the functions carried out by good teams to achieve successful outputs. Although not inferring comprehensiveness, Frame at least provides a starting point in which to teach team competencies to the team members themselves rather than only to team leaders.

Expanding on this output function approach and to further identify characteristics of high-performing teams, Gilson et al. (2000) use an example of a 1995 New Zealand America's Cup syndicate who desired a team with the following characteristics:

We want a small, informed and fully motivated team that:

- *Works in an environment which encourages every member to make a meaningful contribution.*
- *Has a high degree of personal integrity and honesty.*
- *Recognizes personal goals but not hidden agendas.*
- *Continuously monitors and improves its performance.*
- *Is fun to be in* (p.221).

This description is, we believe, an exemplar of high-performance teams and one that can be incorporated into teaching team competencies.

Academics may already be implicitly, mostly unconsciously, engaged as members of teams in addition to their valued and recognized individual roles within higher education environments. To teach team competencies, there must first be a recognition of or a change in their mind set, a concept raised by Senge (1992) to explicitly acknowledge the teams in which they work. For too long, universities have rewarded the individual at, what we would claim, the expense of the effectiveness and possible exemplary outcomes of already existing teams - if only they were acknowledged. We propose that the way forward to teach team competencies is first for educators to adopt the principles themselves, and subsequently apply these in their teaching using a team-centered approach. In this way, educators can truly *model* the team competencies they intend to teach, thus accelerating the understanding of all team competencies in the student population.

Modeling of Team Behavior

When academics also model the team approach, then students have an optimum learning environment in which to learn team competencies. To help educate potential IT project managers, the faculty of IT of one Australian university provides a carefully structured IT project management (ITPM) subject, to provide students with the most appropriate skills and relevant knowledge to prepare them for the workforce, its learning objectives focusing on both project success and the project manager's role in taking responsibility for their projects (Jewels & Bruce, 2003; Jewels & Ford, 2004).

The teaching team responsible for delivering the unit has itself been the recipients of both faculty and university teaching awards for group (team) categories. Usually numbering around eight people, the team is comprised of full-time academic and administrative university staff, postgraduate students and sessional (part-time) staff working in project management fields. Over the 3 years that the unit has been offered, there has yet to be a team member who has willingly given up his or her position; a testament both of their individual commitment to the goals of the unit and to the 'fun' nature of its delivery.

Within a 13-week curriculum for final-year IT students, a single theme is made explicit by continually referring to a 45-year-old quotation:

A project manager's primary tool is the brainpower of other people who are professional specialists in their own fields(Gaddis, 1959).

Most ITPM students will have had earlier experience of group work, but many have difficulty in comprehending (and in many cases, believing) the statements made by the lecturer in the first week, that:

Your most valuable resource in this unit is your fellow students

and

This unit will not be subject to bell-curve marking ... you can all earn grades of 7 (high distinctions), as you are not competing with each other.

In addition to the traditional 'hard' skills, such as methodologies, processes and tools, the content of the unit also includes a number of specific team-related issues:

- Stakeholder analysis
- Team dynamics/group conflict
- Conflict resolution/personality type
- Communication practices
- Organization culture
- Knowledge sharing in projects

Each week, a case incident is drawn from the same single case study used throughout the unit involving a divergent rather than convergent problem (Schumacher, 1977). It requires students to 'dig' around and thus construct their own conception of the possible problems and solutions. Students were originally required to address each week's problem on their own, and their individual solutions were discussed in a tutorial session where the diversity of the problem-solving approaches are made obvious, thus reinforcing the notion that regardless of how well an individual might be able to address an issue, others would be able to provide alternative solutions. More recently, an experimental variation to the unit design was undertaken to allow multiple students to work on weekly problems. Trialing different partners in this manner allowed students to select appropriate partners for the unit's two team assignments, representing 55% of the unit assessment.

Although a formal online discussion forum is available to students in which commonly asked questions can be answered by students themselves, it is not a highly utilized method of interaction; students prefer to engage in more informal discussions, either online or off-line.

Although the teaching team continues to refine its team competency teaching model in terms of modeling its own behavior, in providing appropriate assessments and providing team competency content it is uncertain to what extent students have fully understood the desired learning objectives related to team competency. Unfortunately, the ITPM subject is but one unit in a 36-unit course, and it might be overly ambitious to expect that any single unit could dramatically influence the mind-sets of students regarding the importance of 'teamness.'

Assessment Drives the Learning

It follows that the arguments we have presented for the development of team competencies must also follow through so that team competencies are legitimately included in university assessments. These arguments contributed to the development of the taxonomy described in Table 2, which is a synthesis of works from Katzenbach and Smith, Barnett and Frame.

Table 2. Taxonomy for the assessment of team competency maturity

Levels of Group/ Team Maturity	**Examples of Competency**		
	Individual	Team	Organizational
Working Groups	Individuals are only nominally a group coming together to report on individual progress	Members help each other at a peripheral level in the belief that each member can best perform for the group by working individually	The organization only expects group members to provide individual inputs. Tendency to reward individual, not group, performance
Pseudo Teams	Psychologically, members know they must contribute to team output, but cannot see beyond their own view and perspective	Members know their contribution must interact in the final product and so are prepared to assist each other. Effort is made to listen to and respond to each other	Explicit acknowledgement of value of teams, but no resources or incentives given for team output
Potential Teams	Members recognize their individual responsibilities, but yet to recognize team responsibility	Members recognize personal skills and those of others; are aware how these can contribute to the success of the team project; have nominated strategies; but lack collective accountability	Support given for overall team performance but lack of acknowledgement of the individual's team responsibilities
Real Teams	Prepared to up-skill and do additional work as part of accountability to team	Members hold themselves mutually accountable for the project's direction, development and outcome	Support and resources given to teams and the individuals in them for current work
High-Performance Teams	Members recognize each other's strengths and weaknesses and how the final project can be shaped by these factors	Members all deeply committed to each other's personal growth and success. Contribute so members' contributions are optimized for the collective good	Support provided to teams and individuals for personal growth that is focused toward current and future work

The taxonomy enables group or team work to be defined and, in turn, enables educators to set criteria for assessments in accordance with the expectations of each team description or level.

Establishing this taxonomy begins to tease out what it is we expect students to be actually doing when placed in groups/teams. Currently, it seems that the confusion educators experience about whether to allocate individual or team marks is arbitrary. The taxonomy provides guidelines for establishing the expectations for each type of team, and the design of the assessment should reflect the particular team maturity level expected of students at a particular point in time.

The understanding of the development of teams is one approach enabling students to achieve skills and critical insight required of the future. Barnett proposed in 1997 that higher education needs to dispense with the notions of teaching and learning and acquire a different vocabulary to address a different way of approaching education. The clusters of concepts he proposes are neither singularly student-based/centered or problem-based, but are a rich, complex and dynamic approach capable of proliferating team competencies leading to critical action.

CONCLUSION

A British economist, more than a century ago, said that:

the full importance of an epoch-making idea is often not perceived in the generation in which it is made. ... A new discovery is seldom fully effective for practical purposes until many minor improvements and subsidiary discoveries have gathered themselves around it (Marshall, 1920).

Although introducing team competency principles into curricula appears to be a necessity, the ideas implicit within this teaching may not be fully appreciated by students until there are more units embodying the principles into their own subject matter, making it a course core outcome.

This article has examined the need for teachers to apply team competency principles to address and accommodate the dynamic, fast-paced world in which knowledge management is a feature. A new team-centered pedagogy to team learning has been presented and supported by examples of successful teaching practices. IT educators have a responsibility to their graduates to prepare them to be managers of knowledge in an information-knowledge era.

REFERENCES

Barnett, R. (2004). Learning for an unknown future. *Higher Education Research and Development, 23*(3), 247-260.

Belbin, R. (1981). *Management teams: Why they succeed or fail.* Oxford: Butterworth-Heinemann.

Biggs, J. (2003). *Teaching for quality learning at university* (2nd edition). Buckingham: Society for Research into Higher Education and Open University Press.

Brown, J.S., Collins, A. & Duguid, P. (1989). Situated cognition and the culture of learning. *Educational Researcher, 18*(1), 32-42.

Drucker, P.F. (1964). The big power of little ideas. *Harvard Business Review, 42*(3), 6-8.

Drucker, P.F. (1993). *Post-capitalist society.* New York: Harper Collins.

Dubofsky, M. (1975). *Industrialism and the American worker.* New York: Crowell.

Ferrán-Urdaneta, C. (1999). *Teams or communities? Organizational structures for knowledge management.* Proceedings of SIGCPR 99, New Orleans, LA.

Frame, J.D. (1999). *Project management competence.* US: Jossey-Bass.

Gaddis, P.O. (1959). The project manager. *Harvard Business Review, 32*(May-June), 89-97.

Gilson, C., Pratt, M., Roberts, K. & Weymes, E. (2000). *Peak performance: Business lessons from the world's top sports organizations.* Netley SA: Harper Collins Business.

Grant, R.M. (1997). The knowledge-based view of the firm: Implications for management practice. *Long Range Planning, 30*(3), 450-454.

Jay, R. (1995). *Build a great team.* London: Pitman Publishing.

Jewels, T. & Bruce, C. (2003, June 24-27). *Using a case method approach in an IT project management curriculum: A long look over the shoulder of a practitioner at work.* Proceedings of the Informing Science + IT Education Conference, Pori, Finland.

Jewels, T. & Ford, M. (2004). A single case study approach to teaching: Effects on learning & understanding. *Issues in Informing Science and Information Technology, 1*, 359-372.

Jewels, T. & Underwood, A. (2004). The impact of informal networks on knowledge management strategy. In B. Montano (Ed.), *Innovations of knowledge management* (Chapter 1). Hershey, PA: IRM Press.

Katzenbach, J. & Smith, D. (1993). *The wisdom of teams*. Boston: Harvard Business School Press.

Leonard-Barton, D. (1995). *Wellsprings of knowledge: Building and sustaining the sources of innovation*. Boston: Harvard Business School Press.

Marshall, A. (1920). *Principles of economics*. Retrieved August 2, 2002, from www.econlib.org/library/Marshall/marP0.html

McCann, D. (2005). *Team learning*. Retrieved March 7, 2005, from www.tms.com.au/tms12-2c.html

Nonaka, I. (1991). The knowledge-creating company. *Harvard Business Review, 6*(8), 96-104.

Palloff, R.M. & Pratt, K. (1999). *Building learning communities in cyberspace: Effective strategies for the online classroom*. San Francisco: Jossey-Bass.

Schumacher, E. (1977). *A guide for the perplexed*. USA: Harper & Row.

Senge, P.M. (1992). *The fifth discipline: The art and practice of the learning organization*. Adelaide: Random House Australia.

Simon, H.A. (1976). *Administrative Behaviour: A Study of Decision Making Processes in Administrative Organization 3rd ed.* New York NY: Free Press.

Taylor, F.W. (1967). *The principles of scientific management*. New York: Norton.

Toffler, A. & Toffler, H. (1995). *Creating a new civilization: The politics of the third wave*. Atlanta: Turner Publishing.

Von Krogh, G., Ichijo, K. & Nonaka, I. (2000). *Enabling knowledge creation*. USA: Oxford University Press.

Weick, K.E. (1978). *The social psychology of organizing*. Reading: Addison-Wesley.

Weinberg, G. (1971). *The psychology of computer programming*. New York: Van Rostrand Reinhold.

Wellins, R.S., Byham, W.C & Wilson, A.M. (1991). *Empowered teams: Creating self directed work groups that improve quality, productivity and participation*. San Francisco: Jossey-Bass.

This work was previously published in International Journal of Information and Communication Technology Education, Vol. 3, Issue 1, edited by L. Tomei, pp. 58-69, copyright 2007 by IGI Publishing (an imprint of IGI Global).

Section II
Development Tools

Chapter X
Creating an Interactive PowerPoint Lesson for the Classroom

Lawrence Tomei
Robert Morris University, USA

ABSTRACT

This article helps classroom teachers create an "Interactive Lesson," a self-paced, student-controlled, individualized learning opportunity embedded with assessments. These lessons are offered to learners who need individualized instruction; corrective instruction, additional practice, or enrichment activities. Interactive lessons are not new. However, the practical, sequential methodology offered herein provides a practical design model for creating and integrating Microsoft's PowerPoint for presenting self-paced, personalized lesson content. The presentation can be captured to a floppy diskette, burned onto a CDROM, or sent as an email attachment to students in a classroom, computer lab or at home. The interactive lesson has many practical applications for students needing remedial attention or those attending cyber schools or home-bound students.

INTRODUCTION

Much of the technology used in today's classroom involves or is supported by the Microsoft Office© suite. It has arguably become the integrated software package of choice for many schools and corporate training rooms. Word, PowerPoint, Excel, and Access are the staples for many teachers, trainers, and their students. Complimenting these capabilities, Internet Explorer and Netscape Communicator are the tools of choice for accessing the World Wide Web. Teachers and trainers often opt to utilize these tools to develop text, visual, and Web-based materials for the classroom, preferring to leave the more complex and costly packages to multimedia designers and commercial artists. The success of this practice has been borne out by a blistering growth in applications from K-12 classroom teachers and their corporate training counterparts.

THE INTERACTIVE LESSON DEFINED

Teachers and trainers use Microsoft Word to create text-based class handouts, lesson study guides, and student workbooks based on their own classroom learning objectives. They use Microsoft's Front Page and Netscape's Composer to produce web-based content materials. And, they use Microsoft's PowerPoint to create interactive presentations. Interactive Lessons take the form of a self-paced, student-controlled, individualized learning opportunities embedded with formative and summative assessments to gauge student learning outcomes. Lessons are offered to those who need individualized attention, remedial instruction, additional practice, or enrichment activities. Specifically, an Interactive Lesson:

- Is a **visual-based, behavioral-oriented** teaching strategy appropriate for learners at all levels that may benefit from concrete, graphical learning experiences.
- Contains **self-paced** instructional content appropriate for those who learn best when content is sequenced and delivered at their own pace or who need remedial instruction outside the typical classroom environment.
- Offers **specific, logical, systematic** lessons that foster individualized instruction and sequential learning.
- Is **student-initiated** and **student -managed** learning that places a good deal of the responsibility for mastering the material directly in the hands of the learner.
- Embraces all phases of the **Mastery Learning** instructional technique. It suggests alternatives for presenting the initial mastery objectives, corrective instruction, and enrichment activities.

CREATING AN INTERACTIVE LESSON

The ADDIE instructional system design model guides the creation of the Interactive Lesson.

For each step in the ADDIE Model, a practical, hands-on task is completed as evidence that the skill has been mastered. Here's how it goes:

1. *Analyze*: define the needs and constraints
2. *Design*: specify learning activities, assessment and choose methods and media
3. *Develop*: begin production, formative evaluation, and revise
4. *Implement*: put the plan into action
5. *Evaluate*: evaluate the plan from all levels for next implementation; evaluation is essential after each step.

Each step has an outcome that feeds the subsequent step.

Figure 1. The ADDIE model for instructional system design

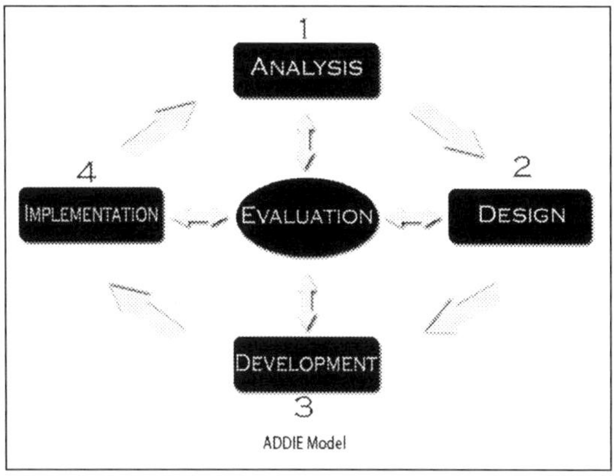

Analysis Phase

Arguably the most crucial step in course design –whether the target learner is a child or an adult, in a traditional classroom or online. At this stage of the process, the instructor sets the scope of the content to be covered by answering the following questions:

- Who are the learners?
- What do they already know (cognitivists call it 'prior knowledge'?
- What learning styles are in play in this classroom or online environment?
- What do the learners need or want to learn, why do they need it, and in what environment will they apply the learning?
- What is the teacher or trainer trying to achieve with the instruction? How would the overall goal or rationale for the course be described?
- What knowledge, skills and attitudes need to be taught during the lesson or must be in place before the lesson begins?

Design Phase

The design identifies the specific learning objectives of the lesson. Specifically:

- What skills, knowledge and attitudes will be developed and/or mastered during the lesson? Higher-order thinking skills, taxonomies (such as Bloom's Cognitive Domain), and multiple intelligences (for example, Gardner's theory) come into play during this phase of instructional design.
- Resources and instructional strategies are selected as teaching aids, learning activities, and, of course, instructional technologies are chosen to complement the learning outcomes.
- The organizational structure and sequence of the content furthers the design phase. of your learning material.
- Assessment of learner outcomes completes the design phase to evaluate the learners' understanding and whether or not they have met the objectives of the instruction.

Often, in the design phase, storyboards are created. Every screen of the course is designed using these storyboards. These storyboards provide the perfect interface for presentations in general and the PowerPoint-based Interactive Lesson specifically.

Development Phase

Development phase prepares the learner, the teacher or trainer, and the support materials (audio, video, and other multimedia), as well as any other technology-based materials. The Interactive Lesson, as well as other instructional materials, is created in the development phase.

Implementation Phase

The implementation phase puts the lesson into practice, first (and most commonly) through a prototype state where the lesson is tested by target learners. There are two stages of testing:

- Alpha-testing provides for in-house experimentation of the lesson. For example, a group of learners reviews the course content, delivery, and evaluation scheme, followed by....
- Beta-testing as the first real implementation of the course with real learners (often, learners who have already successfully completed the lesson, perhaps in a prior course). The feedback from these learners is essential for the improvement of the course.

Evaluation Phase

The evaluation phase constitutes the quality control component of the lesson. The effectiveness of the instructional process and materials is evaluated at this stage. The input from the alpha- and beta-evaluation is collected and the lesson revised based on feedback collected by either formative or summative evaluation.

- Formative evaluation occurs throughout the entire design process, particularly at the completion of each phase of ADDIE.
- Summative evaluations occur at the end of testing and at the completion of each lesson offering.

Lesson design by the numbers...seems fairly simple, right? With the ADDIE Model as a guide, this paper will continue with a practical, hands-on approach for turning a relatively straightforward PowerPoint presentation into a multimedia-rich Interactive Lesson. Any journeyman PowerPoint user can produce a viable Interactive Lesson with the addition of a handful of features no typically explored in most training workshops.

POWERPOINT FEATURES INAN INTERACTIVE LESSON

A menu of options and features make PowerPoint a powerful graphics development and presentation package. Four features in particular make the Interactive Lesson possible:

- Action Buttons
- Hidden Slides
- Kiosk Browser
- Assessment Slide

Figure 2. ActionbButtons (versions 2003 and 2007)

Figure 3. What did you learn?

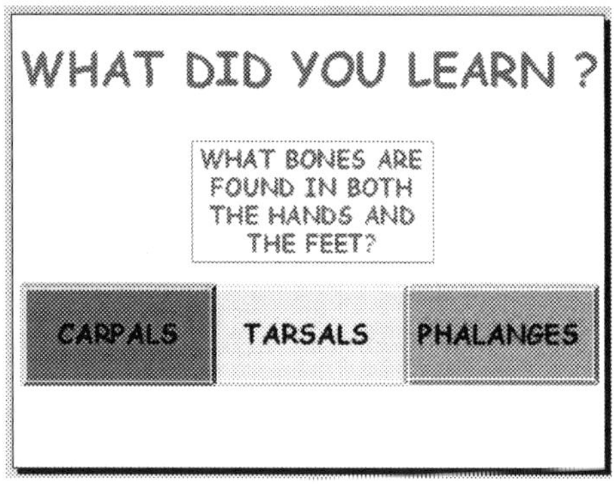

Figure 4. Action setting links to slides (versions 2003 and 2007)

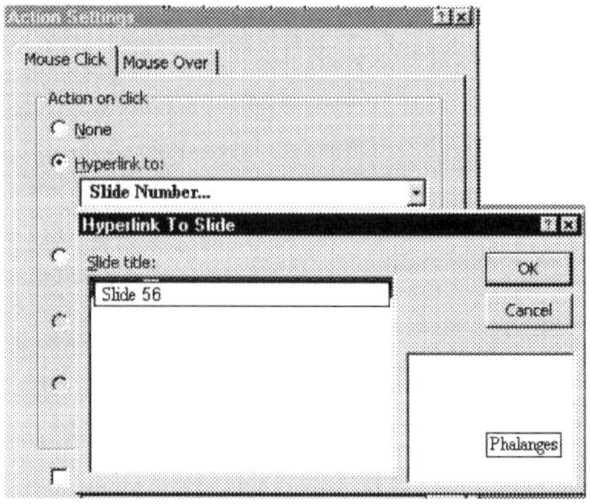

Action Buttons

PowerPoint comes with several built-in responses that are easily inserted into a presentation. There are Action Buttons that go to the next slide, indicate an available movie or sound clip, or request help or information. The Slide Show pop-down menu accesses the Action Button option (Figure 2). The left-side of the figure depicts the path to the Action Buttons in the 2003 version; the right-side shows the 2007 version. Any element in a PowerPoint slide can serve as an Action Button – text, images, even Clip Art.

Even more important, though, is the use of the Action Button to assess student understanding. By creating a simple question with several possible responses, PowerPoint transfers students either to new information (if correct) or to remedial information if additional instruction is necessary. Figure 3 shows an example slide that asks the question, "What bones are found in both the hands and the feet?" Notice the three possible answers. A correct response of "Phalanges" triggers the Hyperlink shown in Figure 4 to advance to the feedback (Slide 56) and from there continue the lesson with Slide 36. If students select one of the incorrect responses "Carpals" or "Tarsals," they advance to Slide 57 (Figure 5) containing negative feedback and, from there, back to Slide 31 to reread the original material. Action Buttons enable this interactive feedback, but they would be confusing to the student without the Hidden Slide feature of PowerPoint.

Figure 5. Hiding a feedback slide (versions 2003 and 2007)

Hidden Slides

In its typical mode, students view PowerPoint slides sequentially from Slide 1 to the final slide at the end of the presentation. There are times, however, when a designer might wish the individual to see certain slides only under particular circumstances, such as the assessment question discussed earlier.

Unless the feedback slides are hidden, they will be viewed in order as the presentation unfolds. This can cause unnecessary confusion for the student, so the feedback slide is hidden using the

Figure 6. Setting a kiosk show (versions 2003 and 2007)

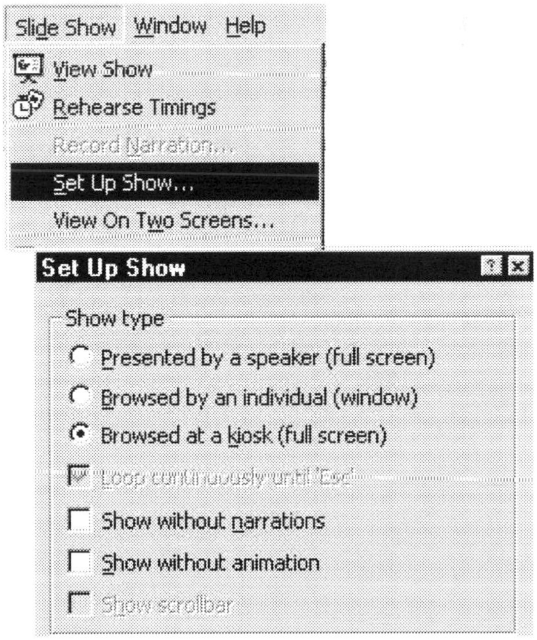

Figure 7. Summative assessment slide

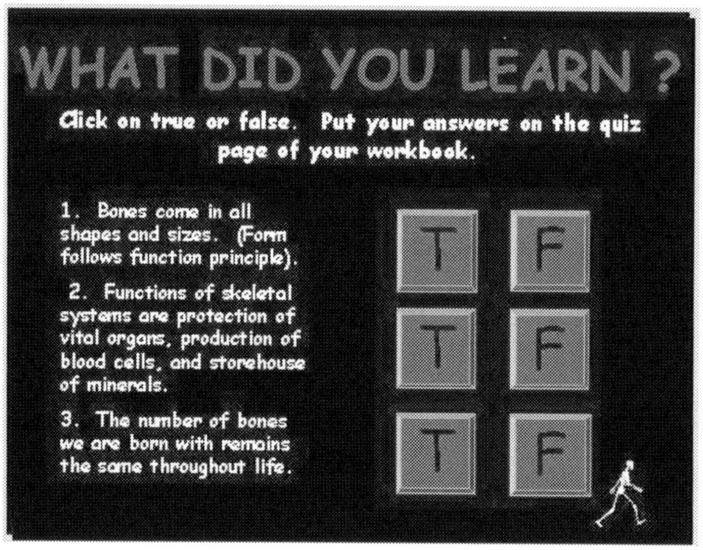

pop-down menu shown in Figure 5. Once hidden, a null icon (a diagonal slash through the slide number) appears when viewing the presentation in the Slide Sorter mode. Now, the only way to view this slide is by directly accessing it using the Action Button—and the Kiosk Browser.

Kiosk Browser

You have seen kiosks before. They are self-running presentations found at many trade shows, amusement parks, and conventions. PowerPoint's Kiosk feature supports unattended slide shows that run continuously unaided, restart automatically after each showing, or require user intervention to advance the slides. It is this last characteristic that makes our lesson Interactive. Figure 6 shows how to Set Up the Show as a Kiosk presentation. The learner must manually advance every slide for this to work properly; that's why each of the slides in the presentation has its own Next slide button on each slide. Otherwise, the presentation would stop dead in its tracks. Also, we need the Kiosk feature to ensure that the student does not skip around the presentation—the teacher alone controls the sequence through the Action Buttons, Hidden Slides, and Kiosk Browser.

Assessment Slide

Earlier in the article, the Interactive Lesson was presented as a mastery learning instructional technique. An important premise with this teaching strategy is its underlying dependence on behavioral psychology. To be successful, the Interactive Lesson must follow a few basic rules. First, it must be logically sequenced; significant time must be spent structuring the progression of information from beginning to end, least important to most, simple to complex. Second, there must be some form of immediate feedback; again, this is accomplished using the hidden slides. And third, there must be a summative (final) assessment.

A final slide (see Figure 7) in the presentation can meet this requirement while ensuring that students have completed the lesson, mastered all the learning objectives, and received some reward for their efforts. In a computer lab environment, this final Assessment Slide, displayed in bold colors on each individual computer monitor, alerts the teacher that the lesson has been completed and the student is ready for the next instructional challenge.

CONCLUSION

Interactive Lessons are not new. They have existed almost since the beginning of instructional technology. But now we offer a structured format for designing such lessons using a popular, highly effective, and relatively easy to use software package, PowerPoint using either version 2003 or, now, its newest version 2007. Once created using the ADDIE Model for instructional system design, the presentation can be captured onto floppy diskettes or burned to CDROMs, copied many times, and provided to learners who can take the lesson in a formal multimedia classroom, informal computer lab, or even on their own home computers.

Chapter XI
Planning Staff Training for Virtual High Schools

Chris Thompson
Elmbrook Schools, USA

Zane L. Berge
University Maryland Baltimore Campus, USA

ABSTRACT

This chapter briefly profiles three virtual schools, each at a different stage of development, yet each dependent upon a successful and sustained distance education program for its professional staff in order to remain viable long into the future. As virtual schools become more accepted by the public and the attention given to the online schools shifts from their sources of funding to their standardized test scores, a model for sustained distance training and education must be in place to deliver quality professional development that can positively impact students' achievement scores on standardized tests for each school's online student population.

INTRODUCTION

Virtual schools are a rapidly growing phenomenon in American elementary and secondary (K-12) education (Berge & Clark, 2005). They are the latest and potentially the most controversial manifestation of the e-learning revolution in schools. As Clark and Else noted, "For the foreseeable future, the World Wide Web is likely to serve as an umbrella technology uniting distance education media for distributed learning...Virtual schooling is the next wave" (Clark & Else, 2003, p. 35-36).

Distance education in today's virtual schools describes not only the education of the students enrolled, but also the professional development programs used to train the faculty and support staff working for each online school.

The three virtual schools used here to explore improvements to staff development includes the *Electronic Classroom of Tomorrow*. This is the largest K-12, electronic school in the United States, servicing almost 7,000 students. The second school is the *iQ Academies* high school, which is in its fourth year of operation at the time of this writing, and claims its 589 (2005-06 school year) students a success. The third school is entirely virtual, and to date imaginary, but ideal in its delivery of distance training and education to its staff and faculty. The first two schools provide examples which collectively will serve as the foundation for a successful professional development program. The *Virtual I.D.E.A.L.* school will serve as a model for future management consideration. Taking into consideration what is known about barriers to online learning, best practices in virtual schools, and how to sustain virtual schools, success of schools that take into account the *Virtual I.D.E.A.L.* program will ultimately have a positive impact on the long-term status and its ability to service each of its students' needs.

The ECOT

The *Electronic Classroom of Tomorrow* (ECOT) was founded in 2000, and today serves as the largest online K 12 school provider in the United States, serving almost 7,000 students (ECOT, 2006). Based in Columbus, Ohio, this school enrolls only Ohio residents, drawing from a student and teacher population that is geographically spread across the state. Like most public, state-sponsored chartered schools, the ECOT high school provides a workstation and internet access for each student and partners with course content providers to offer curriculum through a course management system. Each teacher begins with the provided course content and is able to adapt it to meet the needs of his or her class. Special education teachers are also available, modifying the course content and classroom activities for the students identified with special needs (Hartge, 2005). Like traditional classrooms, each special education student has an Individualized Education Program (IEP), which is meant to guide and document specially designed instruction for each student with a disability based on his or her unique academic, social, and behavioral needs (IEP, 2007). While serving the needs of a very diverse student population and growing at an incredibly rapid rate in its seven years, the ECOT approaches each school year as a work-in-progress, tweaking the system to better serve the changing needs of its online student population.

iQ Academies

The Waukesha *iQ Academies* (*iQ*) was the first virtual high school in the State of Wisconsin, opening its doors September 1, 2004. Wisconsin's open enrollment laws require students wishing to attend a school district other than their home district, to apply to those districts during a three-week window in February of each year. While a student has until August 31st to decide if he/she will attend another district, if the student does not apply in February, the opportunity to enroll somewhere other than the home district is lost (Wisconsin Department of Public Instruction, 2006). The Waukesha *iQ Academies*, a public charter school affiliated with the 13,000 student Waukesha School District (Waukesha School District, 2007) located 20 miles west of Milwaukee, has nearly 500 students apply during the open enrollment process, with 220 enrolling by September 1st (Diener, Interview, 2005).

The *iQ Academies* was created to better meet the needs of students that, for a variety of reasons, did not believe their needs were being adequately addressed. Students requiring an alternative education, traveling the United States, or who were traditionally home-schooled, did not have an educational system that was working for them. The Waukesha School District recognized these needs, and in less than ten months, energized its resources to put into place a complete online virtual high school.

Because of its affiliation with an existing school district and an agreement with the Waukesha Teachers Association Union, all *iQ* teachers are employees of the Waukesha School District and most continue to teach in the classroom in addition to online. In its first year, *iQ* had one-full time instructor, and 27 part-time teachers, teaching as little as 1/12th of their instructional time in the online environment. Each instructor received a computer, but was not given internet access, under the assumption that most instructors continued to teach in a District building with internet access. It is hoped that with the increased enrollment projections, at least five to six teachers can be hired full-time for *iQ*, allowing the online teaching faculty to dedicate more of their time to their online responsibilities. As could probably be expected, in some cases it was believed that in-person classroom needs consistently trumped online classroom needs, which proffers both student and staff frustrations. More full-time online staff will help address the challenge of trying to teach both online and in-person.

Distance Training and Education Efforts for Faculty

Neither *iQ* nor ECOT fit cleanly into any one of Schreiber's (1998) stages of technological maturity indicating an organization's readiness to successfully implement distance training and learning (see Appendix A). Due to the nature of their business, both schools met the stage two requirement of the necessary infrastructure and technological capability to delivery online education and training. *iQ* may fall a little short because of the decision to not provide home internet access to its staff. Nonetheless, all teachers still had access either at home or at school. Furthermore, *iQ*'s full-time staff of two (principal and secretary) simply cannot be expected to train, develop procedures, and evaluate the needs of its staff, also stage two expectations. Hopefully as it grows in size and resources, *iQ* can transition from a stage one school providing separate and sporadic learning events, to an organization that is more established and consistent in its training efforts.

ECOT's size and experience allow it to meet more of the stage two requirements including approaching distance education and training with a more interdisciplinary team and maintaining adaptability and flexibility in its course offerings. By establishing a mentoring program and through its formation of a partnership with the University of North Florida to develop research-based, online faculty training solutions, ECOT approaches stage three, but has yet to fully commit and embrace distance training of its staff as a core belief and philosophy.

Staff Development Efforts

Ironically, both *iQ* and ECOT use on-the-ground instruction as their primary mode of delivery for staff development. The fact that both virtual high schools maintain in-person education and training for most, if not all, of their staff development may not be as much of a contradiction to its focus on online education as it first seems. Rather, a mixed-mode delivery of online and in-person professional development recognizes that each school must first build a learning community among its staff members, most of whom have been trained in an in-person environment, before engaging students in their respective online communities. While not taking advantage of some of the online efficiencies, it seems both institutions are most comfortable with a blended model of professional development delivery.

The *IQ Academies* offered one week of primary instruction to all faculty members before its inaugural school year and, because all teachers currently or formerly taught in the district, it was not difficult to arrange such in-person training. The training consisted of four parts:

- Exploring online teaching and how it compares with in-person teaching;
- a mini-course simulation;
- technical skills training; and
- how to effectively communicate with the online student.

Because most of the course content is provided for the instructor, the training was focused on the supplemental, but very important, aspects of a course such as classroom management, facilitating discussion, course expectations, and school policies. With more full-time staff projected for future school years, it is hoped that *iQ* can at least begin to have virtual faculty meetings as a first step in the distance training process.

The ECOT has an established system of in-person training four times a year, for 2-3 days, held at various locations around the state. ECOT takes advantage of this opportunity to also hold parent-teacher conferences for students in that particular geographic section. ECOT's in-person sessions may include some technical skills training on new products or features, but more often includes training on school initiatives such as competencies and state test scores. Because state test scores are extremely public figures and are oftentimes the measure of a school's success (or failure) in the eye of John Q. Public, it is not uncommon for new charter schools and traditional brick and mortar schools alike, to place great emphasis on achieving higher test scores. Because the state test is not yet online and must be proctored, the challenges presented just in the coordinating of a state standardized test at an online high school are incredible, doubling the need for all staff members' active participation come testing time.

Feedback for Staff Development

ECOT has also established an effective system for collecting feedback on training needs, incorporating suggested education and training opportunities into their quarterly meeting. This supports one of Rosenberg's (2001) five areas of transformation when describing the new era of training and instruction, the ability to move training from cycle time to real time. Historically, training has taken some time to cycle from concept to delivery. But today, real-time turnaround is necessary and critical. ECOT's ability to turnaround its training needs from one quarter to the next demonstrates real-time training and instruction to its faculty members.

VIRTUAL I.D.E.A.L.

Virtual I.D.E.A.L. (IDEAL) school does not exist. It serves here as a model to point toward considerations that need to be made when managing a virtual high school. IDEAL was established five years ago. It maintains manageable growth, its students test high on the required state exams, its staff is well-qualified, and its distance education and training program would be described by Schreiber (1998) as beyond Stage 4. The IDEAL school is just that, yet it somehow strives to be more. What makes its distance education and training program the best? Its practices are based on research, experience, self-reflection, and successful partnerships with industry experts.

A Strong Beginning

A well trained staff is critical for sustaining IDEAL's programs and to meet the needs of students. Training staff at IDEAL begins with a structured program for beginning distance instructors following many of the topics identified in *iQ*'s training. The difference between the two, is that IDEAL takes advantage of the particular strengths of in person and online instruction. Rosenberg (2001) notes, "With all the potential of e-learning, it might be easy to dismiss traditional classroom training as completely antiquated – of no value down the road. Although e-learning has a great deal to contribute, it does not mean the end of a classroom learning" (p.120). Similar in concept to the ECOT, the IDEAL school gives all incoming staff an assessment to determine the individual technical and instructional skills. Tobin (2004) cites ever-changing teacher competencies for graduating teachers to justify the need for the development of competencies for online teachers as well as continuing professional preparation and training for online teachers. This assessment is based on the Educational Technology Standards and Performance Indicators established for all teachers by the International Society for Technology Education (ISTE, 2000). ISTE identifies six areas of focus:

1. Technology operations and concepts
2. Planning and designing learning environments and experiences
3. Teaching, learning, and the curriculum
4. assessment and evaluation
5. Productivity and professional practice
6. Social, ethical, legal, and human issues

Because of its virtual nature, special emphasis is placed on needed technical skills, and any technical remediation is done before the first week of distance training. In addition to the pre-week assessment and technical training, new instructors are also required to complete online learning objects introducing available administrative and support services, tutorials on copyright and policy issues, and an exploration unit on the components of a successful online course. IDEAL also uses its pre-week session to have the instructors introduce themselves, describe their background, and share some of their personal interests. Completing these components online before the actual in-person session demands preparation, encourages camaraderie, and builds skill development, similar to what the instructors will soon be expecting from their students. Finally, the week of training arrives and sessions are presented on the following items:

- Distance learning technology and its impact on learners;
- How distance instruction ties in with the institutional mission;
- Fundamentals of and assistance with course development and adaptation;
- Techniques for encouraging interaction;
- Development of back-up and contingency plans; and
- An opportunity for addressing concerns (Clay, 1999).

Each session builds upon the previous and adequately connects the school's mission and vision with the online classroom teacher's responsibilities and expectations. One unique way the IDEAL insures a connection is made by its new teachers, is through formative and summative assessments. Instructors are quizzed and must complete practice exercises demonstrating formative, or short-term, learning. Instructors must also demonstrate summative learning, evaluated by how effectively they are able to implement training topics into their own virtual classroom, measured by a 3-month, 6-month, and 9-month classroom "visit and evaluation." To help insure their success, trainers teaching the introductory sessions to new teachers can earn financial incentives based on their students' ability to connect session training with students' classroom learning. Success in the classroom is evaluated by the administrators' ability to quantify interactivity, learning, and growing, something that remains quite a challenge in an online school. Sunal, Sunal, Odell, and Sundberg, (2003) offer a *Checklist for Online Interactive Learning* (COIL), which centers on four main topics: student behavior, faculty-student interaction, technology support, and the completeness of the learning environment. Classroom teachers at IDEAL appreciate such an assessment because it is research-based and can provide data to support both successes and future opportunities in their virtual classrooms.

The successful integration of online training and in-person training reinforces Rosenberg's (2001) other indicators necessary to transform distance training and instruction. By using online training both as a precursor to live instruction and as a follow-up to classroom training, deftly blends the strengths of the two modes of instruction. Holding the trainers accountable for their instruction, and subsequently holding the instructors accountable for implementing what they have learned, encourages all participants to make the connection between training and performance. And finally, by maintaining online materials, information can remain current and accessible when needed.

Variety

As the IDEAL's training program moves beyond the first weeks, variety becomes an integral part of its continued success. Just as teachers need to find ways to reach students of all learning styles, distance training programs need a variety of opportunities for learners with varying styles and preferences. Clay advocates for online programs that include mentoring, group sessions, one-on-one labs, printed materials, listservs, regular discussion sessions, and observation of others' courses. Not only does the variety address different learning style needs, but most of the learning options allow educators the opportunity to explore skills and technologies they could ultimately use in their classroom as part of their own instruction. Clay acknowledges the reinforcement of skills as valuable by noting that, "experience shows that training simply won't 'take hold' unless support is ongoing, with job-embedded opportunities for practice."

Knowledge Management

Rosenberg (2001) describes an online training that is supported by expert modeling and stories, learning from others mistakes, and having the opportunity to reuse information after learning. The reuse of information refers to knowledge management and, once again, IDEAL is at the forefront of the virtual training field. Where training and instruction is focused on specific learning outcomes, is sequenced for memory retention, and may contain presentation, practice, feedback, and assessment components, knowledge management focuses on the organization of content, is sequenced for optimum reference, and is centered primarily on effective presentation. IDEAL's knowledge management archives contain many how-to's, alternative instruction and assessment strategies, and many tutorials on school and student policies and procedures. The database is easy to search and tracks hits on various information modules, to help identify future training and instruction opportunities.

Identifying and Assessing Needs

IDEAL's ability to use data from the knowledge management archives is vital to the proper identification of future training needs. While feedback and user input is another valuable collection tool, the data provides not only a basis from which to start, but a means to be able to measure progress over time to determine the success of training and instruction efforts. The power of data in sustaining distance training efforts cannot be overstated. Data can justify a major investment in new software or additional support expenses. In IDEAL's case, data can be used to demonstrate competency and achievement of individual or school goals, and it can also be used to challenge both staff and students to raise the bar higher in all forms of instruction and learning. Identifying and assessing needs for distance training is an important aspect of Rosenberg's Learning Architecture (p.124) and the IDEAL's continued success.

Variety of Teaching Methods

The *Virtual IDEAL School's* successful pre-week assessments, remediation, and topical learning objects, combined with its week of in-person instruction, added to its extensive knowledge management database, and integrated into a system that stresses accountability, data-driven decisions, and constantly assessing learning and training needs, allows its well-trained staff to focus its distance education and training efforts on larger, school-wide initiatives. No longer are instructors learning how to make templates in Microsoft Word, but instead are using discussion board software to consider how to improve literacy, create common assessments, improve communication channels with parents and students, and how to align curriculum to state and national standards. In another chat room instructors are brainstorming at how to integrate more team-building exercises into the curriculum while others are comparing classroom statistics to identify instructor strengths and areas for improvement. IDEAL is able to remain at the forefront of distance education and training through its continued creative use of online learning tools and its desire to tackle challenging issues facing most high schools today.

BARRIERS TO SUSTAINING DISTANCE EDUCATION

Berge and his colleagues (Cho & Berge, 2002; Muilenburg & Berge, 2001) identify ten barrier clusters to establishing successful distance education and training programs:

1. technical expertise
2. administrative structure
3. evaluation/effectiveness
4. organizational change
5. social interaction and quality
6. student support services
7. threatened by technology
8. access
9. faculty compensation and time
10. legal issues

Ohio's ECOT seems to have already addressed many of these barriers. Through its pre-assessment, technical training, and quality internal support, its teachers are technically competent and can adapt to new technologies. It has provided a computer and high-speed internet access for its staff members, and pays its faculty well compared to brick and mortar institutions. ECOT's quarterly meetings encourage social interaction and community, and its willingness to accept feedback and evaluation helps include all staff in the development of the training process. ECOT's needs seem to revolve around an integrated learning architecture and a more substantial investment in online training opportunities for its staff. Like most schools, it struggles with implementing accountability and successfully managing the rapid growth of its student and teaching populations. While it has already started initiatives concerning competencies and state test scores, better use of its online resources will allow it to reach a new level of success when addressing organization issues and goals.

Wisconsin's *IQ Academies* has taken the first step towards sustaining distance education and training just by entering the virtual high school arena, yet there is certainly room for improvement. Faced with a small administrative staff, limited student support services, less provisions for faculty access, and no unique compensation packages, the challenges ahead of *iQ* are formidable, but not impossible. As enrollment increases, more full-time teaching and administrative staff will generate increased opportunities for online collaboration and the development of a knowledge management system. Two more years of experience will solidify its new instructor training initiative and perhaps lead to a pre-assessment to determine both teaching and technical skills.

Increased administrative support may allow for more detailed assessment of classroom instruction and measurable student learning. The key to *iQ*'s future success is continued growth and constant evaluation of the support services and instructional opportunities offered to its teaching faculty.

CONCLUSION

The IDEAL school focuses on the future. While it may be impossible for any one school or district to achieve the comprehensive success of the distance education and training efforts modeled by IDEAL, various components and strategies are attainable and can be matched in terms of quality and sustainability. The IDEAL school may best serve as a reminder that a school's commitment to its teaching staff is constant and unending, and consistent and determined effort is required year in and year out.

These three virtual schools are each at different stages of maturity. Successful and sustained distance education and training for all three schools revolves around continued growth, organizational commitment, quality resources and instruction, and measures of accountability. Not surprisingly, these factors are really not much different than the standards of success identified by brick and mortar institutions. By examining the current status of staff training that exists in a virtual high school, and planning the direction a model program should have, school administration can better plan to close the gap while overcoming common barriers and identifying needs for staff development.

REFERENCES

Berge, Z. (2001). *Sustaining distance training: Integrating learning technologies into the fabric of the enterprise*. San Francisco: Jossey-Bass.

Berge, Z.L. & Clark, T. (2005). *Virtual schools: Planning for success*. Teachers College Press.

Cho, S.K. & Berge, Z.L. (2002). Overcoming barriers to distance training and education. *Education at a Distance, 16*(1). Retrieved November 18, 2007 from http://www.usdla.org/html/journal/JAN02_Issue/article01.html

Clark, T. & Else, D. (2003). Distance education, electronic networking, and school policy. In D.R. Walling & J. F. Jennings (Eds.), *Virtual Schooling*, pp. 31-45. Bloomington, IN: Phi Delta Kappa Educational Foundation.

Clay, M. (1999). Development of training and support programs for distance education instructors. *The Online Journal of Distance Learning Administration* (Fall, 1999). Retrieved on November 18, 2007 from http://www.westga.edu/~distance/clay23.html

Diener, K. (2005). Principal, Waukesha *IQ Academies*. Interview conducted via the phone on April 14, 2005.

ECOT. (2006). High school. Retrieved November 18, 2007 from http://www.ecotohio.org/highschool.html

Hartge, B. (2005). Personal correspondence.

IEP. (2007). L.D in depth. *LD Online*. Retrieved November 18, 2007 from http://www.ldonline.org/ld_indepth/iep/iep.html

iQ Academies. (2006). *iQ Academies* homepage. Retrieved November 18, 2007 from http://www.gotoiq.com/index.asp

ISTE. (2000). Educational technology standards and performance indicators for all teachers. Retrieved November 18, 2007 from http://cnets.iste.org/teachers/pdf/page09.pdf

Muilenburg, L.Y. & Berge, Z.L. (2001). Barriers to distance education: A factor-analytic study. *The American Journal of Distance Education*. 15(2), 7-24.

Rosenberg, M. (2001). *e-Learning: Strategies for delivering knowledge in the digital age*. McGraw-Hill.

Schreiber, D.A. (1998). Organizational technology and its impact on distance training. In D.A. Schreiber and Z.L. Berge (Eds.), *Distance Training: How Innovative Organizations are Using Technology to Maximize Learning and Meet Business Objectives*. San Francisco, CA: Jossey-Bass Inc., Publishers. pp. 3-18.

Sunal, W. D., Sunal, S. C., Odell, R. M. & Sundberg, A. C. (2003). Research-supported best practices for developing online learning. *The Journal of Interactive Online Learning, 2* (1), 1-40. Retrieved November 18, 2007 from http://www.ncolr.org/jiol/issues/PDF/2.1.1.pdf#search=%22Sunal%20%22Research-supported%20best%20practices%20for%20developing%20online%20learning%22%22

Tobin, T. (2004). Assessing educators' online performance. *The Online Journal of Distance Learning Administration* (*Volume VII, Number II, Summer 2004*). Retrieved November 18, 2007 from http://www.westga.edu/%7Edistance/ojdla/summer72/tobin72.html

Waukesha School District. (2007). Retrieved September 15, 2006 from http://www.waukesha.k12.wi.us/District/Default.asp

Wisconsin Department of Public Instruction. (2006, March 3). Public school open enrollment: 2005-2006 timelines for public school open enrollment. Retrieved November 18, 2007 from http://www.dpi.state.wi.us/dpi/dfm/sms/oetime05.html

APPENDIX A

A brief model that describes stages of organizational maturity, or capabilities, with regard to the delivery of distance education (Schreiber, 1998) might be:

- **Stage 1:** Separate or sporadic distance learning events occur in the organization.
- **Stage 2:** The organization's technological capability and infrastructure can support distance learning events. When distance education events occur, they are replicated through an interdisciplinary team which responds to staff and management needs and makes recommendations regarding the organization and management of distance learning among the workforce.
- **Stage 3:** The organization has established a distance learning policy, procedures are in place and planning occurs. This means that a stable and predictable process is in place to facilitate the identification and selection of content and of technology to deliver distance training.
- **Stage 4:** Distance training and education has been institutionalized in the organization as characterized by policy, communication, and practice that are aligned so that business objectives are being addressed. The business unit has established a distance education identity and conducts systematic assessment of distance training events from an organizational perspective.

Chapter XII
Training Prospective Online Instructors: Theories Utilized by Current Online Instructors

MarySue Cicciarelli
Duquesne University, USA

ABSTRACT

Research shows that training prospective online instructors in an online learning environment is advantageous. One effective training topic is on use of theory when designing curriculum. Information in this study reports what empirical research shows about online instructor use of different design theories. It identifies design theories that have not been researched in regard to online instructor utilization of theory, and it illustrates how frequently online instructors use nine of the design theories.

INTRODUCTION

When training instructors to teach a course online, what should be taught, how should they be instructed, and where should the learning experience take place? Researchers argue that teaching instructors how to design and execute courses in an online learning environment is a most advantageous choice because they encounter similar perspectives that their own future students will experience. In addition, they go beyond learning the basic 101 instructional aspects and mechanical abilities needed to manipulate a computer management system and teach a course online (Bird, 2007; Cook, 2007).

Considering that there are many facets to instructing a course online, there are just as many topics that can be presented to prospective online instructors as they participate in training programs aimed at teaching online (Bird, 2007; Dennen, 2007). Dennen (2007) indicated that an instructor persona exists as part of the online discussion in relation to instructor presence (number of posts) and position. Instructors need to learn to be aware of the persona that they present to their students, and they should recognize that student's perceptions of their instructor impacts the learning experience. Bird (2007) presented a design model for e-learning and discussed theoretical underpinnings of utilizing a design model. Mimirinis and Bhattacharya (2007), based on results from their study on deep learning in virtual learning environments, also noted the importance of course design and highlighted the need for students to reflect, inquire, analyze, and synthesize for deep learning.

Research in the field has shown us that an increasing number of individuals have chosen distance education when taking a course or earning a degree because it is a flexible alternative that meets their needs (Chu & Hinton, 2001; Course-Management systems, 2005). We also know from research that there are advantages to taking an online course, and there are challenges that students and instructors must overcome (Figueroa & Huie, 2001; King, 2001; Northrup, Lee, & Burgess, 2002; Prester & Moller, 2001). Another training topic appropriate for prospective online instructors is on use of theory to guide the development of effective courses so that needs can be met, advantages can be recognized, and challenges can be overcome.

In a study conducted by Cicciarelli (2006), information on online instructor use of theory when designing a course is presented. Data from the study was used to answer the three following questions: (a) According to empirical evidence, what does the research show about online instructor use of different design theories? (b) According to lack of empirical evidence, which design theories have not been researched in regard to online instructor utilization of theory? And (c) According to a recent study, how frequently do online instructors use nine of the design theories? Information that answers these three questions can be used when training prospective online instructors on the use of theory to design an effective online course.

REVIEW OF LITERATURE

There are three schools of psychology in which theories are categorized; Behaviorism, Cognitivism, and Humanism. It is believed that one school of theory is not better than the other, and individuals are encouraged to apply the theory that is the most appropriate for the student (Pinar, Reynolds, Slatery, & Taubman, 1996; Joyce, Weil, & Calhoun, 2000; Tomei, 2007).

Behaviorism, Cognitivism, and Humanism

Behaviorism

Experimental psychologists William James and Edward L. Thorndike questioned the use of memorization as a strategy for learning. Experiments that they conducted showed that memory did not increase after the participants had memorized sets of information. These results guided a turn in research toward stimulus-response behavioral psychology. James and Thorndike believed that the environment served as a stimulus, and it could be used to change the way individuals responded. As Behaviorism became more established as a part of the school of psychology, psychologists began to focus on individual's responses to feedback when they performed a task. Other behavioral psychologists that have made contributions to the field of behavior psychology include Ivan Pavlov, B. F. Skinner, Albert Bandura, and Benjamin Bloom (Pinar et al., 1996; Joyce et al., 2000; Cicciarelli, 2007).

Cognitivism

Theories that are cognitive in nature are based on learning tasks that are practical, and they are seen being used in authentic learning environments. Cognitive theorists such as Jean Piaget, Lev Vygotsky, Erik Erikson, and David Ausubel have developed theories that are not only widely accepted, but they have begun the path for the development of other cognitive theories. When instructors utilize theories that are cognitive, they tend to develop learning experiences that help students make connections that are meaningful to themselves (Grabinger, 2004; Cicciarelli, 2006; Tomei, 2007; Cicciarelli, 2008).

Humanism

Theories that focus on a student's affective needs come from the Humanism school of psychology. These theories attend to students' feelings, emotions, values, and attitudes. Some of the earliest work that reflected Humanism came from Colonel Parker who encouraged child centered learning in a democratic school environment. His work later influenced the progressive work of John Dewey. Theorists such as Elliot W. Eisner, Ross Mooney, and Paul Klohr supported the development of learning experiences that focused on self value. Carl Rogers and Abraham Maslow wanted educators to concern themselves less with curriculum development and give more of their attention to understanding curriculum. The work of these psychologists eventually influenced the development of other theories based on Humanism. Collective common factors of theories rooted in Humanism include the attention toward student-centered learning and individualism (Pinar et al., 1996; Cicciarelli, 2007).

ACCORDING TO EMPIRICAL RESEARCH: THEORIES SEEN BEING USED BY ONLINE INSTRUCTORS

Theories of Behaviorism

Social-Cultural Model of Learning

An online course that incorporates the Social-Cultural Model of Learning is reflective of Behaviorism because patterns of communication are utilized. The Social-Cultural Model of Learning uses written and oral dialogue. Threaded asynchronous discussions, synchronous discussions, and e-mail are examples of the tools implemented by instructors during the course design process. The pattern of Behaviorism begins when the instructor poses a question, students respond to the question, and the instructor responds to the student's responses with positive or negative reinforcement comments (Berge & Fjuk, 2006; Lee & Lee, 2006; Schwartzman, 2006).

Researchers found that learners want to interact with each other and their instructor through discussion. However, results show that precautions need to be taken by the instructor to help promote effective discussion (Berge & Fjuk, 2006; Cook, 2007; Dennen, 2007; Tallent-Runnels, Thomas, Lan, Cooper, Ahern, Shaw, & Liu, 2007).

In another study, researchers looked at the behavior of ninety-six individuals who participated in an asynchronous discussion. Results from a Myers-Briggs test were used to divide the participants into twenty-four groups based on personality type. There were eight introverted type groups, eight extroverted type groups, and there were eight mixed groups. Participants from the extroverted and mixed groups posted more asynchronous messages compared to the participants from the introverted groups. However, it was the participants from the mixed groups who showed a greater amount of metacognitive interaction. Online instructors are encouraged to create mixed groups because instant feedback and new ideas presented by extroverted members may stimulate introverted members, and the in-depth discussion prompted by arguments from the introverted members could further stimulate the extroverted members (Lee & Lee, 2006).

Mastery Learning

Mastery learning is a practice originally created by John B. Carroll and Benjamin Bloom. John Carroll's perspective holds that a student's aptitude correlates with achievement. His view of aptitude considers how long it takes for the learner to learn the material as opposed to the learner's ability to master the material. According to this view, every learner can learn as long as the appropriate materials and instruction are provided. Benjamin Bloom's work focuses on organizing the curriculum so that students have the necessary time and ability to benefit from instruction. Now that modern instructional technology has afforded educators with new choices, curriculum developers are encouraged to develop comprehensive curriculum that includes self-administering multimedia units and programmed learning procedures (Pinar et al., 1996; Joyce et al., 2000).

In a study that explores the use of strategies to engage the whole learner, mastery learning is utilized as a strategy to promote significant learning. According to the researchers, an online course should be designed so that different parts of the course encourage deep learning. The syllabus should be detailed, comprehensive, and it should contain direct instruction about interaction. Objectives should be clearly presented so they form the course structure, and the assignments should directly relate to the objectives. Finally, they talk at length about asynchronous discussions and the use of different activities. They indicate that the most effective discussions that lead to deep learning are supported with facilitative directions and rubrics because the learners know what is expected. Interactive discussion activities, discussion tasks, and reflective and exploratory discussions are said to be successful strategies that instructors can use to encourage mastery learning that is significant (Majeski & Stover, 2007; Mimirinis & Bhattacharya, 2007).

Simulations

Learning from simulations through training and self-training is another example of a behavioral learning theory. When simulations are utilized, students take on the role of someone from a real life experience. To succeed when performing the role, students make use of concepts and skills to perform specific tasks. Instructors take the role of explaining, refereeing, coaching, and discussing the simulation experience with the students. They explain the rules, place the students into teams, and they assign roles based on student ability to ensure participation and communication between the students. When coaching, the instructor needs to be supportive, yet avoid interfering with the natural play of the simulation. Students are expected to make mistakes and adjust from those mistakes. Finally, instructors hold a discussion in which students have the opportunity to reflect and identify similarities and differences between the simulation and the real world (Pinar et al., 1996; Joyce et al., 2000).

Online role-playing is not seen as a common practice in online courses. Data collected from a study on role-playing that was part of an online course showed that students who participated enjoyed taking part in the role-plays, and they found that students considered the learning experience to be beneficial. Participants also indicated that the interaction made the classroom experience feel more personal. Students were seen making use of their knowledge to add to the role-play experience (Lebaron & Miller, 2005).

Theories of Cognitivism

Theory of Multiple Representations

Applying multiple representations that connect to content of subject matter is thought to be a valuable practice because students can build mental representations with the information and make information meaningful to themselves. Web environments and computer mediated discussions are said to be conducive to the application of multiple representations during course design (Huang & Liaw, 2004). Researchers provide support and they raise cautions when it comes to using multiple representations during instruction. (Gfeller, Niess, & Lederman, 1999; Moreno, 2002; Ying-Shao & Fu-Kwun, 2002; Huang & Liaw, 2004).

A web-based lesson created to promote situated learning involved one hundred ten high school students from Taipei. Participants in the study were asked to connect a realistic situation with their own life. Social learning theory and multiple representations were used by the students to make connections. Results showed that when multiple representations were used along with situated learning during an asynchronous discussion that the students were able to cultivate and integrate the knowledge (Ying-Shao & Fu-Kwun, 2002). In another study on use of multiple representations, the researcher found that students with stronger technology skills were more successful when it came to using multiple representations. Since their technological skills were stronger, they had a lower amount of cognitive overload compared to students whose technology skills were not as strong (Moreno, 2002).

Bruner's Three-Form Theory

Bruner (1990) states that there are three ways from which individuals see the world; through action, icons, and symbols. They use action to perform or demonstrate what it is they see about the world from their perspective. Icons or mental images are used to present a path, summary, or pattern. Symbolism, which is an abstract way of visualizing reality through the use of words and numbers, is the third form that individuals use. According to Bruner, these three forms of representation are founded on the theory that development must be effectively related to theories of knowledge and instruction.

Studies on Dual-Coding Theory show an influence on learning when visual and aural modalities are combined (Rieber, Tzeng, Trubble, & Chu, 1996; Alty, 2002; Beacham, Elliot, Alty, & Al-Sharrah, 2002). In a study conducted by Rieber et al.(1996), fifty-two college students interacted with a computer simulation created partly with Dual-Coding Theory as a framework. Visual modalities were presented as animated graphics and numeric displays were presented as aural modalities. Results were better for students who were provided with the visual and aural support compared to the students who were provided with one of the two modalities.

Moore's Theory of Transactional Distance

Moore's Theory of Transactional Distance unlike the web-based theories already presented is a distance theory. Many online instructors have applied this theory, because its three dimensions have an affective influence on teaching procedures. Those three dimensions are referred to as interaction, course structure, and learner autonomy (Huang & Liaw, 2004).

Results from a study conducted by Huang (2002) showed that learners do not need to interact with other learners to develop a relationship with an instructor. The researcher found that course structure is easily implemented and adjusted in an online course, and it was found that the more technologically skilled an individual was, the better able the individual was at working independently. Kanuka, Colett, & Caswell (2002) noted from a two year study that observed twelve online university instructors that the instructors were more apprehensive about course structure, dialogue, and autonomy when they first began teaching online. Instructors who participated in the study needed to assess learners' autonomy, they needed to provide students with more feedback when the students did not have enough self-discipline to work independently, and they found that the learners wanted flexibility to be a part of the course structure.

Theories of Humanism

Theory of Immediacy and Social Presence

A model of online learning, which considers social presence during asynchronous discussion to be a significant part of mediated discussion, is presented. Theorists who hold that learning takes place through the interaction of three core components: cognitive presence, teaching presence, and social presence (Rourke, Anderson, Garrison, & Archer, 2001).

After a more in-depth look at social presence responses, three forms of social responses were identified. The three response types are called affective responses, interactive responses, and cohesive responses (Martyn, 2004). These responses were used as indicators by Rourke et al. (2001) when analyzing content during their exploration of computer mediated discussions and affective behaviors among participants. It was found that learners' perceptions were an important factor that instructors kept in mind when designing online courses.

Additional important factors related to social presence were found in other studies. Gunawardena and Duphorne (2002) found in a study that focused on an academic computer conference environment that comfort with participating in discussions, easiness with interacting through text, and assurance with ones self significantly impacted perceptions. Murphy (2004) found that sharing, recognition of group presence, appreciative communication between learners, and the opportunity to express feelings and emotions to be indicators of social presence in an online discussion. It is Murphy's conjecture that learners must be able to contribute their own social presence before moving to higher levels of interaction.

Cooperative Learning Theory

Five facets of the basic elements of cooperative learning help others understand how to design learning experiences that utilize Cooperative Learning Theory. Positive Interdependence takes place when students work together, and they perceive that they are moving toward the same goal. "Direct Interaction" occurs when students discuss what they plan to do and how to go about it. "Individual Accountability" encourages individuals to master learning while sharing and working with others. "Attaining Collaborative Skills" involves individuals working together before they cooperate and learn. Finally, group processing takes place when the individuals in the group discuss and evaluate their work. Upon evaluation, the group members found that they work well together. Instructors who have applied this theory guide their students through each facet of the model. The more students develop, the better they work in a cooperative learning situation (Joyce et al., 2000).

A study that explored the online collaborative experiences and attitudes of twelve graduate students found after assessing asynchronous posts that poor communication, conflicts between members, and poor attitudes challenged the success of the online collaboration. Groups that did not have these challenges produced projects of higher quality. The researchers suggest implementing strategies to reduce the challenges so that chances for effective collaboration can increase (Thompson & Hing-Yu, 2006). Another study that investigated the impact of cooperative learning showed that cooperative learning did impact cognitive learning outcomes. Researchers found that learners become more involved with the online learning experience when they work cooperatively with others compared to when they worked independently (Riley & Anderson, 2006).

ACCORDING TO EMPIRICAL RESEARCH: THEORIES NOT YET RESEARCHED

Theories of Behaviorism

Elaboration Theory

Elaboration Theory, which is reflective of Behaviorism, is a practice that is concerned with the organization of materials for a course. While cognitive aspects to this theory exist, it is also considered reflective of Behaviorism because the instructor adjusts the learning environment to meet student needs. This theory holds that new learning should be presented first in the simplest form and carefully move to more complex forms of content and learning. For this reason, online instructors tend to begin with knowledge that students are already familiar with. Then, they transition to the exploration of new knowledge which helps students make the appropriate connections to help them understand the content (Ludwig, 2000; Huang & Liaw, 2004).

Dircct Instruction

Direct Instruction is, referred to by behaviorists as, "modeling with reinforced guided performance." The focus of this model of learning involves dividing performance into goals and tasks, breaking the tasks into smaller tasks, creating training activities that directly target the objectives and ensure mastery of each task, and the inclusion of prerequisites that students have to achieve before they can go on to more advanced concepts. Critics of Direct Instruction Theory note that the application of this theory should be used with caution because it is not appropriate for all educational objectives and all students (Joyce et al., 2000). A study that focused solely on use of direct instruction was not available in the literature, however there is information that suggests using direct instruction as part of an online course when presenting specific directions to students and when providing information needed to participate (Bellefeuille, 2006).

Theories of Cognitivism

Cognitive Flexibility Theory

Jonassen (2003) explains that much research looks at the presentation of problems to learners and identifies two conflicts with how problems that need to be solved are presented. First, it is a conflict when the problems are presented as structured problems because real life problems are not structured. The other conflict is that students do not transfer problem solving skills very well. Jonassen suggests using cognitive flexibility theory to prohibit the conflicts.

Active learning through discussion and exchange of ideas are important aspects of learning (McAlpine & Ashcroft, 2002). For effective distance learning to take place, constructivism and cognitive flexibility needs to be present. Learners should be active participants and instructors cannot be distributors of information since students process information differently. According to the researchers, students should solve problems in ways that are best for themselves (Notar, Wilson, and Montgomery, 2005).

Gagne's Conditions of Learning

Huang and Liaw (2004) identify Gagne's Conditions of Learning as an instructional learning process that is methodical and logical. Gagne's Conditions are a descriptive theory of knowledge that contain five separate categories of outcomes labeled as intellectual skills, verbal information, cognitive strategies, motor skills, and attitudes. Having the ability and knowledge to categorize and use materials are characteristics of intellectual skills. Abilities that allow individuals to show "what" something is or means are verbal information abilities. Cognitive strategies have to do with the learning skills that individuals own. Simple and complex movements make up an individual's motor skills, and attitudes are the feelings that we develop as a result of interactions that are either constructive or unconstructive. Researchers note that Gagne's work has grown into a system of nine practices: gaining attention, informing learners of the objective at hand, stimulating recall of prior learning, presenting the content, providing learning guidance, eliciting performance, providing feedback, assessing performance, and finally, enhancing retention and transfer (Gagne, 1985; Smith & Ragan, 1996; Molenda, 2002; Gagne, Wager, Golas, & Keller, 2005).

Merrill's Instructional Transaction Theory

This theory holds that learners can be motivated by processes of transactions that help them make connections. This theory has a set of conventions to which objects of knowledge are selected and sequenced (Huang & Liaw, 2004). Identifying relationships between educational and technical factors are possible with Instructional Transaction Theory. Instructional Transaction Theory consists of two facets: schemes of knowledge and procedures for applying the knowledge. Merrill's position states that for learning to take place, the learner needs to have more than one knowledge structure illustrated for anything to make sense. According to the researchers, Instructional Transaction Theory learning consists of the object that is to be learned or the content that is to be taught. It is possible to combine the different facets of content that need to be taught and group them into one structure of knowledge. Individuals have internal representations of knowledge and structures of knowledge are external. The theory utilizes transactions as a way to categorize the content that is to be taught (Buendia et al., 2002).

It is believed that there are three data types used when a transaction of knowledge takes place. There is a knowledge base, a resource base, and there are instructional boundaries. These three facets of instructional transaction are then subdivided into more descriptive categories. A knowledge base is for example, divided by entities, activities, and processes. Resource databases among other possibilities are subdivided by mediated representations of the knowledge field, presentation techniques, and communication techniques. Instructional boundaries, of which vary by situation, can be divided according to population, learning task, and the environmental situations. So when an online instructor applies Instructional Transaction Theory to course design empirical research is used to help set the categories in a knowledge base, build resource database classes, and define the parameters that are used to set the boundaries (Zwart, 1992).

Theories of Humanism

Phenomenal Field Theory

A humanistic theorist named Arthur Combs presented his Phenomenal Field Theory with psychologist Donald Snygg. According to this theory, to understand human behavior, the time must be taken to consider the point of view of another. They believed that if one wanted to change another person's behavior that they must first modify his or her beliefs or perception. One had to "walk in their shoes" if they wanted to understand and guide change. By taking this line of thinking, educators had to recognize that the learner needed to find meaning and understand the learning as opposed to learning and understanding the strategies (Boeree, 2007; Tomei, 2007).

Combs and Snygg felt that if they were to understand and foresee the behavior of another that they had to reach into the person's phenomenal field. Since it was impossible for them to physically look into another person's mind, they had to make inferences from what was observed. When educators utilize this theory, they cannot choose a topic of instruction, implement a strategy, and expect every child to be motivated by what has been placed before them because the information does not connect to their own lives. Instead, the educators have to get to know the learner's phenomenal self and create learning experiences that have meaning to the learner. Once instructors take this path, the student that was not motivated to learn at one time will become connected to the learning experience (Boeree, 2007; Tomei, 2007).

Self-Actualization Theory

Maslow believed that strong beliefs about ones self was connected to the thought of self-actualization. According to his thinking, individuals with strong self-actualization interacted well with others, and they found ways to develop and contribute to the world around them fairly easily. Those who did not have strong self-actualization choose to live within their environment and accept what comes their way instead of reaching into their environment and making new opportunities happen for themselves (Joyce et al., 2000; Pinar et al., 1996; Tomei, 2007).

For a person to reach the level of self-actualization, he or she has to be fulfilled at each level of what Maslow referred to as the Hierarchy of Needs. The first level is the biological level. At this level, an individual's need for food and shelter must be met before the individual can move to another level. At the next level, the individual has to feel secure. Level three of the hierarchy of needs demands that the individual feel as though he or she belongs and is loved. Needs for self-respect, achievement, attention, and recognition must be fulfilled if an individual is to move past the esteem level of the hierarchy. When an individual has past each of those levels, he or she has reached the final level, the level of self-actualization. At this point, the individual's ability to reach potential can take place. While each level has to be fulfilled, they do not have to stand alone and one behavior can satisfy more than one level on the hierarchy. Instructors who utilize this theory when designing and conducting a course look to see if their students needs have been met to help them understand student behavior (2000; Pinar et al., 1996; Joyce et al.; Tomei, 2007).

Frequency of Design Theories Utilized by Online Instructors

In a recent study, online instructors were asked how often they utilized nine design theories when creating an online course. Questions about the design theories were presented as part of a larger study on online instructor support for the practice of telementoring (Cicciarelli, 2006).

METHODOLOGY

During the quantitative exploratory study based on descriptive research design, 2000 online instructors were sent a link to an anonymous contingency survey. Nine of the survey questions asked the online instructors how often, according to a Likert scale, they utilized each theory when designing an online course. Once 323 responses to the survey had been submitted, access to the survey was turned off and the data was collected for analysis. A univariate, descriptive level analysis of frequency distributions was run for each variable. Bivariate relationships between the independent and dependent variables were examined using Spearman Rho tests, and cross-tabulations were calculated to provide a deeper look at the data.

Results According to Theory

Theory of Multiple Representations

Spearman rank order of coefficient of correlation (Spearman rho) was computed to measure the strength between the two variables. Results showed that the correlation at a .01 level was significant for a two-tailed test (r = .113, p = .045). According to cross-tabulation results, 32.6% of the online instructors said that they always utilize Theory of Multiple Representations, 32.9% indicated that they used the theory more often than occasionally, and 20.8% indicated occasional use. There were 7.3% who said they utilized the theory less often than occasionally, and 6.4% said they never made use of the Theory of Multiple Representations.

Theory of Immediacy and Social Presence

Spearman rho results for a .01 level two-tailed test that correlated online instructor use of Theory of Immediacy and Social Presence were significant (r = .187, p = .001). Cross-tabulation results showed that 36.8% always use Theory of Immediacy and Social Presence, 36.1% utilize it more often than occasionally, and 15.2% use it occasionally. There were 8.1% who said they utilize the theory less often than occasionally and 3.9% said they never utilize Theory of Immediacy and Social Presence when designing an online course.

Elaboration Theory

Results from a .05 level two-tailed Spearman rho test that correlated online instructor use of Elaboration Theory was significant (r = .146, p = .010). Cross-tabulation results showed that 31.5% of the online instructors always use Elaboration Theory, 41.2% said that they use the theory more often than occasionally, and 16.2% indicated that they use it occasionally. There were 6.8% who said that they utilize the theory less often than occasionally and 4.2% who said that they never use Elaboration Theory when designing an online course.

Dual-Coding Theory

Spearman rho results for a two-tailed test that correlated online instructor use of Dual-Coding Theory was not significant (r = .076, p = .179). Cross-tabulation results showed that 23.5% always utilize Dual-Coding Theory when designing an online course, 23.5% more often than occasionally use the theory, and 18.3% occasionally utilize the theory. There were 17.7% who said that they utilize the theory less often than occasionally and 17.0 percent said that they never use Dual-Coding Theory when designing an online course.

Moore's Theory of Transactional Distance

Results from a .01 level two-tailed Spearman rho test that correlated online instructor use of Moore's Theory of Transactional Distance was significant (r = .233, p = .000). Cross-tabulation results showed that 39.5% always utilize Moore's Theory of Transactional Distance when designing an online course, 37.5% use the theory more often than occasionally, and 14.9% indicated that they utilize the theory occasionally. There were 5.5% who said that they utilize the theory less often than occasionally and 2.6% said that they never utilize Moore's Theory of Transactional Distance when they design an online course.

Merrill's Instructional Theory

Spearman rho results for a .01 level two-tailed test that correlated online instructor use of Merrill's Instructional Theory was significant (r = .181, p = .002). Cross-tabulation results showed that 22.4% of the online instructors said that they always utilize Merrill's Instructional Theory when designing an online course, 39.3% indicated that they utilize the theory more often than occasionally, and 21.8% said that they utilize it occasionally. There were 8.6% who said that they utilize the theory less often than occasionally and 7.9% said that they never utilize Merrill's Instructional Theory when designing an online course.

Gagne's Conditions of Learning

Results from a .01 two-tailed test that correlated online instructor use of Gagne's Conditions of Learning was significant (r = .257, p = .000). Cross-tabulation results showed that 28.9% of the online instructors indicated that they always utilize Gagne's Conditions of Learning when designing an online course, 42.8% said that they utilize the theory more often than occasionally, and 13.8% said that they occasionally utilize the theory. There were 9.0% who indicated that they utilize the theory less often than occasionally and 5.5% said that they never utilize Gagne's Conditions of Learning when designing an online course.

Cognitive Flexibility Theory

Spearman rho results for a .01 level two-tailed test that correlated online instructor use of Cognitive Flexibility Theory was significant (r = .226, p = .000). Cross-tabulation results showed that 30.1% of the online instructors indicated that they always utilize Cognitive Flexibility Theory when they design an online course, 36.9% said that they utilize the theory more often than occasionally, and 22.4% said that they occasionally use the theory. There were 5.4% who indicated that they utilize the theory less often than occasionally and 5.1% said that they never utilize Cognitive Flexibility Theory when designing an online course.

Bruner's Three Form Theory

Results from a .01 level two-tailed test that correlated online instructor use of Bruner's Three Form Theory was significant (r = .169, p = .003). Cross-tabulation results showed that 14.6% of the online instructors indicated that they always utilize Bruner's Three Form Theory when designing an online course, 27.3% said that they utilize the theory more often than occasionally, and 26.0% said that they utilize the theory occasionally. There were 17.9% who said that they utilize the theory less often than occasionally and 14.3% who said that they never utilize Bruner's Three Form Theory when designing an online course.

CONCLUSION

Information presented in this paper focused on answering three questions that individuals could utilize when training online instructors to use theory when designing a course. The first question asked, "According to empirical evidence, what does the research show about online instructor use of different design theories?" The evidence showed that there is empirical evidence available on eight of the fifteen theories presented in this paper. Researchers found positive experiences, and they made suggestions for online instructors to follow when utilizing these theories. According to what has been presented, it would be an effective choice to frame course development with the different theories based on the course and the needs of the students.

The second question presented asked, "According to the lack of empirical evidence, which design theories have not been researched in regard to online instructor utilization of theory?" Considering that evidence on use of seven of the theories is not available, a signal has been made for the need for research. It would be of benefit to the field, if online instructors decided to present how they utilize these theories as part of their online course, and it would be especially useful for those training prospective online instructors.

Finally, the third question asked, "According to a recent study, how frequently do online instructors use nine of the design theories?" Of the nine theories presented there were significant findings for all but one theory. This indicates that online instructors are making use of theory to support course design, and it points to which theories they tend to use and how often. Other online instructors may find this helpful when they question the use of the different theories.

Further research on this topic should investigate to learn if implementing use of theory when designing a course as part of a prospective online instructor training program was beneficial or not. From a different angle, it would be of interest to find a description of the online instructors who indicated how often they utilized nine of the different theories when designing an online course. Researchers could also conduct a study that asked online instructors to provide a qualitative explanation of how they use design theory when creating a course. This would be a favorable way for instructors to share their work so that others could learn and possibly develop other unique ways to make use of the theories for the purpose of creating effective online courses.

REFERENCES

Alty, J. (2002). Dual coding theory and computer education: Some media experiments to examine the effects of different media on learning. In P. Kommers & G. Richards (Eds.), *Proceedings of World Conference on Educational Multimedia, Hypermedia, and Telecommunications 2002* (pp. 42-47). Chesapeake, VA: AACE.

Beacham, N. A., Elliot, A. C., Alty, J. L., & Al-Sharrah, A. (2002). Media combinations andlearning styles: A dual coding approach. In P. Kommers & G. Richards (Eds.), *Proceedings of World Conference on Educational Multimedia, Hypermedia, and Telecommunications 2002* (pp. 111-116). Chesapeake, VA: AACE.

Bellefeuille, G. L. (2006). Rethinking reflective practice education in social work education: A blended constructivist and objectivist instructional design strategy for a web-based child welfare practice course. *Journal of Social Work Education, 42*(1), 85-103.

Berge, O., & Fjuk, A. (2006). Understanding the roles of online meetings in a net-based course. *Journal of Computer Assisted Learning, 22*(1), 13-23.

Bird, L. (2007). The 3 'c' design model for networked collaborative e-learning: A tool for novice designers. *Innovations in Education and Teaching International, 44*(2), 153-167.

Boeree, C. G. (2007) Donald Snygg and Arthur Combs: Personality theory. Retrieved January, 14, 2007 from http://www.social-psychology.de/cc/click.php?id=40

Bruner, J. (1990). *Acts of meaning.* Cambridge, MA: Harvard University Press.

Buchanan, E. A., Myers, S. E., & Hardin, S. L. (2005). Holding your hand from a distance: Online mentoring and the graduate library and information science student. *The Journal of Educators Online, 2*(2), 1-18.

Buendia, F., Diaz, P., & Benlloch, J. V. (2002). A framework for the instructional design of multi-structured educational applications. In P. Kommers & G. Richards (Eds.), *Proceedings of World Conference on Educational Multimedia, Hypermedia & Telecommunications 2002* (pp. 210-215). Chesapeake, VA: AACE.

Bush, G. (2007). Learning about learning: from theories to trends. *Teacher Librarian, 34*(2), 14-18.

Chu, H.C., & Hinton, B. E. (2001). *Factors affecting student completion in a distance learning mediated HRD baccalaureate program.* Paper presented at the Academy of Human Resource Development 2001 Conference, Fayetteville, AR. (AHRD Reference No. 021)

Cicciarelli, M. (2006). Telementoring and computer mediated discussions: A description of online instructors' support. *Dissertation Abstracts International,* (UMI No. 3238537)

Cicciarelli, M. (2007). Behavioral theories that guide online course design. *Encyclopedia of Information Technology Curriculum Integration,* Pennsylvania: Idea Group Publishing.

Cicciarelli, M. (2007). Cognitive theories that guide online course design. *Encyclopedia of Information Technology Curriculum Integration,* Pennsylvania: Idea Group Publishing.

Cicciarelli, M. (2007). Humanistic theories that guide online course design. *Encyclopedia of Information Technology Curriculum Integration,* Pennsylvania: Idea Group Publishing.

Cicciarelli, M. (2008). A description of online instructors use of design theory. *International Journal of Information & Communication Technology Education, 4*(1), 25-32.

Collins, M., & Berge, Z. (1996). *Facilitating interaction in computer mediated online courses.* Paper presented at the FSU/AECT Distance Education Conference. Retrieved June 29, 2005, from http://www.emoderators.com/moderators/fleehtml.

Cook, K. C. (2007). Immersion in a digital pool: Training prospective online instructors in online environments. *Technical Communication Quarterly, 16*(1), 55-83.

Course-Management Systems. (2005). *Library technology reports, 41*(3), 7-11.

Dennen, V. P. (2007). Presence and positioning as components of online instructor persona. *Journal of Research on Technology in Education, 40*(1), 95-108.

Figueroa, S., & Huie, C. (2001). *The use of blackboard in computer information systems courses.* New York: City University of New York Hostos Community College. (ERIC Document Reproduction Service No. ED469883)

Gagne, R. M. (1985). *The conditions of learning and theory of instruction* (4th ed.). New York: Holt, Rinehart, and Winston.

Gagne, R. M., Wager, W. W., Golas, K. C., & Keller, J. M. (2005). *Principles of instructional design* (5th ed.). Belmont, CA: Wadsworth/Thomson Learning.

Gfeller, M. K., Niess, M. L., & Lederman, N. G. (1999). Preservice teachers' use of multiple representations in solving arithmetic mean problems. *School Science and Mathematics, 99,* 250-257.

Grabinger, S. (2004). Design lessons for social education. In Duffy, T. M., & Kirkley, J. R., *Learner-Centered Theory and Practice in Distance Education: Cases from Higher Education 2004* (pp. 49-60).

Gunawardena, C. N., & Duphorne, P. L. (2000). Predictors of learner satisfaction in an academic computer conference. *Distance Education, 21*(1), 101-117.

Huang, H. M. (2002). Student perceptions in an online mediated environment. *International Journal of Instructional Media 29,* 405-422.

Huang, H. M., & Liaw, S. S. (2004). Guiding distance educators in building web-based instructions. *International Journal of Instructional Media 31*(2), 125-137.

Jonassen, D. (2003). Using cognitive tools to represent problems. *Journal of Research on Technology in Education, 35,* 362-381.

Joyce, Weil, & Calhoun. (2000). *Models of Teaching (6th Ed.).* Needham Heights, MA: Allyn & Bacon.

Kanuka, H., Collett, D., & Caswell, C. (2002). University instructor perceptions of the use of asynchronous text-based discussion in distance courses. *The American Journal of Distance Education, 16*(3), 151-167.

King, K. P. (2001). Playing out the realities of web-based bulletin boards: Enhancing face to face learning. *New Horizons in Adult Education, 15*(1).

Lebaron, J., & Miller, D. (2005). The potential of jigsaw role playing to promote the social construction of knowledge in an online graduate education course. *Teachers College Record, 107*(8), 1652-1674.

Lee, J., & Lee, Y. (2006). Personality types and learners' interaction in web-based threaded discussion. *Quarterly Review of Distance Education, 7*(1), 83-94.

Ludwig, B. (2000, August). *Web-based instruction: Theoretical differences in treatment of subject matter.* Paper presented at the Annual Meeting of the Psychological Association, Washington, DC. (ERIC Document Reproduction Service No. ED453708)

Majeski, R., & Stover, M. (2007). Theoretically based pedagogical strategies leading to deep learning in asynchronous online gerontology courses. *Educational Gerontology, 33*(3), 171-185.

Makrakis, V. (1998). *Guidelines for the design and development of computer mediated collaborative open distance learning courseware.* Paper presented at the 1998 World Conference on Educational Multimedia and Hypermedia & World Conference on Educational Telecommunications. Frieburg, Germany. (ERIC Document Reproduction Service No. ED428694)

Martyn, M. A. (2004). *The effect of online threaded discussion on student perceptions and learning outcomes in both face-to-face and online courses.* Doctoral dissertation, University of Akron, 2004. (UMI No. 3123389)

McAlpine, I., & Ashcroft, B. (2002). Turning points: Learning from online discussions in an off-campus course. In P. Kommers and G. Richards (Eds.), *Proceedings of World Conference on Educational Multimedia, Hypermedia& Telecommunications 2002* (pp. 1251-1257), Chesapeake, VA: AACE.

Mimirinis, M. & Bhattacharya, M. (2007). Design of virtual learning environments for deep learning. *Journal of Interactive Learning Research, 18*(1), 55-64.

Molenda, M. (2002). *A new framework for teaching in the cognitive domain.* Syracuse, NY: ERIC Clearinghouse on Information and Technology (ERIC Document Reproduction Service No. ED470983)

Moreno, R. (2002). Who learns best with multiple representations? Cognitive theory implications for individual differences in multimedia learning In P. Kommers and G. Richards (Eds). *Proceedings of World Conference on Educational Multimedia, Hypermedia & Telecommunications 2002* (pp. 1380-1385). Chesapeake, VA: AACE.

Murphy, E. (2004). Recognizing and promoting collaboration in an online asynchronous discussion. *British Journal of Educational Technology, 35*, 421-431.

Nortar, C. E., Wilson, J. D., & Montgomery, M. K. (2005). A distance learning model for teaching higher order thinking. *College Student Journal, 31*(1), 17-26.

Northrup, P., Lee, R., & Burgess, V. (2002). Learner perception of online interaction. In P. Kommers & G. Richards (Eds.), *Proceedings of World Conference on Educational Multimedia, Hypermedia, and Telecommunications 2002* (pp. 1462-1467). Chesapeake, VA: AACE.

Pinar, W. F., Reynolds, W. M., Slattery, P., & Taubman, P. m. (1996). Understanding curriculum. New York, NY: Peter Lang Publishing, Inc.

Prestera, G. E., & Moller, L. A. (2001, April). *Facilitating asynchronous distance learning: Exploiting opportunities for knowledge building in asynchronous distance learning environments.*

Paper presented at the Annual Mid-South Instructional Technology Conference, Murfreesboro, TN. (ERIC Document Reproduction Service No. ED463723)

Rieber, L. P., Tzeng, S. C., Tribble, K., & Chu, G. (1996). *Feedback and elaboration within a computer-based simulation: A dual-coding perspective.* Paper presented at the 1996 National Convention of the Association for Educational Communications and Technology, Indianapolis, IN. (ERIC Document Reproduction Service No. ED397829)

Riley, W., & Anderson, P. (2006). Randomized study on the impact of cooperative learning. *Quarterly Review of Distance Education, 7*(2), 129-144.

Rourke, L., Anderson, T., Garrison, D. R., & Archer, W. (2001). Assessing social presentation asynchronous text-based computer conferencing. *Journal of Distance Education, 14*(2), 50-71.

Schwartzman, R. (2006). Virtual group problem solving in the basic communication course: Lessons for online learning. *Journal of Instructional Psychology, 33*(1), 3-14.

Simonson, M., Smaldino, S., Albright, M., & Zvacek, S. (2000). *Teaching and learning at a distance: Foundation of distance education.* Upper Saddle River, NJ: Merrill.

Smith, P. L., & Ragan, T. J. (1996). *Impact of R. M. Gagne's work on instructional theory.* Paper presented at the 1996 National Convention of the Association for Educational Communications and Technology, Indianapolis, IN. (ERIC Document Reproduction Service No. ED397841)

Tallent-Runnels, M. K., Thomas, J. A., Lan, W. Y., Cooper, S., Ahern, T. C., Shaw, S. M., Liu, Xiaoming. (2007). Teaching courses online: A review of the research. *Review of Educational Research, 76*(1), *93-135.*

Thompson, L., & Hing-Yu, K. (2006). A case study of collaborative online learning. *Quarterly Review of Distance Education, 7*(4), 361-375.

Tomei, L. A. (2007). *Learning theories: A primer exercise.* Retrieved January 14, 2007, from http://www.acacemics.rmu.edu/~tomei/ed711psy/human.htm

Williams, S. W. (2001). *Experiences of web-based instruction among African-American students enrolled in training and development graduate courses.* Paper presented at the Academy of Human Resource Development 2001 Conference, Raleigh, NC. (AHRD Reference No. 059)

Ying-Shao, H., & Fu-Kwun, H. (2002, June). *The use of multiple representations in a web-based and situated learning environment.* Paper presented at the ED-MEDIA 2002 World Conference on Educational Multimedia, Hypermedia & Telecommunications, Denver, CO. (ERIC Document Reproduction Service No. ED477029)

Zwart, W. J. (1992, June). *Instructional transaction theory applied to computer simulations.* (ERIC Document Reproduction Service No. ED352945)

Chapter XIII
The Impact of PowerPoint Presentations on Student Achievement and Student Attitudes

Michael Fedisson
Bellefonte Area Middle School, USA

Silvia Braidic
California University of Pennsylvania, USA

ABSTRACT

Seventh grade students were tested on their knowledge of sentences and nouns in a language arts classroom. This study was conducted over a two-year time frame. In the first year, classes consisted of twenty-eight (28) males and thirty-one (31) females. All students are Caucasians with the exception of two African American males. In year two, the classes consisted of thirty-two (32) females and thirty-two (32) males. All students are Caucasians with the exception of one African American female and one Nicaraguan-American female. Students are predominantly from middle class families. All three classes are grouped heterogeneously. During instruction for two units, classes were taught with the following approaches: 1) using traditional methods of book work and handouts for one unit, and 2) using technological aids such as Microsoft PowerPoint for a second unit. Test results from three classes during both units were compared. The data indicates that when using technological aids as teaching tools, student test grades increased in year one, especially for low-achieving students or for those with learning disabilities. In year two, those same results were not achieved. A technology survey was also used to establish each student's comfort level with technology and their attitudes towards the use of technological aids in the classroom.

INTRODUCTION

The Bellefonte Area School District recently acquired ceiling-mounted LCD projectors for use in all school classrooms. The projectors have the capability to project images and video from teacher computers, VCR's, DVD players, and television sets. Teachers also have been given a remote to turn on the projectors, as well as a wireless remote to change between computer and video sources. This remote can also be used as a wireless means of operating PowerPoint presentations and as a "mouse" for searching the internet.

Seventh grade language arts teachers have been plagued by a lack of grammar textbooks for our students. There are not enough books for each student to have his/her own copy. As a result, many teachers utilize overhead projectors with transparencies and homework packets to teach grammar. Many students find this means of teaching as stoic and ineffective. Therefore, the purpose of this research was to measure the attitudes and achievement of students when comparing traditional methods of teaching versus the use of the overhead LCD projector in conjunction with PowerPoint presentations.

REVIEW OF THE LITERATURE

Technology is everywhere in today's schools and larger society. For the current youth generation, the Internet has *always* existed. Online technologies have profoundly contributed to a dramatic technocultural shift in contemporary society, transforming how we learn, work, play, and socialize (Steinkuehler, C., University of Wisconsin–Madison) For those who have grown up with such technologies, this heterogeneous, networked, online global, "flat" (Friedman, 2005) world is the unremarkable mainstream.

Technology is available in our classrooms, and it is changing the way educators think about teaching and the way students think about learning. Yet, students will not make significant gains on their own. Students spend countless hours at home playing games on their PC's or surfing the internet. This does not necessarily transfer to an increase in student achievement. Furthermore, 45 to 90 minutes a week in the computer lab does not foster the type of learning that will improve student achievement (Kozlowski, 2000). Research reminds us that technology generally improves performance when the application directly supports the curriculum standards being assessed (Cradler, McNabb, Freeman, & Burchett, 2002). A review of studies conducted by the CEO Forum (2001) emphasizes that "technology can have the greatest impact when integrated into the curriculum to achieve clear, measurable educational objectives."

Extreme advances in its use as a teaching tool are more apparent every day. Many researchers point to the great value that technology brings in motivating students and increasing achievement. However, others still find that technological resources are misused and abused and can create more problems than good. It must be made clear that technology in the schools will not change motivation and achievement alone. The introduction of technology in the classrooms may produce enthusiasm from both teachers and students, but having rooms full of computers, projectors, software, and handhelds are useless dust collectors if not productively accessed by trained professionals. Students cannot reach their full potential with technological tools without a well-trained staff of professionals to guide them. Technology should be used when it is the most appropriate tool for the lesson or activity, not because it is simply available. Time and resources must be allocated to help teachers acquire the expertise necessary to feel comfortable using technology to create student-centered learning environments. Teachers must be given the proper training and support to integrate technology into their classrooms for the positive effects of technology on student engagement to last (Sandholtz, Ringstaff,

& Dwyer, 1994). The technological debate will continue to ignite the flames of controversy among school teachers across the country. It is clear that technology will continue to change the way students learn for decades to come. It is up to the teachers and school districts of this country to decide how effective we want to be in using technology to reach our students. The International Society for Technology in Education (ISTE) is an excellent reference for teachers currently in the field. Although designed to focus on preservice teacher education, the National Educational Technology Standards (NETS) for teachers provides a framework to identify fundamental concepts, knowledge, skills, and attitudes for applying technology in educational settings (NETS, 2000). The standards also serve as guidelines for teachers currently in the classroom. The six standards (NETS, 2000) are:

- Technology operations
- Planning and designing learning environments and experiences
- Teaching, learning and the curriculum
- Assessment and evaluation
- Productivity and professional practice
- Social, ethical, legal, and human issues

In reviewing the vast field of information related to technology, most authors highlight the many positive uses of technology. Today, many new programs exist that were mere dreams ten years ago. Students now use laptops, reading programs, countless word-processing software programs, and the internet to gather, learn, and present information. Most proponents of technology point to the great increase in student motivation over the past ten years. Students at this point in education have grown up with computers and are more than happy to show others how to use them. Students today are those of the millennial generation. The millennial generation are those born between 1982-2002. "When Generations Collide" (Butterfield, The Forbes Group, 2005) shares the following about millennial: The millennials are realistic, optimistic, progressive, loyal, inclusive, collaborative, and scheduled. They have always been consulted. Millennials have always known diversity. They want constant feedback and they want it to be timely and two way. They are great collaborators and see leadership as participative. Millennials were born with technology." Butterfield's comments clearly illustrate the need for students to "branch out" from the more traditional assessment methods of yesteryear. They have a tremendous mastery of surfing the net and creating extraordinary presentations. The applications for students seem endless. The American educational system has done its best to keep pace, providing Internet connections to virtually all schools (99% in 2001), 87% of which are accessible to students via classrooms, libraries, computer labs, and other regulated spaces (Kleiner & Farris, 2002). Still, the *culture* of schooling carries on with business as usual – as it was ten or twenty years, ago, that is. As a Pew Internet & American Life Report (Levin & Arafeh, 2002) on the digital disconnect between children and their schools details with excruciating clarity, what students do with online technologies *outside* the classroom is not only markedly different from what they do with them in schools (e.g. instant messaging, blogging, sharing files, consuming and producing media, engaging in affinity spaces, gaming, building social networks, downloading answers to homework, and researching for school projects and assignments), but it is also more goal-driven, complex, sophisticated, and engaged. (Steinkuehler).

These endless possibilities are most apparent in motivating students to learn and increasing achievement. Students often feel more comfortable completing assignments with technology than with the traditional paper and pencil method. Our students are coming to school having already developed diverse learning style preferences. The processes teachers' employ for engaging student in the learning process, once assigned to a classroom, is just as diversified. New technologies are being introduced into classrooms to address this dynamic. They are the masters of PowerPoint, word processing, and creating videos of which they are most eager to share and demonstrate. Technology allows students to more easily make corrections and add creative touches to showcase their individuality. Furthermore, this generation's fast-paced attitude is supplemented by the use of the internet versus using print materials in a library. Technology has given students a whole new vehicle for learning. The technology provides opportunities for teachers to meet the needs of students with various learning styles through the use of multiple media (Bryant & Hunton, 2000).

Other studies go beyond making a point just about student motivation and find that technology greatly increases achievement (Andrews, 2006). The use of laptops and programs such as Accelerated Reader or Reading Counts has helped students to maximize their full potential. In Andrews' article, Tony Sambunjak states, "The motivation level of the students rose astronomically. The students could learn about a new tool or jig or fixture in class, over lunch or in the evening, design a part on Mastercams, and come into the lab and actually create the part the next day by plugging the laptop into a milling machine. Interest level skyrocketed." These programs, and similar ones, have helped teachers to organize, deliver, and test information. Students are learning more in less time and retaining more than ever. Specifically, as in *Laptops + Challenging Curriculum = Student Success*, Andrews cites that on average students master 159 of the 347 competencies of the PMT program in their junior year and a total of 279 competencies after their senior year. However, the first group of students to use the online curriculum and laptops averaged a mastery of 295 competencies in their junior year alone; more than the previous classes did in two years. As a result, students are becoming more self-sufficient and willing to share knowledge.

Of course, not everyone will agree with any issue. Those that oppose the overuse of technology point to the misuse and abuse of technology. These researchers contend that certain programs like Reading Counts or Accelerated Reader are all a cover-up for a larger problem. Instead of helping students increase reading ability, the programs just focus on moving students from level to level and to receive "feel good" awards (Chenoweth, 2001). Also, these researchers point to the fact that students are misusing tools such as the internet. Instead of using it for valuable learning, students are more interested in playing games, sending email, or downloading pornography (Borja, 2006). Along with constant technical difficulties and hardware maintenance, technology is creating a distraction among our school students.

RESEARCH QUESTION

How will the use of technological aids in the classroom increase student achievement and/or attitudes?

Method and Design

Over a two-year time period, two different seventh grade classes participated in this study. Three language arts classes participated in two separate grammar units. The first was a unit focused on sentences. This unit included the types of sentences (imperative, exclamatory, declarative, interrogative, simple, compound, complex, and compound/complex). The second unit focused on nouns. This included common nouns, proper nouns, nouns as subjects, direct objects, possessive, and plural nouns. All classes were taught the exact same material and given the exact same tests. All students completed the same homework assignments, as well.

The three language arts classes all received similar, but different methods of instruction. The first class, Language Arts "A", was taught the sentences unit in a traditional method. They received instruction from handouts, book work, and chalkboard instruction. For the second unit on nouns, the class experienced these same methods, but with a PowerPoint presentation as a supplement. The PowerPoint presentation was presented on a ceiling-mounted LCD projector. The presentation expanded each lesson by providing extra examples and examples from the homework. Students were able to solve sample problems and then instantly see the answers on a large screen in the classroom. This presented them with instant feedback.

The second class, Language Arts "B", received similar instruction. In their case, the sentences unit was taught with the same methods of the nouns unit for Language Arts "A". The sentences unit used conventional methods, plus a PowerPoint presentation related to the material on sentences. During the second unit, on nouns, this class received the traditional instructional methods that Language Arts "A" received during the sentences unit. Essentially, both classes received similar instruction, just during opposite units.

The third class, Language Arts "C", experienced the traditional instructional methods plus the PowerPoint presentation for both units (sentences and nouns). This served to act as a control group for Language Arts classes "A" and "B" and is summarized in Table 1.

After each instructional unit, students participated in a review session prior to each examination. The class reviewed each section that was present on the test and the teacher answered any questions of the students. Each test mimicked the homework assignments. Each test section was in the same sequence as the topics discussed in class. Furthermore, the problems on the test were taken directly from the homework assignments. Students were asked to complete the same tasks on the test as on the homework assignments. This included identifying types of sentences/nouns, changing sentence/nouns forms, and correcting mistakes in writing.

Each test was scored by awarding the number of points correct out of the total. All questions on both tests were weighted equally. Each of the test scores was then doubled. In the first year of the study, the sentences test had a total of 50 questions, for a total of 100 points. The nouns test had 42 questions, for a total of 84 points. It is imperative to note that total points did not serve as the final data evaluation. In the second year, the tests changed slightly. After the first year's research, a concern arose that the differences in test format between the "sentences test" and "nouns test" could skew the results. In an effort to eliminate this factor, the tests in the second year of research were more consistent. During the second year of research, both tests consisted of 35 multiple choice questions. Again, the scores were doubled for a total possible score of 70 points. It is important to note that both tests were compared using the final score percentage to maintain consistency with year one's research. Data was then analyzed to see if any changes were detected in final scores across the two units within each class. (I.e. scores of the sentences test

Table 1.

Class	Sentences Unit	Nouns Unit
Language Arts "A"	Traditional methods No PowerPoint	Traditional methods plus PowerPoint
Language Arts "B"	Traditional methods plus PowerPoint	Traditional methods No PowerPoint
Language Arts "C"	Traditional methods plus PowerPoint	Traditional methods plus PowerPoint

versus the nouns test for Language Arts "A"). The point was to analyze the effect of the PowerPoint presentation and the use of technology in varying student scores. Scores were listed for all students, in separate classes, in percentage form. Then, the percent change was included for all students and as a class average. Furthermore, the data was analyzed to determine the total percent increase and decrease for the class from test one to test two. Lastly, the number of students who had a positive grade change and/or a negative grade change from test one to test two was noted. Also, both tests were compared separately across all three classes. It was important to document the degree of similarity on either test between all three classes. Again, were there any patterns associated with the addition of instructional technology?

Lastly, students were asked to complete a brief technology survey. The survey was comprised of three main parts. First, a simple questionnaire measured the students' attitudes and comfort levels related to the use of technology. The second questionnaire focused on the use of the PowerPoint presentations and the use of technology in the language arts classroom. In year one, these two parts asked students to respond to a number of statements. Students were to check one of five boxes to describe their level of comfort with each statement. The categories of the Likert scale were labeled as, "Strongly Agree", "Agree", "Same", "Disagree", and "Strongly Disagree." After the data from year one was analyzed, a pattern appeared. A large number of students checked the "Same" category for many of the statements. This left an unanswered question. Did those students have a stronger feeling towards agreeing or disagreeing with the statement? In an effort to make this distinction more decisive, in year two, the "Same" category was eliminated from the survey. Therefore, a clear distinction would be formed as to whether students agreed or disagreed with each statement. Finally, six open-ended questions sought to gain specific information about student attitudes related to technology in general and its specific use in the language arts classroom. All surveys were collected. Answers for all sections of questionnaire parts I and II were tallied to create an overview for all three classes. In year two, students were given the same survey twice, once before the introduction of the use of technology as a teaching tool and once after the units were presented with technology. Those numbers are presented in a pre/post form. Open-ended responses were evaluated for particularly thoughtful responses.

Data Analysis

After collecting the data from all three classes related to test scores, it is apparent that the use of the LCD projector played a significant impact on the grades of lower-achieving students during the first year of this study. For other classes, results were mixed. While the other classes did show numerical improvements, they may not be significant enough to determine that the sole reason for the increase was due to the LCD projector and PowerPoint presentation. Consider Table 2 as an illustration of the data:

Table 2. Average Class Scores Year 1 / Year 2

Class	Average Score: Sentences	Average Score: Nouns
LA "A"	75% / 73%	*77% / *74%
LA "B"	*67% / *75%	60% / 80%
LA "C"	*76% / *76%	*80% / *79%

**denotes the use of the LCD and PowerPoint presentations in addition to traditional instructional methods*

As the above table indicates, during year one of the study there was very little noticeable change in LA "A" from the first test average to the second. Only a two (2) percent increase was not significant enough to determine that the increase was due solely to the use of the LCD and PowerPoint presentation. A similar case also occurred during the second year of the study. Again, there was not a significant increase in the total class average of LA "A". Also, in looking at the scores of LA "C", a similar increase was noted, but both units were taught with the addition of the LCD and PowerPoint presentations.

However, there was a dramatic difference in LA "B". In the first year of the study, the second test illustrates a drop of seven (7) percentage points when not using the LCD with PowerPoint presentations. It should be noted that this class is comprised of seven (7) learning disabled students. This may suggest that these students benefit from the use of the technological instructional methods. During the second year of the survey, there was also a significant difference between test one and test two of LA "B." However, the result was the complete opposite of the previous year's research. In the absence of the PowerPoint instruction, student test scores actually increased. This data might suggest that the nouns test was considerably easier than the sentences test. The data from the control group, LA "C", would support this conclusion. It should be noted, however, that the students who had the greatest percentage increases in scores all scored 60% or less on the first test (sentences).—*See Table 3: Year 2.* Hence, these students had the most potential for increase. That particular data would misrepresent the fact that an equal number of students both increased and decreased percentage scores from test one to test two. Therefore, while there was a significant difference in total percentage increase and total percentage decrease of LA "B" in year two, the majority of the disparity can be contributed to the scores of five students. Without considering the scores of those five students, the difference between the total percentage increase and total percentage decrease of LA "B" is very minimal. One should not jump to conclusions that the data would indicate the decreased value of using technological teaching tools in the classroom.

Upon the continued analysis of individual scores, there appeared a more significant change in the use of the LCD and PowerPoint from the first year of the study. LA "B" showed the most significant change in class average. Also, when looking closely at the data, a correlation between achievement and the use of technology in the classroom was noticeable. Of the fifteen (15) students in the class, more than half had lower scores on the nouns test than on the sentences test. This was without the use of technology during instruction. Eight (8) students earned lower grades without the PowerPoint instruction as compared to the first test (sentences). Only four (4) students actually achieved a higher grade without the PowerPoint. One student recorded no change. Three students either did not take the first test or the second test. Their data was incomplete and not factored into class averages for the particular test missed.

Similarly, LA "C" had a more direct correlation when observing individual data. Students in this class had more students achieve higher scores on the second test (nouns) than on the sentences test. Eleven (11) students achieved higher scores, while only seven (7) received lower scores. In year two, nine (9) students achieved higher scores, and seven (7) received lower scores, respectively. Since LA "C" had the use of technology in their instruction for both tests, this may suggest that test two (nouns) was easier than test one (sentences). Therefore, one would not expect LA "B" to have such a dramatic decrease in scores from test one to test two of year one. This would confirm the advantage of using technology in instruction to increase scores. Lastly, when comparing the total percentage increase and decrease for LA "C", students had a total percentage increase of 222% on test two (nouns) compared to only a 54% decrease on the same test. Again, this would confirm the effectiveness of the technological instruction for LA "B".

Due to the data above, LA "B" from year one, which is comprised of seven (7) learning disabled students, would benefit more from the use of technology in instruction. Research also supports that "being sensitive to learning profile provides benefits for instruction. (Tomlinson, 2000). Numerous researchers over a period of decades have concluded that addressing an individual's learning styles through flexible and compatible teaching results in increased academic achievement. In a meta-analysis of research on the effect of learning-style accommodation, Sullivan (1993) found that accommodating learning style through complementary teaching or counseling interventions resulted in significant academic and attitude gains for students from all cultural groups. When looking at data from LA "A", no direct conclusions could be drawn from the data. Students in LA "A" experienced technologically supported instruction on test two (nouns). From test one to test two, an equal number of students increased their scores as did receive lower scores. Eleven students showed an increase with the PowerPoint instruction, while an equal eleven students showed a decrease with the PowerPoint instruction. One student had identical percentage scores on both tests. During year two of this experiment, similar results were achieved. Seven (7) students achieved

Table 3. Year One

Language Arts "B" Name	Sentences Test PowerPoint (T1)	Nouns Test No PowerPoint (T2)	Percent Change (T2-T1)/ T1
AB	22	31	+41
BB	62	45	-27
CB	68	38	-44
DB	68	60	-12
EB	50	60	+20
FB	64	64	0
GB	72	69	-4
HB	68	INC	INC
IB	72	69	-4
JB	INC	40	INC
KB	88	INC	INC
LB	92	79	-14
MB	60	48	-20
NB	64	83	+30
OB	76	81	+7
PB	82	71	-13
Class Average	67%	60%	8 neg. change 4 positive change 1 no change 3 incomplete

Total % increase = 98
Total % decrease = 138

Table 3. Year Two

Language Arts "B" Name	Sentences Test PowerPoint (T1)	Nouns Test No PowerPoint (T2)	Percent Change (T2-T1)/T1
AB	100	100	0
BB	94	94	0
CB	69	74	+7
DB	97	91	-6
EB	86	83	-3
FB	80	69	-14
GB	41	66	+61
HB	47	INC	INC
IB	53	66	+25
JB	91	94	+3
KB	60	69	+15
LB	66	69	+5
MB	94	100	+6
NB	91	91	0
OB	41	74	+81
PB	71	66	-7
QB	60	71	+18
RB	94	91	-3
SB	74	71	-4
TB	63	66	+5
UB	94	94	0
Class Average	75%	80%	6 negative change 10 positive change 4 no change 1 incomplete

Total % increase = 226
Total % decrease = 37

higher scores, while twelve (12) students earned lower scores. The class average was a mere one percent different. Furthermore, the total percent change was insignificant. This class, LA "A" has no learning disabled students in year one or year two of this study. This would suggest that more impact can be shown when technology is used with LD students Table 3 - 5 provides a detailed breakdown of this information.

After exploring the numerical data, the technology questionnaire was examined to find more evidence to support or deny the use of technology in the classroom. In relation to general technology questions, the majority of students find technology easy to use and useful in completing school assignments. Furthermore, students agreed that technology allows them to improve the quality of their work and understand their classes better. On the other side of the coin, students disagreed with the statement that others know more about technology than they do personally. Also, the majority of students feel that they do not need

Table 4. Year One

Language Arts "A" Name	Sentences Test No PowerPoint (T1)	Nouns Test PowerPoint (T2)	Percent Change (T2-T1)/ T1
AA	72	90	+25
BA	74	93	+26
CA	80	93	+16
DA	58	83	+43
EA	84	90	+7
FA	96	86	-10
GA	92	86	-7
HA	88	71	-19
IA	60	36	-40
JA	54	INC	INC
KA	80	79	-1
LA	88	88	0
MA	68	81	+19
NA	72	71	-1
OA	80	81	+1
PA	50	48	-4
QA	84	93	+11
RA	78	69	-12
SA	72	55	-27
TA	72	88	+22
UA	76	74	-3
VA	84	88	+5
WA	66	48	-27
XA	62	71	+15
Class Average	75%	77%	11 positive change 11 negative change 1 no change

Total % increase = 189
Total % decrease = 151

to learn many more skills to use technology effectively. Individual responses are noted below in Table 6, Year One.

Similar results were found during year two of the research. It should be noted that the survey was changed slightly during the second year of data collection. During the first year, a large number of students checked the "Same" column. This made the data confusing. It was not clear whether students were more in favor or less in favor of each statement. In order to clarify the students' responses, the "Same" column was eliminated from the survey during year two. Furthermore, students who participated during the second year

Table 4. Year Two

Language Arts "A" Name	Sentences Test No PowerPoint (T1)	Nouns Test PowerPoint (T2)	Percent Change (T2-T1)/ T1
AA	97	80	-18
BA	71	80	+13
CA	74	80	+8
DA	69	57	-17
EA	66	49	-28
FA	57	54	-5
GA	54	49	-9
HA	91	91	0
IA	91	89	-2
JA	71	97	+37
KA	77	71	-8
LA	89	66	-26
MA	69	86	+25
NA	77	57	-26
OA	54	80	+48
PA	89	89	0
QA	86	60	-30
RA	80	89	+11
SA	91	89	-2
TA	57	60	+5
UA	80	77	-4
Class Average	73%	74%	7 positive change 12 negative change 2 no change

Total % increase = 147
Total % decrease = 175

of the research were given a pre-survey and a post-survey. The pre-survey was administered before the introduction of PowerPoint as a means of an instructional aid. The post-survey was administered after finishing the "nouns test"; hence, after the introduction of PowerPoint as an instructional aid. The data was analyzed for any changes in attitudes before and after experiencing technology in instruction. Table 6, Year Two illustrates the data.

In response to particular methods used in language arts class, responses favored using technology as a teaching tool. Answers were mixed in response to the students' thoughts about the effectiveness of using the LCD projector and PowerPoint presentations to increase achievement. While a large number of students agreed or strongly agreed, a large number of students were also undecided or disagreed. In the last section

Table 5. Year One

Language Arts "C" Name	Sentences Test PowerPoint (T1)	Nouns Test PowerPoint (T2)	Percent Change (T2-T1)/ T1
AC	96	81	-16
BC	56	83	+48
CC	40	52	+30
DC	70	83	+19
EC	80	74	-8
FC	62	83	+34
GC	88	79	-10
HC	60	48	-2
IC	70	83	+19
JC	74	90	+22
KC	64	69	+8
LC	80	88	+10
MC	96	93	-3
NC	76	93	+22
OC	86	76	-12
PC	84	90	+7
QC	86	83	-3
RC	92	95	+3
Class Average	76%	80%	7 positive change 11 negative change

Total % increase = 222
Total % decrease = 54

Table 5. Year Two

Language Arts "C" Name	Sentences Test PowerPoint (T1)	Nouns Test PowerPoint (T2)	Percent Change (T2-T1)/T1
AC	86	86	0
BC	69	69	0
CC	77	80	+4
DC	86	94	+9
EC	83	94	+13
FC	89	89	0
GC	80	77	-4
HC	71	80	+13

continued on following page

Table 5. continued

IC	69	80	+16
JC	77	71	-8
KC	97	91	-6
LC	31	46	+48
MC	91	91	0
NC	66	60	-9
OC	83	86	+4
PC	31	58	+87
QC	86	80	-7
RC	100	94	-6
SC	80	89	+11
TC	INC	51	INC
UC	74	69	-7
	76%	79%	9 positive change 7 negative change 4 no change

Total % increase = 205
Total % decrease = 47

of open-ended questions, students noted that the instruction alone would not dictate test scores. Students observed that other factors contribute to their scores. Such items like time spent studying, talking to friends, and distractibility were listed. It is clear, however, that students do favor such technology tools as teaching methods. Students found technology to be more motivating in regards to learning. They also noted that it was used effectively in the language arts classroom. There was resounding disagreement to the statement that technology is used too much in the classroom. Individual responses are noted below.

Lastly, the students responded to six open-ended questions to get specific responses to the use of technology in the classroom. Students were asked the following questions:

1. How did you feel about the use of the projector and PowerPoint presentations to teach grammar?
2. Did you feel the use of the projector and PowerPoint presentations were helpful in learning the material about sentences and nouns? Explain.
3. Do you think your grade on the sentences or nouns test was higher simply because of the use of the projector? Explain.
4. What factors helped you to get a better grade on one test, rather than on the other?
5. Should we use the projector more often for PowerPoint presentations of grammar? Why?
6. Should we use the projector more often for showing examples of writing when working

Table 6. Year One

Category	Strongly Agree	Agree	Same	Disagree	Strongly Disagree	Failed to Answer
Computers make school work easier to do.	26	22	6	1	0	0
I prefer to use computers to do school-work rather than paper and pencil.	20	13	18	4	0	0
Using computers to do school-work can also have disadvantages.	5	32	15	2	0	1
Most days, I look forward to attending school.	6	17	20	6	6	0
Computers make school-work more fun/ interesting.	23	12	19	0	0	1
Computers help me to improve the quality of my school-work.	15	25	15	0	0	0
Computers help me understand my classes better.	8	9	27	9	2	0
I need to learn many new skills to use computers for my school-work.	3	8	9	22	13	0
I generally enjoy school-work.	1	10	20	8	16	0
Compared to other students, I really enjoy being in class.	3	17	24	5	6	0
Many of my classmates know more about computers than I do.	1	3	22	17	11	1
Having a computer in class is an advantage to learning.	19	22	12	1	1	0
Students should each have a laptop to do school-work on.	35	10	7	2	1	0
I think my abilities with computers will affect the grades I get.	15	13	21	4	2	0

Table 6. Year Two

Category	Strongly Agree	Agree	Disagree	Strongly Disagree	Failed to Answer
Computers make school work easier to do.	33 / 37	22 / 18	3 / 1	0 / 2	0 / 0
I prefer to use computers to do school-work rather than paper and pencil.	37 / 36	15 / 13	3 / 6	2 / 3	1 / 0
Using computers to do school-work can also have disadvantages.	7 / 10	34 / 32	9 / 7	8 / 7	0 / 2
Most days, I look forward to attending school.	4 / 5	24 / 25	13 / 13	17 / 15	0 / 0
Computers make school-work more fun/ interesting.	34 / 39	18 / 11	5 / 3	0 / 4	1 / 1
Computers help me to improve the quality of my school-work.	19 / 24	27 / 24	12 / 9	0 / 1	0 / 0
Computers help me understand my classes better.	19 / 27	19 / 16	16 / 13	4 / 2	0 / 0

continued on following page

Table 6. continued

I need to learn many new skills to use computers for my school-work.	**6 / 6**	**19 / 16**	**21 / 27**	**11 / 8**	**1 / 1**
I generally enjoy school-work.	**2 / 1**	**18 / 14**	**16 / 17**	**22 / 26**	**0 / 0**
Compared to other students, I really enjoy being in class.	**2 / 4**	**21 / 22**	**23 / 18**	**12 / 13**	**0 / 1**
Many of my classmates know more about computers than I do.	**2 / 3**	**15 / 15**	**28 / 29**	**12 / 11**	**1 / 0**
Having a computer in class is an advantage to learning.	**25 / 33**	**29 / 20**	**3 / 3**	**0 / 2**	**1 / 0**
Students should each have a laptop to do school-work on.	**43 / 43**	**12 / 12**	**0 / 1**	**2 / 1**	**1/ 1**
I think my abilities with computers will affect the grades I get.	**21 / 20**	**18 / 22**	**11 / 13**	**8 / 3**	**0 / 0**

**Pre-survey results / Post-survey results*

Table 7. Year One

Category	Strongly Agree	Agree	Same	Disagree	Strongly Disagree	Failed to answer
The use of the projector helps me to learn better than handouts or writing on the board.	**10**	**21**	**19**	**5**	**0**	**0**
PowerPoint presentations help me learn better than worksheets or writing on the board.	**14**	**19**	**21**	**1**	**0**	**0**
PowerPoint presentations on the projector are more motivating than handouts.	**18**	**24**	**12**	**1**	**0**	**0**
I would rather see information on the projector than on the overhead projector.	**28**	**18**	**6**	**3**	**0**	**0**
Using the projector as a teaching tool for grammar would increase my test scores.	**6**	**23**	**24**	**3**	**0**	**0**
More teachers should use projectors or other forms of technology for teaching.	**13**	**31**	**11**	**1**	**0**	**0**
Many teachers use technology effectively in the classroom.	**8**	**24**	**17**	**4**	**2**	**0**
More assignments should allow students to use the computer to complete the assignment (PowerPoint, Word, Excel, etc.)	**19**	**19**	**13**	**3**	**1**	**0**
Teachers use technology too much in classroom instruction.	**1**	**2**	**13**	**21**	**18**	**0**
The use of technology in Language Arts class as a teaching tool is very effective.	**16**	**24**	**12**	**2**	**0**	**1**
Technology use in Language Arts should be increased.	**14**	**19**	**15**	**6**	**0**	**0**

Table 7. Year Two

Category	Strongly Agree	Agree	Disagree	Strongly Disagree	Failed to answer
The use of the projector helps me to learn better than handouts or writing on the board.	21 / 29	20 / 18	10 / 9	7 / 2	0 / 0
PowerPoint presentations help me learn better than worksheets or writing on the board.	18 / 31	25 / 20	12 / 4	3 / 2	0 / 1
PowerPoint presentations on the projector are more motivating than handouts.	28 / 35	22 / 18	4 / 2	3 / 2	1 / 1
I would rather see information on the projector than on the overhead projector.	34 / 29	16 / 19	5 / 9	3 / 1	0 / 0
Using the projector as a teaching tool for grammar would increase my test scores.	14 / 20	16 / 21	26/ 14	2 / 2	0 / 1
More teachers should use projectors or other forms of technology for teaching.	32 / 31	17 / 21	5 / 4	2 / 1	2 / 1
Many teachers use technology effectively in the classroom.	11 / 11	32 / 31	10 / 9	3 / 5	2 / 2
More assignments should allow students to use the computer to complete the assignment (PowerPoint, Word, Excel, etc.)	37 / 34	16 / 17	5 / 4	0 / 3	0 / 0
Teachers use technology too much in classroom instruction.	2 / 4	6 / 6	20 / 24	30 / 23	0 / 1
The use of technology in Language Arts class as a teaching tool is very effective.	21 / 25	29 / 27	6 / 3	2 / 2	0 / 1
Technology use in Language Arts should be increased.	23 / 27	20 / 21	11 / 5	4 / 4	0 / 1

**Pre-survey results / Post-survey results*

on a writing piece (ex. narrative, persuasive, etc.)? Why?

Many of the students favored the use of technology in the classroom. They cited specifically that the PowerPoints were more motivating and made learning easier. Also, they liked being able to connect the examples on the PowerPoint with the examples in the homework. Furthermore, they found that students paid closer attention to the PowerPoints and enjoyed working through the examples together as a class. The students had this to say:

"I think that they (PowerPoint presentations) helped most of us, including me, to learn. It's a more fun and interesting way to learn."

"I think that it catches your eye. It also gets you interested. It was easier to understand, too."

"I think it is a good way because worksheets you do by yourself and projectors you have practice with the whole class."

"I though they really helped me and they were a lot more fun. They actually made me want to learn."

Other students also mentioned that PowerPoint presentations alone may not increase test grades. Many students wrote that students are individuals and everyone learns differently. Also, just because the information is presented differently doesn't mean that students will necessarily pay attention or study more. Students had the following comments:

(Q: Do you think your grades would be higher simply because of the use of the projector? Explain.)

"No, it doesn't matter how the information is presented to me just so it's told to me."

"It depends on what kind of information you put on the projector."

"If the teachers don't explain everything that well, then we wouldn't know what to study."

PLAN OF ACTION

Now that both the hard data of test scores and students' individual responses about technology have been considered, it is clear that its use is worthy. PowerPoint presentations will continue to be used to teach grammar units. It is probably the one part of language arts that is the most confusing and least motivating. "Motivation means having the desire and willingness to do something. Teachers who want to motivate students to stay on task, increase their knowledge and skills and improve their ability to process information, must guide the initiation, direction, intensity and persistence of learning behavior." (Business Wire, 2007). But how should educators do that? Although nothing will replace a highly qualified teacher, technology can help to motivate students in their learning. It will be nice to create these PowerPoints because it will be easy to make changes as necessary and store them in an easily-accessible fashion. Also, there will be an attempt to use technology in other areas of language arts. Currently, the Reading Counts program is used in the school. It would be interesting to explore its effect on student achievement and motivation. All in all, these data have provided a stepping stone to increasing the use of technology as a teaching tool and as a means of assessment.

CONCLUSION

The data suggests that students are highly motivated by the use of technology in the classroom. Students feel confident in their abilities with technology and seek to use it whenever possible. Furthermore, scores indicate that technology can have an impact on student achievement, specifically for those students with learning difficulties. However, the data also shows that the use of PowerPoint as an instructional method does not alone increase test scores. Despite its motivating factor for students, the students themselves are still responsible for completing homework and preparing for tests. Therefore, the use of technological aids in the classroom will continue to increase student attitudes towards learning. Yet, the effectiveness of technology in increasing achievement is largely debatable. Schools and teachers must make a conscious effort to balance technology and proven methods of instruction. Technology is a powerful tool if used correctly. Resources must be used in a way that prepares teachers to use technology effectively and engage students in a learning environment.

The questions on how technology impacts achievement are continuing to grow. Obviously, students are chomping at the bit to use technology in the classroom. Multiple programs and software provide all sorts of avenues for students to be creative and demonstrate active learning. Still, a conclusion about using technology as a teaching tool to increase test scores is still inconclusive. Allowing students to take tests that are more similar in nature did not confirm results. Perhaps using tests on different material miscued the results. Students may have found one test to be easier than the other. Also, a number of social and personal factors could contribute to dramatic changes in scores. For example, students may be having problems at home, have other activities that affect their study time, or may have been absent for portions of the instruction. In addition, revising the survey to obtain more specific answers from the students was helpful. By eliminating the category marked as "same", students were more decisive in their answers.

Overall, there are many connections with the review of the literature. Research exists that both supports and denies the foundation of technology in achievement. This experimental data suggests similar responses depending on the interpretation of the data. However, it is apparent that technology can have a profound positive effect on some students and greatly influence many. If the use of technology is helping to motivate students and helping at least one student achieve higher mastery of skills, its use is well worth the effort.

REFERENCES

Alfaro, R. (1999). The Technology-Reading Connection. *Educational Leadership, March.* 48-50.

Andrews, G. (2006). Laptops + Challenging Curriculum = Student Success. *Techniques, May.* 42-44.

Borja, R. (2006). Researchers Weigh Benefits of One Computer Per Lap. *Education Week, 25(36).* 10.

Bryant, S. M. and Hunton, J. E. (2000). The Use of Technology in the Delivery of Instruction: Implications for Accounting Educators and Education Researchers. *Issues in Accounting Education, 15*(1), 129-163.

Chenoweth, K. (2001). Keeping Score. *School Library Journal, 47(9).* 48-51.

Chou, S.W. and Liu C.H. (2005). Learning Effectiveness in a Web-based Virtual Learning Environment: A Learner Control Perspective. *Journal of Computer Assisted Learning, 21.* 65-76.

Cradler, J., McNabb, M., Freeman, M., & Burchett, R. (2002). How does technology influence student learning? *Learning and Leading with Technology, 29, 46-52.*

Cramer, S. and Smith, A. (2002). Technology's Impact on Student Writing at the Middle School Level. *Journal of Instructional Psychology, 29 (1).* 3-14.

Friedman, T. (2005). *The World is Flat.* New York.

Kleiner, A. & Farris, E. (2002). *Internet access in U.S. public schools and classrooms: 1994–2001* (NCES 2002-018). Washington, DC: U.S. Department of Education, National Center for Education Statistics.

Kozlowski, S. (2000). Better Learning with Technology? *The School Administrator, April.*

Kramarski, B. and Feldman, Y. (2000). Internet in the Classroom: Effects on Reading Comprehension, Motivation and Metacognitive Awareness. *Educational Media International, 37(3).* 149-55.

Levin, D. & Arafeh, S. (2002). *The digital disconnect: The widening gap between Internet-savvy students and their schools.* Washington DC: Pew Internet & American Life Project.

National Education Teacher Standards (NETS, 2000). International Society for Technology in Education.

Sandholtz, J., Ringstaff, C., & Dwyer, D. (1994). Student engagement revisited: Views from Technology Rich Classrooms. Retrieved October 4, 2007, from http://www.apple.com/education/k12/leadership/acot/pdf/rpt21.pdf.

Snyder, L., Caccamise, D. and Wise, B. (2005). The Assessment of *Reading* Comprehension: Considerations and Cautions. *Topics in Language Disorders, 25*(1), 33.

Sullivan, M. (1993). *A meta-analysis of experimental research studies based on the Dunn & Dunn learning styles model and its relationship to academic achievement and performance.* Doctoral dissertation, St. John's University, Jamaica, New York.

"Technology and Motivation: Can Computers Motivate Students to Read?" Business Wire. April 13, 2007. FindArticles.com. 25 Nov. 2007. http://findarticles.com/p/articles/mi_m0EIN/is_2007_April_13/ai_n19001667

Tomlinson, C. & Allan, S. (2000). *Leadership for Differentiating Schools and Classrooms.* ASCD.

Chapter XIV
Teaching Java™: Managing Instructional Tactics to Optimize Student Learning

Henry H. Emurian
University of Maryland—Baltimore, USA

ABSTRACT

Information systems students in a graduate section and an undergraduate section of an introductory Java graphical user interface course completed the following initial assignments to learn a simple program: (1) automated programmed instruction tutoring, (2) hands-on learning with a lecture, and (3) collaborative peer tutoring. Tests of knowledge transfer and software self-efficacy were administered before students began the first assignment and following completion of each one. The results showed progressive improvement in rule test performance and software self-efficacy across the several instructional events. Taken together, the results of these classroom observations extend the generality of previous work to an updated set of instructional materials and assignments, and that outcome shows the reliability of the learning processes with new groups of students. Students who are new to Java had the privilege of exposure to an initial repertoire of teaching tactics that are synergistic and cumulative.

INTRODUCTION

The research reported here is part of an ongoing stream of formative evaluations of instructional tactics that are intended to help novice, college-level students acquire skill and confidence in computer programming by means of an integrative approach to curriculum development (Emurian, in press: a). Direct mastery of the core knowledge in a discipline is recognized as a fundamental requirement to apply and extend that knowledge to solve novel problems, and that implies consideration of an instructional design to overcome the empirically verified shortcomings of teaching tactics that provide minimal guidance during a student's learning experiences (Kirschner, Sweller, & Clark, 2006). The integrative tactics adopted in our classrooms are in furtherance of helping all of our students to succeed.

Our previous work consistently confirmed the value of programmed instruction in teaching introductory information systems students a simple Java applet as a first technical training exercise in preparation for advanced learning (Emurian, 2004, 2005, 2006a,b). A Web-based, programmed instruction tutoring system to accomplish that objective was presented in Emurian, Hu, Wang et al., (2000), and behavior principles supporting the design and implementation of the system were described by Emurian, Wang, and Durham (2003) and Emurian and Durham (2003). Similar value of programmed instruction is evident in its applications within other symbol intensive disciplines, such as chemistry (Kurbanoglu, Taskesenligil, & Sozbilir, 2006), and its training effectiveness in fostering parent-teacher communications has been demonstrated (Ingvarsson & Hanley, 2006). The objectives of our work are to apply programmed instruction and to assess its effectiveness as a tactic to promote a common level of mastery by all students for a designated learning objective in Java programming. An optimal outcome of such a direct mastery approach is taken to reflect a *true gain* in learning (Anderson, Corbett, Koedinger et al., 1995).

Among several recommendations for effective learning principles to foster retention and transfer of knowledge is repeated practice with different instructional modalities (Halpern & Hakel, 2003) and with socially supported interactions (Fox & Hackerman, 2003). The modalities that have been adopted in our most recent classroom applications include: (1) programmed instruction, (2) lectures with hands-on learning, and (3) collaborative peer tutoring (Emurian, 2006b; in press:b). These tactics are demonstrably effective in promoting programming skill, software self-efficacy, and generalizable knowledge, the latter reflecting *far transfer* of learning (Barnett & Ceci, 2002). The benefits on student learning of a somewhat different, "blended" instructional approach to teaching introductory Java have been reported by Boyle, Bradley, Chalk et al. (2003), where repetition of similar topics occurred throughout the course syllabus. Our assessments of student learning, however, sometimes showed room for improvement in the goal of achieving maximal performance by all students on a far transfer test that was administered immediately following collaborative peer tutoring (Emurian, 2006b; in press:b).

To potentiate the effectiveness of the collaborative peer tutoring, the present classroom studies undertook a modification to the instructions and materials that made available to students to prepare them for collaborative peer tutoring and to use during the collaboration session. The modified procedure allowed the collaborating students to view and discuss together the questions that constituted the test of far transfer. Collaborating students also had direct hypertext access to instructional frames that were otherwise presented sequentially and contingently within the Java programmed instruction tutoring system. Finally, the Java program to be learned by students, as the first technical exercise in a course, contained more items of code to be mastered in comparison to the previous work in this area of classroom applications and research.

METHOD

Subjects

Subjects were as follows: (1) 13 graduate students, four females and nine males, taking IS 613 (*GUI Systems Using Java*) during a four-week summer session (summer 2006), and (2) 14 upper-level undergraduate students, six females and eight males, taking the equivalent undergraduate course (IS 413) during a 14-week fall session (fall 2006). There were more students enrolled in each class than are represented in the data analysis, which was based only on data collected on all assessment occasions by the students. If a student missed any data collection class or assignment, that student was not included in the analysis. The summer 2006 class met three times each week, and each class lasted three hours. The fall 2006 class met once each week for 2.5 hours. The course was designed for information systems students, and the prerequisite was one prior programming course for both classes. The technical content was identical for both classes, but there were more presentation and writing assignments, based upon reviews of journal articles, for the graduate students in comparison to the undergraduate students.

Prior to using the tutor, demographic information was collected, including age, number of prior programming courses taken, rated Java experience, and rated programming experience. The rating scales were 10-point ordinal scales where ***1 = No experience. I am a novice*** to ***10 = Extensive experience. I am an expert.*** Appendix A presents the scales that were administered during the pre-tutor and post-tutor assessments.

For the summer 2006 class, the background characteristics of the students were as follows: age (median = 28 yrs, range = 23 to 33), number of prior programming courses taken (median = 3, range = 1 to 15), rated prior Java experience (median = 2, range = 1 to 5), and rated prior programming experience (median = 5, range = 2 to 8).

For the fall 2006 class, the background characteristics of the students were as follows: age (median = 22 yrs, range = 21 to 32), number of prior programming courses taken (median = 5.5, range = 3 to 8), rated prior Java experience (median = 2, range = 1 to 7), and rated prior programming experience (median = 5, range = 2 to 8). A Welch robust test (Maxwell & Delaney, 2004, p. 134) showed a significant difference only for the age variable (W = 11.231, p = .003).

The research protocol was exempt from informed consent by the IRB, and the course syllabus clearly indicated that questions both embedded in the Java tutor and administered during several assessment occasions in class were eligible to appear on a graded quiz. The course description and syllabus provided information about the Java tutor and the collaborative peer tutoring, and they presented the rationale for the repetition of initial learning using the several different instructional modalities under consideration.

Materials[2]

Java Program and Tutor

The instructional tactics in this study are based upon teaching students a JApplet program that would display a JLabel object within a browser window on the World Wide Web. The program was arbitrarily organized into 11 lines of code (*e.g.*, ***JLabel myLabel;***) and 37 separate items of code (*e.g.*, ***getContentPane()***). The 37 items (1 item per cell), and the 11 lines of code are presented in Table 1.

The rationale supporting the tutor's design is based upon the learn unit formulation of Greer and McDonough (1999). In the tutoring system, each successive component, or learn unit, within eight tutor stages, required accurate responding for the learner to transition from one component to the next. The occasion and events supporting such a transition constitute a ***natural fracture of instruction***, which is "a unit of a compound that separates naturally from other components as a result of lawful conditions" (Greer, 2002, p. 18). Each cell and each line in Table 1 constituted a learn unit, and there were other learn units in the tutor.

The Web-based Java tutor consists of the following eight stages: (1) introduction and example of the program running in a browser (learn units = 1), (2) learning to copy an item of code (learn units = 37), (3) learning to recognize an item of

Table 1. The Java program

import	javax.swing.JApplet	;			
import	javax.swing.JLabel	;			
import	java.awt.Color	;			
public	class	MyProgram	extends	JApplet	{
JLabel	myLabel	;			
public	void	init()	{		
myLabel	=	new	JLabel ("Java")	;	
getContentPane()	.	setBackground (Color.yellow)	;		
getContentPane()	.	add(myLabel)	;		
}					
}					

code in a list (learn units = 37), (4) learning the semantics of an item of code (learn units = 37) and learning the syntax by typing the item by recall (learn units = 37), (5) learning to type a line of code (learn units = 11), (6) learning to recognize a line of code in a list (learn units = 11), (7) learning the semantics of a line of code (learn units = 11) and learning the syntax by typing the line by recall (learn units = 11), and (8) writing the entire program by recall (learn units = 1). Thus, the minimum number of learn units to complete the tutor was 194. If a learner answered incorrectly at any point, the components of the learn unit were repeated iteratively until the correct answer was produced. Some learn units, such as Stage 1, only required a button click to initiate a transition. Those learn units did not iterate because the correct response was simply to follow the instruction to click the button.

Multiple-choice tests for items and lines of code were embedded in the tutor, and each question had five answer choices. For an incorrect items answer, there was a 5-sec delay or "time-out" in the tutor's interaction with the learner. For a correct items answer, a confirmation window appeared stating a general rule associated with the correct answer or an elaboration of the explanation of the meaning of the item. The lines Stage 7 had no delay interval or confirmation window. Experience suggested that most students in our courses could complete the tutor within two to three hours. The tutor transitioned automatically between stages, and students were able to take breaks between and within stages. The instructions, however, encouraged students to complete each stage before taking a break.

Questionnaires

Java software self-efficacy was assessed by requesting a rating of confidence, for each of the 23 unique items of code (*e.g.*, ***import***) in the program, in being able to use the Java item to write a program that displays a text string, as a JLabel object, in a browser window. The scale anchors were ***1 = No confidence*** to ***10 = Total confidence.*** Twelve multiple-choice questions were also administered that required applying a general concept (*i.e.*, rule) of Java object-oriented programming to solve. Appendix B presents the

12 rule questions. These 12 rule-based questions did not appear within the Java tutor, and they intended to assess far transfer or meaningful learning (Mayer, 2002). Each question had five choices, and for each question, a rating of confidence was made that the selected choice was the correct choice. The scale anchors were ***1 = Not at all confident*** to ***10 = Totally confident.*** Ratings of classification and functionality learning for eight pairs of Java symbols were obtained, as given in the online material, but they are beyond the scope of this paper. The questionnaire version that was first presented (pre-tutor questionnaire) also solicited demographic information.

The post-tutor questionnaire omitted the demographic information, and it additionally assessed evaluations of the tutor for: (1) overall effectiveness, (2) effectiveness in learning Java, and (3) usability. The anchors were ***1 =Totally negative*** to ***10 = Totally positive.*** Questionnaires presented after the lecture and after the interteaching omitted evaluations of the tutor.

Procedure

Java Tutor

At the first class meeting, students completed the pre-tutor questionnaire. Students next completed the Web-based Java tutor. The tutor taught a JApplet program that displays a text string, as a JLabel object, in a browser window on the Web. The Java code and a brief description of the eight stages of the tutor are presented as part of the open source material. When a student finished the tutor, he or she completed a post-tutor questionnaire, which duplicated the software self-efficacy ratings and multiple-choice rule questions and confidence ratings. The student next accessed a set of questions and guidelines (Appendix C), posted on Blackboard, that were to be used to structure the collaborative peer tutoring session during a subsequent class. This material also presented a link to access the textual explanations of the items and lines of code presented in the Java tutor. The instructions with this material indicated that the questions presented were eligible to appear on a quiz.

Lecture

At the second class meeting, the instructor (HHE) gave a lecture on the program taught in the Java tutor. The students wrote the code in a Unix™ text editor during the lecture, which repeated the information presented in the tutor. The students were also taught the HTML file, used to access the Java bytecode file, as a URL on the Web. Support was provided so all students successfully ran the JApplet program at the conclusion of this lecture-based exercise.

This lecture required approximately one hour to complete, and the remaining class time was spent on the next unit of material, which related to the life cycle of an Applet. Students were encouraged to help each other during the subsequent classes in the semester, which combined lectures and hands-on demonstrations, with the understanding that files were not to be copied without prior permission of the instructor.

Interteaching

At the third class meeting, a collaborative peer tutoring session occurred based upon the dyadic "interteaching" model (Boyce & Hineline, 2002). Students formed dyads on their own for the session, which lasted one hour. If there were an odd number of students, one three-person group was formed. The assignment was for the students to discuss the set of questions and guidelines made available at the conclusion of the Java tutor work undertaken at the first class meeting. Also presented was the questionnaire, to include the rule questions, and students were encouraged to discuss the questions together prior to answering individually. The interteaching questionnaire instructions stated that the 12 rule questions were eligible to appear on a quiz, but the remaining items were there only to assess instructional effectiveness of the interteaching session. The interteaching questionnaire also requested ratings of the effectiveness of the session for: (1) learning the material and (2) readiness to be tested on the material, where ***1* = *Not effective*** to ***10* = *Totally effective***.

During the interteaching session, students also had access to a hypertext version of the Java program that returned the textual frames of information that were embedded within the tutor[3]. These, then, were the major innovations in the current study: (1) providing the opportunity for students to discuss the rule questions together, (2) and providing direct access to information embedded within the Java tutor. During the interteaching session, students posted questions on a Blackboard discussion forum, and the instructor provided feedback.

Later on that day as the interteaching session, the instructor posted an announcement on Blackboard giving the single question that was answered incorrectly by two of the students in the summer 2006 class. The announcement was as follows: "Some students answered 'c' below for this question (also presented in the announcement). The 'c' choice is not correct because JScrollPane is a class, not an object. An object name begins with a lowercase letter. If you have a question about this, please send me email." All student inquiries were answered privately in a way to promote understanding of the principle involved. The correct answer was not given.

For the fall 2006 class, nine of the 14 students made at least one incorrect choice on the rule questions, and 11 of the 12 questions were answered incorrectly across these students. Accordingly, later on the same day as the interteaching session for this class, these 11 questions were posted on Blackboard along with the correct answer. Students' inquiries about these questions and answers could be posted on an anonymous Discussion forum on Blackboard.

The two approaches to providing feedback were based upon our intention to facilitate optimal learning in relationship to the students' performances observed within and between classes. The tactic was adjusted in accordance with our perceived needs of the students as they pursued mastery of this challenging material. This tactic is consistent with design-based research (Wang & Hannafin, 2005) as a method to improve instructional effectiveness and student performance over successive offerings of a course. In all cases, the instructor bears responsibility for providing what are considered optimal tools of learning for the students.

Graded Quiz

At the fourth class meeting, a quiz was administered that included questions embedded within the Java tutor and the 12 rule questions as indicated above. The graded quiz did not include any rating assessments.

RESULTS

Figure 1 presents boxplots of correct answers on the rule test over the five assessment occasions for students in the summer 2006, and fall 2006 classes. For each of the 12 questions answered during the pre-tutor assessment, one student in the summer 2006 class did not select any answer, but instead indicated being unprepared to answer. The figure shows graphically that the median total correct answers increased over the first four occasions and reached the ceiling of 12 on the interteaching occasion for the summer 2006 students and on the quiz occasion for the fall 2006 students.

For the summer 2006 students, a Friedman test (Conover, 1971, p. 264) was significant (Chi-Square = 42.259, p = 0.000). The figure shows that the greatest change for these students occurred between the pre-tutor and post-tutor occasions,

Figure 1. Boxplots of total correct answers on the rule test for students in the summer, 2006, and fall, 2006 classes across the five assessment occasions. Circles are outliers and triangles are extreme values.

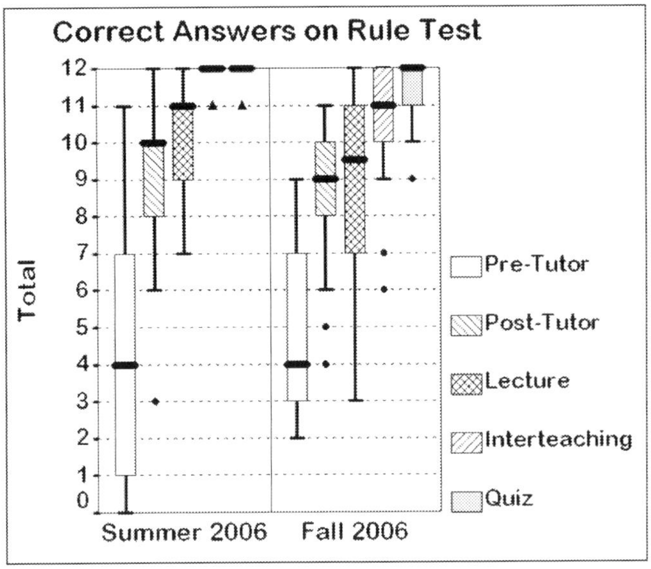

and both medians were 12 for the interteaching and quiz occasions. A Welch test, based on the differences, Di, in correct answers between successive pairs[4] of occasions over the five occasions, was significant (W = 10.889, p = 0.000). Planned pairwise comparisons were significant[5] for D1 and D2 (W = 10.145, p = 0.005), not significant for D2 and D3 (W = 1.513, p = 0.231), and significant for D3 and D4 (W = 12.295, p = 0.003).

For the fall 2006 students, a Friedman test was significant (Chi-Square = 44.000, p = 0.000). A Welch test based on the differences, Di, in correct answers between successive pairs of occasions over the five occasions, was significant (W = 8.950, p = 0.000). Planned pairwise comparisons were significant for D1 and D2 (W = 24.870, p = 0.000), not significant for D2 and D3 (W = 1.125, p = 0.301), and not significant for D3 and D4 (W = 0.207, p = 0.654).

The improvement process was somewhat different between the two classes, but the outcome for both classes reached the intended ceiling for the quiz, at least with respect to the median. With respect to individual student performance on the quiz, in the summer 2006 class two students made one error on the rule test. In the fall 2006 class, two students made one error, one student made two errors, and one student made three errors.

Figure 2 presents boxplots, over four successive occasions, of the ratings made by the students regarding confidence that the selected answer on the rule test was correct for answers that were right (R) and for answers that were wrong (W). Ratings were not obtained during the graded quiz. The number below each boxplot reflects the number of students who answered right and/or wrong over the four assessment occasions, and that is the reason that the frequency for a boxplot is sometimes less than 13 or 14 (*e.g.*, number of students giving incorrect answers for the interteaching occasion). The Welch robust test was used for both classes because of unequal sample

Figure 2. Boxplots of confidence ratings in the correctness of the rule test answers for students in the summer, 2006, and fall, 2006 classes across the four assessment occasions: 1 = Pre-Tutor, 2 = Post-Tutor, 3 = Lecture, and 4 = Interteaching. The scale anchors were ***1 = No confidence*** *to* ***10 = Total confidence****. The figure shows ratings for answers that were right (R) and for answers that were wrong (W). The N reflects the total number of students who answered correctly and/or incorrectly across the assessment occasions. Circles are outliers and triangles are extreme values.*

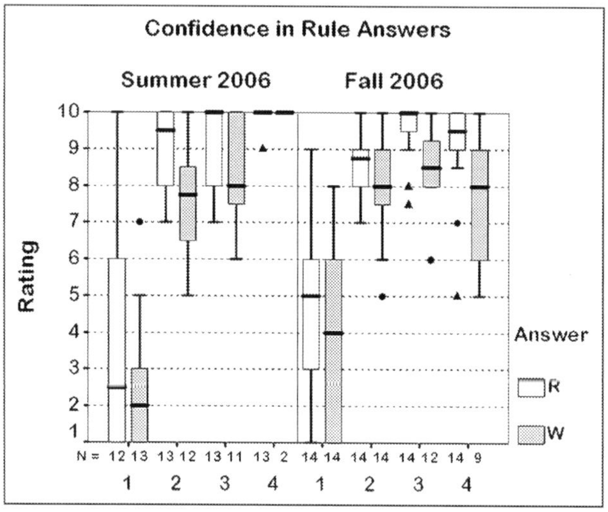

sizes, although the summer 2006 class did show all 14 students consistently making correct answers across the four assessment occasions.

For the summer 2006 students, the Welch test was significant for right answers (W = 16.632, p = 0.000) and for wrong answers (W = 40.864, p = 0.000). The latter test was based on the first three occasions because the variance for the interteaching occasion was zero. For right answers, planned pairwise comparisons were significant for pre-tutor and post-tutor (W = 27.398, p = 0.000), not significant for post-tutor and lecture (W = 0.108, p = 0.745), and not significant for lecture and interteaching (W = 4.959, p = 0.044) occasions. For wrong answers, planned pairwise comparisons were significant for pre-tutor and post-tutor (W = 55.646, p = 0.000) and not significant for post-tutor and lecture (W = 1.220, p = 0.282) occasions. An overall comparison of confidence ratings between right and wrong answers was significant (W = 9.481, p = 0.003).

For the fall 2006 students, the Welch test was significant for right answers (W = 16.231, p = 0.000) and for wrong answers (W = 13.477, p = 0.000). For right answers, planned pairwise, comparisons were significant for pre-tutor and post-tutor (W = 27.955, p = 0.000), significant for post-tutor and lecture (W = 9.512, p = 0.005), and not significant for lecture and interteaching (W = 1.265, p = 0.274) occasions. For wrong answers, planned pairwise, comparisons were significant for pre-tutor and post-tutor (W = 29.141, p = 0.000) not significant for post-tutor and lecture (W = 2.009, p = 0.169), and not significant for lecture and interteaching (W = 1.943, p = 0.190) occasions. An overall comparison of confidence ratings between right and wrong answers was significant (W = 4.690, p = 0.033).

Figure 3. Boxplots of ratings of the interteaching session for students in the summer, 2006, and fall, 2006 classes. Ratings were obtained for effectiveness of the session in understanding the material and for confidence in being tested on the material. The scale anchors were 1 = Lowest effectiveness or confidence to 10 = Highest effectiveness or confidence. The circle is an outlier and the triangle is an extreme value.

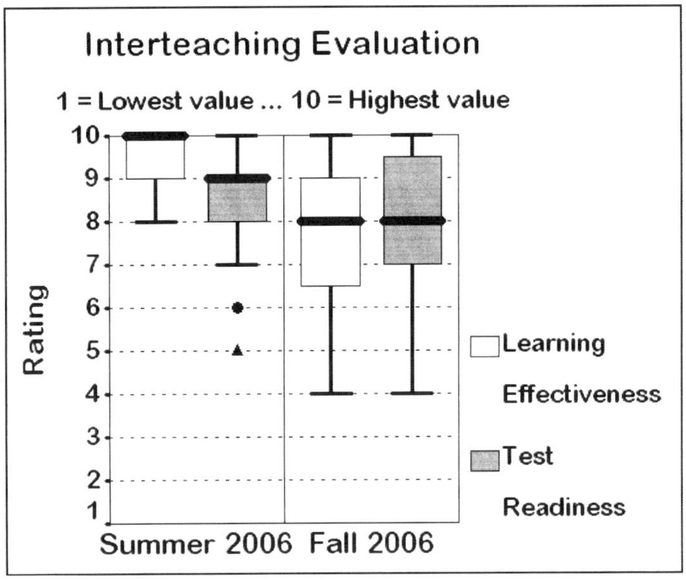

Figure 4. Boxplots of ratings of software self-efficacy for students in the Summer 2006 and Fall 2006 classes across the four assessment occasions. The ratings are based on the 23 unique items of code in the program. The scale anchors were 1 = No confidence to 10 = Total confidence. The triangles are extreme values.

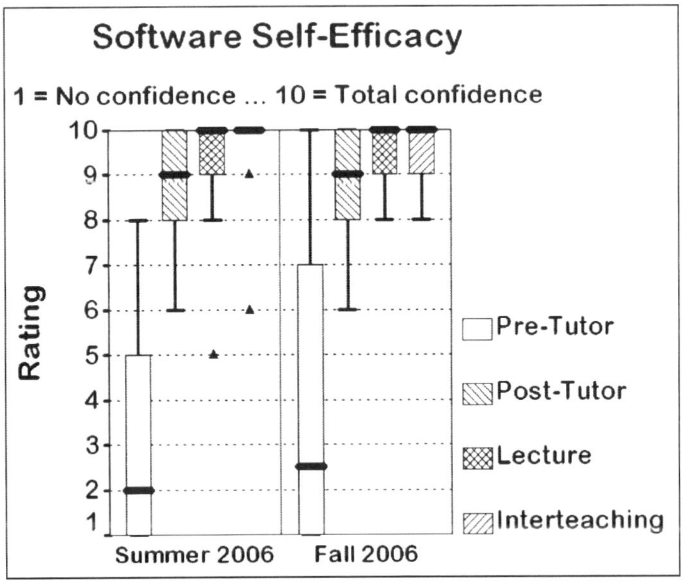

For both classes, confidence generally increased over the assessment occasions, reaching the ceiling for correct answers after the lecture. However, confidence increased for both correct and incorrect answers, although an overall comparison favored the correct answer choices.

Figure 3 presents boxplots of ratings on the interteaching evaluation, which was administered at the conclusion of the interteaching session, for students in both classes. Only 11 of the 14 students in the fall 2006 class provided an evaluation, although all 14 students participated in the interteaching session. The figure shows graphically the students' reported value in the interteaching session even when it occurred after using the Java tutor and after running the program on the Web. For the summer 2006 students, the median rating of learning impact reached the scale's ceiling of 10, with eight being the lowest rating observed. The rating of test readiness was only slightly less, with a median of nine. A Friedman's test was significant (Chi-Square = 5.444, p = 0.020). For the fall 2006 students, both scales showed median ratings of eight, and a Friedman's test was not significant for this class (Chi-Square = 0.667, p = .414). Although the median ratings for the fall 2006 students were comparatively lower than the summer 2006 students, taken together, these data show that almost all students reported value in the collaborative peer tutoring even when the session followed several other instructional experiences. No rating below a value of four was observed by any student.

Figure 4 presents boxplots of software self-efficacy ratings across the first four assessment occasions for students in the summer 2006 and fall 2006 classes. These ratings were not obtained during the graded quiz. Each boxplot is based upon the median rating over the 23 unique items of code in the program for the 13 students in the

*Figure 5. Boxplots of ratings of the tutor for students in the summer, 2006, and fall, 2006 classes for three scales. The scale anchors were **1 = Totally negative** to **10 = Totally positive**. The circle is an outlier, and the triangle is an extreme value.*

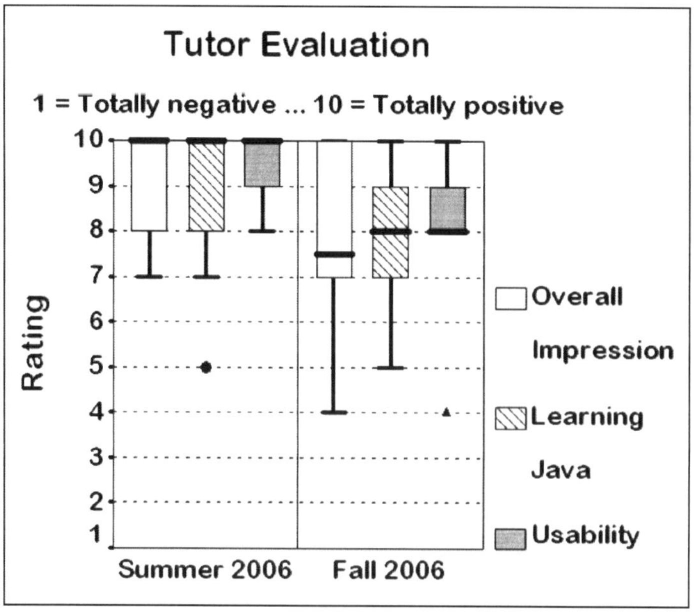

summer 2006 class and for the 14 students in the fall 2006 class. For ratings across all occasions for both classes, Cronbach's alpha reliability of the ratings within each assessment exceeded 0.90, and all values were significant ($p < .05$).

For the summer 2006 class, a Friedman test was significant (Chi-Square = 32.614, p = 000). A Welch test, based on the differences in ratings between successive pairs of occasions, was significant (W = 30.222, p = 0.000). Planned pairwise, comparisons of the differences, Di, were significant for D1 and D2 (W = 60.215, p = 0.000) and not significant for D2 and D3 (W = 1.330, p = 0.260). Software self-efficacy increased over the assessment occasions, and it reached the ceiling following the lecture.

For the fall 2006 class, a Friedman test was significant (Chi-Square = 32.741, p = 000). A Welch test, based on the differences in ratings between successive pairs of occasions, was significant (W = 18.450, p = 0.000). Planned pairwise comparisons of the differences, Di, were significant for D1 and D2 (W = 29.911, p = 0.000) and not significant for D2 and D3 (W = 3.452, p = 0.075). Similar to the summer 2006 students, software self-efficacy increased over the assessment occasions, and it reached the ceiling following the Lecture.

Figure 5 presents boxplots of ratings of evaluation of the tutor taken during the post-tutor assessment for students in both classes. Ratings on the following three scales were requested: (1) overall impression of the tutor, (2) effectiveness of the tutor in learning Java, and (3) usability of the tutor interfaces. The scale anchors on each 10-point scale were ***1* = *Totally negative*** to ***10* = *Totally positive***. For students in the summer 2006 class, median ratings for all three scales reached the scale ceiling of ten, with only a single outlier observed for Java Learning. For students in the fall 2006 class, the medians were comparatively lower, but all medians were higher than seven. Since ordinal data are problematic for between-group comparisons, these differences will not be interpreted statistically. However, the evaluation ratings from both classes together suggest that students reported value in their use of the tutor, despite an occasional extreme value toward the lower end of a scale.

DISCUSSION

The results of this study show the value of applying several different instructional modalities in furtherance of having information systems students achieve skill and understanding with respect to a simple Java applet, presented as a first technical exercise in a semester-long course. The data support the utility of this approach as reflected in students' rule test performance and software self-efficacy, which progressively improved over the successive assessment occasions. Rehearsal is an intuitively obvious and well-researched factor in knowledge and skill acquisition (*e.g.*, Salas & Cannon-Bowers, 2001), and the present study shows how structured rehearsal may be managed using the several modalities under consideration. Principles underlying such managed skill acquisition with different instructional modalities are presented elsewhere (Fox & Hackerman, 2003; Halpern & Hakel, 2003). Finally, although the predictive influence of self-efficacy on future performance has been questioned (Heggestad & Kanfer, 2005), self-efficacy assessments continue to be viewed as an important indicator of the effectiveness of training programs that are intended to produce both skill and motivation to learn (*e.g.*, Johnson, 2005).

Despite the apparent benefits of applying different instructional modalities to support student learning, however, the research base in instructional design typically compares one modality or instructional method with others with respect to student performance assessed at only one point in time. Even the U.S. Department of Education's What Works Clearinghouse[6] favors such an approach. Related to the present study, for example, Harrington (1999) reported that graduate social-work students with relatively high grade-point averages did not differ in final grades when a statistics course was taught either by a traditional lecture format or by "programmed instruction" in a distance learning setting. Saville, Zinn, Neef et al. (2006) reported that quiz scores for graduate and undergraduate students were higher after an interteaching session in comparison to scores observed after a lecture.

With respect to teaching computer programming to college-level students, Williams, Wiebe, Yang et al. (2002) reported that the percent of undergraduate students passing an introductory Java course was higher for a pair-programming laboratory section in comparison to students whose laboratory section involved solo programming. The benefits of collaborative learning, in comparison to solitary learning, when applied to computer programming were also shown in college students' program generation abilities using LISP-LOGO (Jehng, 1997). Although experimental designs that compare average student performances between and among conditions may have value in identifying an optimal technique to use when there is only a single and time-limited occasion to teach or to learn, such studies have little to offer in the engineering of instructional tactics when the objective is to have each individual student reach a criterion of mastery (cf Perone, 1999). Meyer (2004) questioned the value of old-fashioned experimental "horse-race" designs in another context, but the argument seems relevant within the current context as well.

This study constitutes a systematic replication (Sidman, 1960). A set of teaching tactics was revised with the expectation that student learning would be improved. The methodology reflects design-based research, which is a type of formative evaluation (Collins, Joseph, & Bielaczyc, 2004) that is emerging as an alternative methodology in support of developing and assessing improvements in instructional design within the context of the classroom (Bell, Hoadley, & Linn, 2004; Design-Based Research Collective, 2003). In that regard, the order of presenting the several instructional tactics was determined by anecdotal observations of student performance over the several classroom evaluations that were previously undertaken in this stream of work. It was decided that a hands-on lecture would benefit from students' prior rehearsal with the Java code and that collaborative peer tutoring would benefit from the cumulative learning obtained from the programmed instruction and the lecture.

Since the components in the current ordering are well received by students and since a desired learning outcome was achieved, we have the view that it is worthwhile now to direct our attention to developing advanced instructional material, rather than to "prove" the optimal ordering under conditions of a traditional "effect-size" experiment. Support for that view is implicit within designed-based research and has been discussed by educational scholars and training designers (*e.g.*, Mayer, 2004; Sackett & Mullen, 1993). Importantly, students reported value in the Java tutor and in the collaborative peer tutoring, and taken together with the lecture, these approaches to managing rehearsal in the classroom environment converge on what are increasingly recognized as vital ingredients to facilitate science education, in general (DeHaan, 2005).

The content and functionality of the Web-based programmed instruction tutoring system have been upgraded and continuously revised since the initial report (Emurian et al., 2000), and the system has been demonstrably effective and well received by our students. However, it is to be understood that other approaches to automated tutoring systems offer advantages in meeting the needs of the individual learner. For example, Butz, Hua, and Mcguire (2006) reported the application of Baysian networks to determine instructional events at the level of the individual learner in a Web-based intelligent tutoring system for computer programming. That and similar artificial intelligence (*e.g.*, Zhang, 2004) and multi-media applications (*e.g.*, Zhang, Zhou, Briggs et al., 2006) have obvious promise in improving the capabilities of the current programmed instruction orientation to automated instructional design.

Having students discuss rule questions together may have enhanced understanding and retention in the present context as indicated in subsequent rule test performance. However, an obvious challenge for collaborative peer tutoring, in general, and for interteaching, in particular, is to insure that participating students are, indeed, teaching one another and to make certain that they are sufficiently informed to know when their solutions to questions are correct. Boyce and Hineline (2002) and Saville et al. (2006) suggest several approaches to oversee and to evaluate interteaching to assure a beneficial session, such as the awarding of "quality points" by a monitor of the session. Similar to our previous observations, however, students showed "overconfidence" in incorrect rule answers, and that issue requires exploration in the design of future work. Tactics to be explored to improve the effectiveness of interteaching include the adoption of vignettes and rubrics to facilitate higher-order thinking and academic achievement (Kish, 2006).

The list of approaches to teaching and learning computer programming continues to grow. In this article, reported techniques include (1) a "blended" instructional approach (Boyle et al., 2003); (2) an emphasis on mathematics and algorithms (Hu, 2006); (3) supportive programming environments such as BlueJ (Kolling, Quig, & Rosenberg, 2003), DrJava (Hsia, Simpson, Smith et al., 2005), and PigWorld (Lister, 2004); (4) Problem-Based Learning (Tsang & Chan, 2004); (5) the Environment for Learning to Program (Truong, Bancroft, & Roe, 2005); (6) collaborative peer tutoring (Williams et al., 2002) and collaborative learning (Jehng, 1997); (7) a Traffic Light System Simulator (Yuen, 2006); a Computer Clubhouse learning environment (McDougall & Boyle, 2004), and (9) a Web-based personalized system of instruction (Koen, 2005). With the possible exception of Boyle et al. (2003), research studies in this domain typically emphasize a student's singular exposure to a task within the context of a single instructional modality.

As an alternative to the aforementioned approaches, the instructional tactics adopted in the classroom at the start of a semester's work are based initially upon *programmed instruction*, which is a form of structured and optionally automated instruction, as discussed by Emurian and Durham (2003) and Emurian et al., (2003) with respect to teaching computer programming. They also include a *lecture* with hands-on learning. They also include *interteaching*, which is a form of collaborative peer tutoring (Boyce & Hineline, 2002). As implemented in the present context to foster repeated practice with different instructional modalities and with socially supported interactions, these tactics originated from behavior analysis. The Cambridge Center for Behavioral Studies[7] provides fundamental definitions and a wealth of information regarding the philosophical underpinnings and applications of this approach to science, in general, and to education, in particular. Finally, these tactics are to be understood as providing only an initial series of learning experiences to students in preparation for subsequent learning with other instructional and program development tools and techniques, to include the use of an integrated development environment (IDE) such as Eclipse.

Although educators might have the success of their students as a primary goal of teaching, it is less certain that what happens in the classroom is based on empirical evidence of effectiveness: a rational pedagogy (Emurian, 2001). In addition, it is sometimes the case that expecting students prematurely to solve general computer programming problems and to understand complex control structures and algorithms neglects the skills that students must possess to undertake such higher-order learning. Too often, perhaps, educators may view an introductory course in science, engineering, and mathematics (STEM) as an occasion to eliminate marginally prepared students rather than as an opportunity to teach them the skills necessary to succeed. Although we also have the goal of helping students to learn the syntax and semantics of advanced programming such as recursion, we argue that our approach is deliberately and constructively designed to meet the needs of novice students, those *ineffective novices* who lack experience and self-efficacy in this domain (Robins, Rountree, & Rountree, 2003).

In furtherance of providing those skills to our students, techniques derived from behavior analysis have been demonstrably effective in promoting skill, confidence, and meaningful learning by novitiate students regarding an object-oriented programming language. Behavior analysis is one promising approach in identifying the ontogenetic instructional learn units (Greer & McDonough, 1999) whose mastery provides the textual tools essential for advanced understanding, thinking, and problem solving in the domain of computer programming. Teachers facing the difficult challenge of providing effective instruction to the diversity of students who enroll in introductory computer programming courses need to be mindful of all approaches to helping their students succeed. The present study represents a reconfirmation of one set of instructional tactics that are effective for information systems students and well received by them. All students deserve to have access to such evidenced-based tactics.

REFERENCES

Anderson, J.R., Corbett, A.T., Koedinger, K.R., & Pelletier, R. (1995). Cognitive tutors: Lessons learned. *Journal of the Learning Sciences, 4*(2), 167-207.

Barnett, S.M., & Ceci, S.J. (2002). When and where do we apply what we learn? A taxonomy for far transfer. *Psychological Bulletin, 128*, 612-637.

Bell, P., Hoadley, C.M., & Linn, M.C. (2004). Design-based research in education. In M.C. Linn, E.A. Davis, & P. Bell (Eds.). *Internet environments for science education* (pp. 73-88). Laurence Erlbaum Associates.

Boyce, T.E., & Hineline, P.N. (2002). Interteaching: A strategy for enhancing the user-friendliness of behavioral arrangements in the college classroom. *The Behavior Analyst, 25*, 215-226.

Boyle, T., Bradley, C., Chalk, P., Jones, R., & Pickard, P. (2003). Using blended learning to improve student success rates in learning to program. *Journal of Educational Media, 28*(2-3), 165-178.

Butz, C.J., Hua, S., & Mcguire, R.B. (2006). A web-based bayesian intelligent tutoring system for computer programming. *Web Intelligence & Agent Systems, 4*(1), 61-67.

Collins, A., Joseph, D., & Bielaczyc, K. (2004). Design research: Theoretical and methodological issues. *Journal of the Learning Sciences, 13*(1), 15-42.

Conover, W.J. (1971). *Practical nonparametric statistics.* New York, NY: John Wiley & Sons, Inc.

DeHaan, R.L. (2005). The impending revolution in undergraduate science education. *Journal of Science Education and Technology, 14*(2), 253-269.

Design-Based Research Collective (2003). *Educational Researcher, 32*(1), 5-8.

Emurian, H.H. (2001). The consequences of e-learning (Editorial), *Information Resources Management Journal*, April-June, 3-5.

Emurian, H.H. (2004). A programmed instruction tutoring system for Java: Consideration of learning performance and software self-efficacy. *Computers in Human Behavior, 20*(3), 423-459.

Emurian, H.H. (2005). Web-based programmed instruction: Evidence of rule-governed learning. *Computers in Human Behavior, 21*(6), 893-915.

Emurian, H.H. (2006a). A Web-based tutor for Java™: Evidence of meaningful learning. *Journal of Distance Education Technologies, 4*(2), 10-30.

Emurian, H.H. (2006b). Assessing the effectiveness of programmed instruction and collaborative peer tutoring in teaching Java™. *International Journal of Information and Communication Technology Education, 2*(2), 1-16.

Emurian, H.H. (in press:a). Applications of behavior analysis to ICT education: Teaching java with programmed instruction and interteaching. *Encyclopedia of Information Technology Curriculum Integration*. Hershey, PA: IRM Press.

Emurian, H.H. (in press:b). Managing programmed instruction and collaborative peer tutoring in the classroom: Applications in teaching Java™. *Computers in Human Behavior.*

Emurian, H.H., & Durham, A.G. (2003). Computer-based tutoring systems: A behavioral approach. In J.A. Jacko & A. Sears (Eds.), *Handbook of human-computer interaction* (pp. 677-697). Mahwah, NJ: Lawrence Erlbaum & Associates.

Emurian, H.H., Wang, J., & Durham, A.G. (2003). Analysis of learner performance on a tutoring system for Java. In T. McGill (Ed.), *Current Issues in IT Education* (pp. 46-76). Hershey, PA: IRM Press.

Emurian, H.H., Hu, X., Wang, J., & Durham, A.G. (2000). Learning Java: A programmed instruction approach using applets. *Computers in Human Behavior, 16*, 395-422.

Fox, M.A., & Hackerman, N. (2003). *Evaluating and improving undergraduate teaching in science, technology, engineering, and mathematics.* Washington, DC: The National Academies of Science Press.

Greer, R.D. (2002). *Designing teaching strategies: An applied behavior analysis systems approach.* New York: Academic Press.

Greer, R.D., & McDonough, S.H. (1999). Is the learn unit a fundamental measure of pedagogy? *The Behavior Analyst, 22*, 5-16.

Halpern, D.F., & Hakel, M.F. (2003). Applying the science of learning to the university and beyond: Teaching for long-term retention and transfer. *Change, 35*(4), 37-41.

Harrington, D. (1999). Teaching statistics: A comparison of traditional classroom and programmed instruction/distance learning approaches. *Journal of Social Work Education, 35*(3), 343-352.

Heggestad, E.D., & Kanfer, R. (2005). The predictive validity of self-efficacy in training performance: Little more than past performance. *Journal of Experimental Psychology: Applied, 11*(2), 84-97.

Hsia, J.I., Simpson, E., Smith, D., & Cartwright, R. (2005, February 23-27). Taming Java for the classroom. *SIGCSE'05,* St. Louis, MI, 327-331.

Hu, C. (2006). It's mathematical after all: The nature of learning computer programming. *Education and Information Technologies, 11*(1), 83-92.

Ingvarsson, E.T., & Hanley, G.P. (2006). An evaluation of computer-based programmed instruction for promoting teachers' greeting of parents by name. *Journal of Applied Behavior Analysis, 39*(2), 203-214.

Jehng, J-C. J. (1997). The psycho-social processes and cognitive effects of peer-based collaborative interactions with computers. *Journal of Educational Computing Research, 17*(1), 19-46.

Johnson, R.D. (2005). An empirical investigation of sources of application-specific computer-self-efficacy and mediators of the efficacy-performance relationship. *International Journal of Human-Computer Studies, 62*(6), 737-758.

Kirschner, P.A., Sweller, J., & Clark, R.E. (2006). Why minimal guidance during instruction does not work: An analysis of the failure of constructivist, discovery, problem-based, experiential, and inquiry-based teaching. *Educational Psychologist, 41*(2), 75-86.

Kish, M.H.Z. (2006). Overview of using vignettes to develop higher order thinking and academic achievement in adult learners in an online learning environment. *International Journal of Information and Communication Technology Education, 2*(3), 60-74.

Koen, B.V. (2005). Creating a sense of "presence" in a web-based PSI course: The search for Mark Hopkins' log in a digital log. *IEEE Transactions on Education, 48*(4), 599-604.

Kolling, M., Quig, B., Patterson, A., & Rosenberg, J. (2003). The BlueJ system and its pedagogy. *Journal of Computer Science Education, 14*(Dec), 1-12.

Kurbanoglu, N.I., Taskesenligil, Y., & Sozbilir, M. (2006). Programmed instruction revisited: A study on teaching stereochemistry. *Chemistry Education Research and Practice*, 7(1), 13-21.

Lister, R. (2004). Teaching java first: Experiments with a pigs-early pedagogy. *Proceedings of the 6th Conference on Australian Computing Education* Vol. 30 (pp. 177-183), Dunedin: Australian Computer Society, Inc.

Mayer, R.E. (2002). *The promise of educational psychology. Volume II. Teaching for meaningful learning.* Upper Saddle River, NJ: Pearson Education, Inc.

Mayer, R.E. (2004). Should there be a three-strikes rule against pure discovery learning? *American Psychologist, 59*(1), 14-19.

Maxwell, S.E., & Delaney, H.D. (2004). *Designing experiments and analyzing data: A model comparison perspective (2nd Ed).* Mahwah, NJ: Lawrence Erlbaum Associates.

McDougall, A., & Boyle, M. (2004). Student strategies for learning computer programming: Implications for pedagogy in informatics. *Education and Information Technologies, 9*(2), 109-116.

Perone, M. (1999). Statistical inference in behavior analysis: Experimental control is better. *The Behavior Analyst, 22*(2), 109-116.

Robins, A., Rountree, J., Rountree, N. (2003). Learning and teaching programming: A review and discussion. *Computer Science Education, 13*(2), 137-172.

Sackett, R.R., & Mullen, E.J. (1993). Beyond formal experimental design: Towards an expanded view of the training evaluation process. *Personnel Psychology, 46*, 613-627.

Salas, E., & Cannon-Bowers, J.A. (2001). The science of training: A decade of progress. *Annual Review of Psychology, 52*, 471-499.

Saville, B.K., Zinn, T.E., Neef, N.A., Norman, R.V., & Ferreri, S.J. (2006). A comparison of interteaching and lecture in the college classroom. *Journal of Applied Behavior Analysis*, 39, 49-61.

Sidman, M. (1960). *Tactics of scientific research.* New York: Basic Books.

Truong, N., Bancroft, P., & Roe, P. (2005, June 27–29). Learning to program through the web. *Proceedings of the 10th annual SIGSCE conference on innovation and technology in computer science education*, ITiCSE'05. Monte de Caparica, Portugal: ACM Press.

Tsang, A.C.W., & Chan, N. (2004). An online problem-based model for the learning of java. *Journal of Electronic Commerce in Organizations, 2*(2), 55-64.

Wang, F., & Hannafin, M.J. (2005). Design-based research and technology-enhanced learning environments. *Educational Technology Research and Development, 55*(4), 5-24.

Williams, L.A., Wiebe, E., Yang, K., Ferzli, M., & Miller, C. (2002). In support of pair programming in the introductory computer science course. *Computer Science Education*, *12*(3), 197-212.

Yuen, A.H.K. (2006). Learning to program through interactive simulation. *Educational Media International*, *43*(3), 251-268.

Zhang, D. (2004). Virtual mentor and the LBA system—towards building an interactive, personalized, and intelligent e-learning environment. *Journal of Computer Information Systems, XLIV*, (3), 35-43.

Zhang, D, Zhou, L., Briggs, B., & Nunamaker, J. F. (2006). Instructional video in e-learning: Assessing the impact of interactive video on learning effectiveness. *Information & Management*. *43*(1), pp. 15-27.

ENDNOTES

1 A portion of the summer, 2006, data was accepted for presentation at the 2007 convention of the Information Resources Management Association.

2 All materials used in this study are freely available. They include the online Java tutor, the open source code for the tutor, the course materials, and all assessment instruments: http://nasa1.ifsm.umbc.edu/learnJava/tutorLinks/TutorLinks.html

3 http://userpages.umbc.edu/~emurian/learnJava/swing/tutor/v2/explanations/Explanations.html

4 In the present study, the difference values for respective assessment variables were computed as follows: D1 = (Post-Tutor – Pre-Tutor); D2 = (Lecture – Post-Tutor); D3 = (Interteaching – Lecture); and D4 = (Quiz – Interteaching). The Welch test applied to these differences is similar to the multivariate approach for within-subjects designs recommended by Maxwell and Delaney (2004, p. 624). Planned pairwise comparisons were to detect possible differences in effect magnitude over the successive conditions.

5 To control for the experimentwise error rate, the significant *p* value for each planned comparison must be less than 0.05/number-of-planned-comparisons.

6 http://www.whatworks.ed.gov/

7 http://www.behavior.org/index.cfm

This work was previously published in Managing Worldwide Operations and Communications with Information Technology, edited by M. Khosrow-Pour, pp. 9-12, copyright 2007 by IGI Publishing (an imprint of IGI Global).

Section III
Collaborative Tools

Section III.a
Asynchronous Tools

Chapter XVI
Toward an Increase in Student Web Portfolios in New York Colleges and Universities

John DiMarco
St. John's University, USA

ABSTRACT

This research project investigated the existence of web portfolios on academic websites in New York State. It cites disappointing results when surveying the websites of New York State Colleges and University for web portfolios. Recognizing the problem of a lack of web portfolios, this chapter also provides a syllabus sample that can be used in technology based classroom environments across disciplines to integrate web portfolios into curriculums. The goal of this project was to promote web portfolios, provide interpretation of the current level of student web portfolio usage and activity within all New York colleges and universities, and suggest a sample syllabus to build web portfolios into curriculums. Major findings were that there is a low quantity of web portfolios in relationship to overall student enrollment, thus providing impetus to study a new phenomenon, lack of web portfolios. The study yielded data providing a breakdown of where and how many web portfolios were found. This study provides a basis for further research by scholars into web portfolios within academic settings.

INTRODUCTION

Understanding what a web portfolio is and is not is sometimes not easy. Is a web portfolio a course website? No it is not. Is a web portfolio a non-professional personal website used for posting personal data related to social outcomes? No it is not. A web portfolio is a personal website that provides evidence of your skills and expertise in the form of artifacts (photos, professional documents, artwork, and multimedia content including audio, video, and animation) from any discipline or field. The ideas behind the web portfolio as a tool for assessment, lifelong learning, and skill building have not yet been proven nor have they been embraced by society or academia at large. Those who exude confidence and passion for the notion of universal access to web portfolios and web portfolio skills cannot yet say that the web portfolio has become an accepted, professional cyber identity. This paper yields conclusions that uncover some intriguing dialog surrounding web portfolios. There are not very many student web portfolios found through college and university websites. The new phenomenon that has emerged from this research is lack of web portfolios.

Goldsby and Fazal note that student created portfolios are (2001, 607-608):

[Commonly] used in teacher preparation programs to demonstrate teaching skills and expertise. This practice was introduced as test scores alone lack the comprehensive scope needed for effective assessment and evaluation, portfolios can be implemented to interpret/make decisions regarding learning of teaching competencies.

The case for the student portfolio in any discipline can be made on the same basis; electronic portfolios provide a new level of assessment that cannot be measured by traditional methods such as standardized tests, applications, and resumes. Electronic portfolios and web portfolios provide assessment of competency within a discipline as well as a marketable tool for graduates. The web portfolio has promise as a tool, platform, and impetus for worldwide learning and growth in technological skills. The objective of this research project is to provide an accurate interpretation of the level of web portfolio usage within the colleges and universities of New York.

As we move towards more fluent, ubiquitous platforms for web media such as internet ready phones, web based television, and wireless personal digital devices, the web portfolio, and its place as an assessment tool, a learning tool, and a vehicle for lifelong learning has been recently scrutinized by scholars. Scholarly definitions of the electronic portfolio vary from discipline to discipline. To define the web portfolio, we must first define the e-portfolio, also known as the electronic portfolio. DiMarco put forth this definition (2005, 13):

The electronic portfolio is a collection of artifacts, project samples, cases, and focused content presenting the messages and professional and public appearance of an individual or a company through electronic media (web, DVD, CD-Rom). The e-portfolio provides evidence of skills, experience, and learning. I define the web portfolio as: an electronic portfolio that is an internet delivered, interactive, mass communication used to persuade users.

Greenberg (2004, 28-29) writes:

Ideally, all work in an electronic portfolio not only is digital but also is available on the Internet. Yet even though materials may be visible on the Web, the ePortfolio is not simply a personal home page with links to examples of work. In addition, unlike a typical application program, such as word processing, an ePortfolio is a network application that provides the author with administrative functions for managing and organizing work (files) created with different applications and for controlling who can see the work and who can discuss the work (access).

This definition presents several items for closer analysis. First, Greenberg makes a distinction that the electronic portfolio is not only digital but also available on the Internet. Development of the electronic portfolio and delivery are typically centered on using the Web. By using the Internet for delivery, electronic portfolios become less effective and more prone to failure. Also, an electronic portfolio is not just a home page. Any portfolio needs to be filled with work samples and evidence of growth and learning throughout a career, an amount of information that cannot be delivered effectively in only one page. An electronic portfolio must be a narrative that gives perspective to the viewer. The perspective of the viewer is shaped by the content and structure of the web portfolio. Greenberg also describes the e portfolio as having a network function. The web portfolio specifically is a content container that allows dynamic storage capabilities, as well as obvious delivery features. Content management is essential to bringing web portfolios into use across jobs and disciplines. The creator of a web portfolio will gain technical skills by acting as an administrator for his or her own web site, which will be an electronic portfolio. The electronic portfolio allows a student to manage his or her work throughout an academic and a professional career.

Either creation of an electronic portfolio is fostered within a learning environment, or, the skills are gained through self learning. The electronic portfolio provides opportunity for both. The electronic portfolio is a tool for lifelong learning and will be part of learning and growing throughout college and professional life. Gathering materials and creating web pages provides a learning experience that will carry over into a professional skill set. As network computers and the Internet become standards in every aspect of our lives, the skills and abilities needed to present creative and intellectual capital will become paramount to success in a technological marketplace.

Greenberg (2004) says there are three types of electronic portfolios. Each is defined by the author's assumed goals. This can be helpful in developing content management structure in a simplified manner. The structure of each type of e-portfolio differs based on the point of origin of the work. Greenberg believes this results in three types of e-Portfolios (Greenberg 2004, 29):

1. The *showcase ePortfolio:* organization occurs after the work has been created.
2. The *structured ePortfolio:* a predefined organization exists for work that is yet to be created.
3. The *learning ePortfolio:* organization of the work evolves as the work is created.

Greenberg's three portfolio types are effective for establishing types of portfolios based on content. A more appropriate structure might focus on the audience for the electronic portfolio, because it is important that the electronic portfolio is user and audience-centered. Greenberg's eportfolio types fit into an author-based definition of electronic portfolio types, which is a broad approach to classifying electronic portfolios. However, referring specifically to web-based electronic portfolios, three types can be defined:

1. The personal web portfolio for students or individuals
2. The teacher web portfolio
3. The business web portfolio

Scholars including Sanders (2000), Moonen and Tulner (2003), have praised the virtues of web portfolios. These scholars agree that there is a need to embrace the web portfolio as a tool, regardless of their discipline. These scholars also agree that the web portfolio is a tool and should be mastered by teachers and taught to students within the appropriate contexts of their disciplines. Specifically, one example would be that of an art portfolio. This type of portfolio has a structure and presentation style that will focus on the artwork and the skills of the artist. The same method can be adapted for a student in the discipline of English. In this case, the portfolio should focus on the writings and literature aptitude of the creator. In his personal case study on web-based portfolios for technology education, Professor Mark E. Sanders (2000, 11) states that:

"The information age is not just a cliché-we're living it! Global networked information systems such as the World Wide Web are changing nearly every aspect of our lives. These technologies should be prominent within our curriculum. Often, they are not. Web-based portfolios offer a meaningful way for technology students to gain a thorough understanding of these critical new technologies beyond mere web research. Web-based portfolios provide benefits that can never be realized with conventional portfolios."

The web portfolio is growing well beyond the boundaries of education and technology fields and is finding its way outside of educational institutions and into human resources and other corporate directions. This idea is supported by Moonen and Tulner (2004) who reported:

"But also beyond regular education, interest in electronic **portfolio** *is growing. EifEL [7] that commencing in May 2004, EIfEL (European Institute for E-Learning) is going to provide all of its Members with an electronic* **portfolio**, *the most innovative and fastest growing technology in the field of education, training and human resource development."*

The literature reveals variations of web portfolio usage. Several educational themes which are seen are building information literacy skills, branding teachers, applying PowerPoint in the web portfolio process, and performing assessments using web portfolios. Sharma (2007) found that web-based research portfolios effectively enable instructors of information literacy courses to assess objective based skills concretely. The portfolio assessment process is used to attempt to view and grade a collection of authentic evidence of student learning over time. Keller and Stuve (2006) explained that web portfolios are empowering teachers to have new strategies for demonstrating their effectiveness and instructional acumen and thus these methods resemble the concept of "brand" used in economic contexts. The thought of a brand for teachers can be a productive thought and is definitely a focus of the web portfolio. The pragmatic ability of the web portfolio is to enable a professional self and the use of web presence technologies ultimately influences how teacher quality is cultivated for mutual benefit to teachers, schools, and society (Keller and Stuve 2006).

D'Ambrosio (2003) demonstrated how classroom teachers and library media specialists can work together to develop educational Web sites and student Web portfolios using Microsoft PowerPoint as a tool for Web site construction, especially for those with little or no experience. This approach has possibilities as instructors across curriculums attempt to embrace web portfolio usage in their curriculums. Scholars, knowingly understand the lack of evidence behind the value of web portfolios, however, the surge toward web portfolio integration into curriculums is highly discussed.

A lack of systematic research evidence on the effect of performance assessment on teacher learning has not stopped proponents of electronic portfolios laud the benefits of using them as assessment tools for teacher candidates (preservice teachers) and as tools for student assessment (Stansberry, Kymes, 2007, Pecheone, Pigg, Chung, and Souviney, 2005). Callison (2007, 4) found that "Portfolio experiences for students and teachers are most effective when they not only document accomplishments, but also focus on a future course of achievement based on improved communication and information management skills".

Mainstream integration of the use of web portfolios in colleges and universities is a noble and promising endeavor for scholars, administrators, and students. The case for web portfolios is one that seems to be emerging slowly. With a growing body of evidence supporting web portfolios in existence, one might assume that searching for and finding web portfolios on college websites would be an easy task. However, finding a multitude of web portfolios and rhetoric about web portfolios on New York college and university websites was not a fruitful endeavor. Ultimately, this research found small evidentiary pockets of web portfolios on New York college and university websites. The data from this research helped identify a new phenomenon--lack of web portfolios.

PROJECT GOALS

The goal of this project was to promote web portfolios and the main objective was to provide interpretation of the current level of student web portfolio usage and activity within colleges and universities. For the present study I examined the number of web portfolios posted to the internet by students through a web portal or academic web site from each of the 294 private and public colleges and universities in New York State. These were located using the website http://www.nymentor.edu. This study is exploratory and will be used to enable subsequent action research into the development of a comprehensive state-wide web portfolio program in New York. This study also provides a basis for further research into web portfolios within academic settings.

To promote the use of web portfolios, I provide an syllabus model that may be used or modified. The course outline is interdisciplinary in nature and can be modified to fit any curriculum.

RESEARCH QUESTION

A personal web portfolio serves as a self-selected, self-developed multimedia presentation of work that offers multiple views of a person's learning and development (DiMarco, 2005). Driven by creative expression and college learning experiences, web portfolios provide tangible evidence of growth and accomplishment. Web portfolios also allow students to present research papers, essays, and academic projects that incorporate text, images, audio, and video. The disciplines of computer science, art, and education have embraced web portfolio development most frequently. However, students in all disciplines need a web portfolio when they leave college. Having tools to help illustrate one's skills and experience is critical to a college graduate. A resume is one such tool. However, web portfolios go beyond the traditional resume in scope of content and presentation by providing a broader platform for visual and interactive communication. After college, the web portfolio can become a personal hub for professional communication with potential employers and the public. It can serve as a platform for publishing career accomplishments and presenting skills and experiences through content.

There is evidence that the use of web portfolios in society is surging. This notion is present when reviewing the scholarly literature on academic web portfolios, as well as content analysis of Google based searches on web portfolio programs. For example, web portfolio programs are campus-wide at Penn State University (DiMarco, 2005), and even state-wide in Minnesota (http://www.efoliominnesota.com). Web portfolio programs are becoming an important component of academe and are helping build technical communication and presentation skills within today's information society.

The primary research questions guiding this study was:

First, what is the level of web portfolio usage (as measured by the presence of a web portfolio) and activity (as measured by content and artifact incorporation) by students within the 294 colleges and universities of New York State?

Second, what actions can be taken in academia to increase the population of web portfolios in New York State Colleges and universities?

METHODOLOGY

This study is a basic survey of the websites of the 294 New York Colleges and Universities, undertaken to determine the level of web portfolio existence or non-existence. Data gathering methods included recorded observations of college and university websites combined with content analysis and rigorous note-taking during website observations. This data gathering method pointed to specific usage or non-usage of web portfolios by the students of each institution.

Survey research methods provided this research project with nominal quantitative data from an unknown population of national and international web portfolios. The study used a sample frame of websites (294 out of 294) from the colleges and universities of New York.

Web portfolio analysis of 294 cases (college websites) began with a list that was obtained from nymentor.edu. This included the demographic and enrollment data on all known colleges and universities in New York State. The nymentor.edu website is managed by the New York State Department of Higher Education. The nymentor.edu site was used for data gathering because it provides topical, credible NYS education data. The site has an .edu domain extension which means that it is an educational domain and provides current data on the colleges and universities of every city and county in New York State. After studying initial cases of websites, the data revealed no web portfolios at college sites for nursing and medical degree programs. The process of data gathering consisted of the following steps:

- **Step 1:** Went to the school website based on the website address published on nymentor.edu.
- **Step 2:** Used the site search feature (if it was available) to perform limited keyword searches using the following six terms…
    - "web portfolios"
    - "digital portfolios"
    - "student portfolios"
    - "electronic portfolios"
    - "student websites"
    - "student web portfolios"
- **Step 3:** Performed searches looking for clues to web portfolios on the school site. After a minimum of 10 minutes of exhaustive searching, the next case was studied.
  Step 3: Observation Methods:
    - Specific website sections that were investigated:
    - Observed technology services menus
    - Observed home pages
    - Observed sitemaps
    - Observed search features within sites
    - Observed technology departments
    - Observed education departments
    - Observed graphic arts departments
    - Observed computer science depts.
    - Observed career services depts.
- **Step 4:**
    - Entered observation notes into Excel using the insert comment function.
    - Entered notes for websites that had web portfolios or evidence of articles, press releases or faculty papers that referred to the use of web portfolios or the planning of future use of web portfolios art the institution.
- **Step 5:** Counted total number of web portfolios for each college or university and entered the data into MS excel spreadsheet containing the listed NYS colleges and universities. Generated percentages of web portfolios against enrollments.
- **Step 6:** Read the field notes, reviewed sites again, and generated open-coding content analysis about the websites of New York State colleges and universities.

Table 1.

College and University Web Portfolio Totals

CASE #	School Name	Region	Setting	Total Enrollment	Web Portfolio link on homepage	Web space provided by School (~ accts)	Total # of Working Student Web sites/ Portfolios found	% of web portfolios to enrollment
10	Bard College	Mid-Hudson	Rural Setting	1,726	no	yes	15.00	0.86%
	Bard has a student server that allows students to voluntarily post personal web pages and web sites Found 19 web portfolio/personal websites listed -Only 15 active with content. Several blogs.							
88	Hobart and William Smith Colleges	Genessee Valley	Small Town	1,847	no	yes	10.00	0.53%
	10 web pages from a computer graphics class found. Did not fit typical web portfolio criteria-class projects only.							
115	LaGuardia Community College	New York City	Urban	12,875	no, but one link away!	yes	17.00	0.13%
	Notes: Found 17 web portfolios posted. Entire web portfolio program. Found some scattered samples of other student portfolios. Summary: The college with the most prominent, pedagogically integrated e-portfolio (web portfolio) program is LaGuardia Community College. The program is highly visible throughout the college website.							
122	Long Island University - C.W. Post	Long Island	Suburban	8,421	no	yes	4.00	0.04%
	4 Portfolios found.-Doctoral students in Information Studies. Personal URLs. Not colleges sponsored space.							
224	State University College at Buffalo	Western	Urban	11,072	no	yes	1.00	0.01%
	1 found. Art student.							
225	State University College at Cortland	Central	Small Town	7,331	no	yes	47.00	0.64%
	47 found. Educational technology students. Class project.							

continued on following page

Table 1. continued

252	SUNY College of Technology at Farmingdale	Long Island	Small Town	6,250	no	yes	7.00	0.11%
	7 found. Graphic design students							
263	SUNY Institute of Technology at Utica/Rome	Northeast	Suburban	2,432	no	yes	30.00	1.23%
	Notes: 30 found. Technical communication students. Well done work. Summary: The web portfolios that exhibited the highest quality work and design were located at SUNY Institute of Technology at Utica/Rome. The work was from bachelors and associates degree students in Technical Communication.							
267	Syracuse University	Central	Urban	16,753	no	yes	4439.00	26.49%
	Notes: 4439 found. Some were just used as space. Some had home pages. ~ accounts on SU servers. Unix editors or FTP. Summary: The college with the most web portfolios: Syracuse University with 4,439. This institution was by far and away the most populated web portfolio gallery. The space was an account off of the SU server. The web portfolios are kept active by the university for ten years after students graduate. Student web addresses have ~ in the URL.							
Totals				68,707 enrollment			4570.00 web portfolios	6.60% % of web portfolios against enrollment.

LIMITATIONS

As a researcher with a high level of experience in web portfolio research, I entered this study with a level of bias that initiated a loosely shaped hypothesis that very few web portfolios actually existed. However, during data gathering I remained objective and attempted to uncover web portfolios even when my intuition led me to think that there would be none at the particular case web site.

FINDINGS AND CONCLUSIONS

As this study progressed, it was evident that there are a small percentage of web portfolios in comparison to total enrollment. The cumulative enrollment total reported by www.nymentor.edu for New York State Colleges and Universities is 1155042 students. The total number of web portfolios found was 4,570, which represents .39% of the enrollment population. 9 of the 294 schools had active retrievable web portfolios. This points to a very small percentage of web portfolios against student enrollment within the colleges and universities of New York. The same level of involvement seems to continue throughout the colleges and universities where web portfolios were found. Among the nine schools where evidence of web portfolios was found, the highest percentage of web portfolios was found at Syracuse University where 4,439 alphabetically listed web portfolios existed among a student enrollment of 16,753. Syracuse University had the highest web portfolio usage found with 26.49% of the

Figure 1.

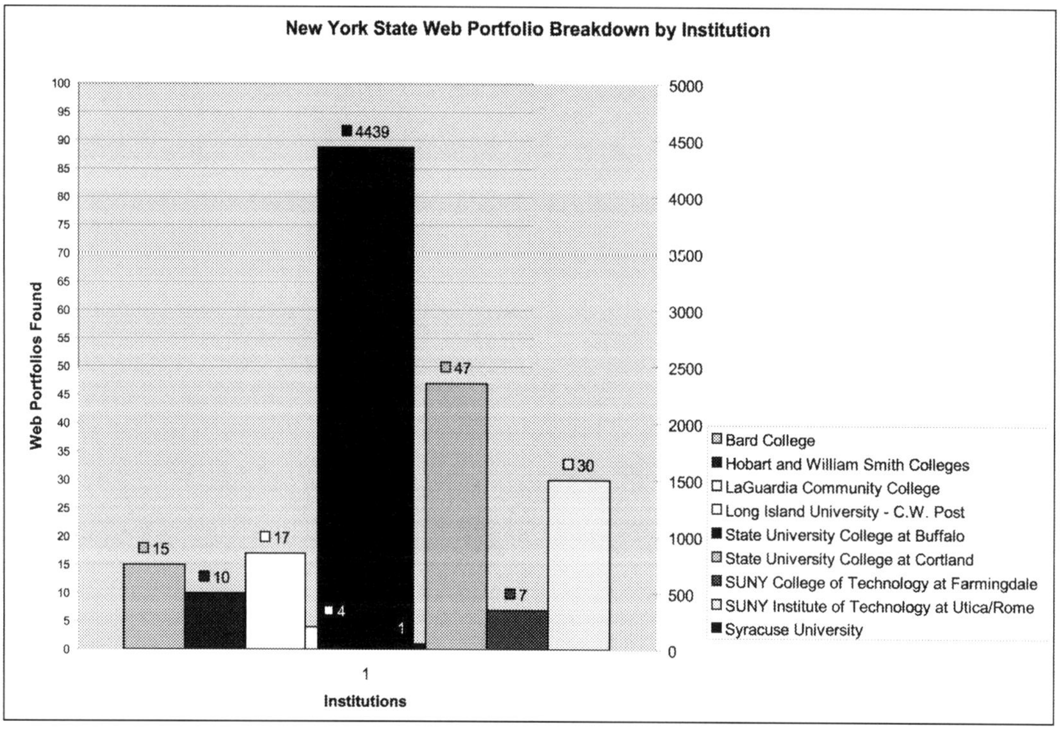

Table 2.

Web Portfolio Course Syllabus Sample
(See DiMarco, 2007)

Interdisciplinary Portfolio Design and Development
3 Credits
Graduate or Undergraduate Level
Submitted by:

COURSE DESCRIPTION
This course provides students across disciplines with a creative opportunity to develop a digitally designed print-based portfolio book and a multimedia-based web portfolio. Instruction includes conceptualization and categorization of assets and artifacts for portfolio development. Laboratory exercises will include technology lessons in the use of digital imaging, MS Office output to web pages, and tutorials in industry standard print and web development software applications. Portfolio-based interview techniques and professional presentation will also be covered. Emphasis and final grading is assessed on completion of an effective print portfolio book and a web based electronic portfolio.

RATIONALE
Creating a personal electronic/web portfolio helps initiate students into professionally focused lifelong learning. The constructivist approach to teaching portfolio development challenges students to perform self-assessment and teaches them how to develop works of guided self-expression for use throughout their careers. The skills learned while developing a portfolio include analysis, inquiry, and reflection, and design for new media. Students learn concepts and applied skills in creating a personal information system and utilizing interactive communications technology.

PREREQUISITES
• none

LEARNING OBJECTIVES:
-Understand why portfolios are important tools for lifelong learning and career communications.
-Conceptualize, plan, design, and output a print based career portfolio.
-Conceptualize, plan, design, and output a web based career portfolio.
-Evaluate and execute artifact content collection decisions and processes.
-Develop assets and thematic content.
Apply industry standard software packages (MS Office, PDF, Photoshop, Dreamweaver) for design, print layout, content development, web authoring, and multimedia.
-Critically review and evaluate created portfolios to insure they meet specific disciplinary criteria and career goals.
-Perform reflective writing during the portfolio development process.
-Perform an interview presentation using print and web portfolios.

MATERIALS
-256 or larger USB removable media, also known as a clip drive to save projects (or IPod, external HDD, or Flash memory Card with adapter)
-1 Ebony Sketching Pencil and a set of fine point and super fine point sharpies (black only)
-Strathmore sketch pad-9 x 12
-3 Blank CDROMS
-University web space account or commercial web space account (www. portfoliovillage.com)
-1 package of laser color copy paper/ cover stock sheets

GRADING
- Participation 10 pts
- Assignments 20 pts
- Midterm* 20 pts
- Final Portfolios and Presentation* 50 pts

COURSE OUTLINE

Table 3.

Weekly Learning Modules	Topic
One	*Introductions* *Review Syllabus* ---------------------------------- *Print Portfolio Definitions* *Electronic/Web Portfolio Definitions* *Defining the portfolio within your discipline and context.* **Activity one:** -Based on class discussion, in a one page paper, describe how the print and web portfolio fits into your specific academic discipline and career goals. -Find three URL's of web portfolios from your discipline -Assigned reading schedule -Start your journal
Two	*Conceptualize/Brainstorm the Portfolio products you will create.* *Defining the audience.* *Explain how the web portfolio will be used to persuade the audience.* **Activity two:** -Prepare a written concept statement that defines the concept, images, messages, and themes that may be part of the web portfolio. -Write a journal entry on your experiences
Three	*Portfolio Content* *Content Evaluation Methods* *Writing the Content List* *Writing project/work/artifact descriptions* **Activity three:** Research web portfolios within your discipline and others to determine possible categories of artifacts and visual themes. -Create your written content list -Create your content Outline -Write two project descriptions using the format presented in class
Four	Information Design Navigation issues Developing a Flowchart Page counts and scope Combining the scope documents (concept statement, content list, content outline, and flowchart) **Activity four:** Develop a flowchart of your web portfolio site Submit complete scope report -Write a journal entry on your experiences
Five	Visual Design Developing storyboards Content development and digital capture techniques Screen resolution and graphical sizing issues Web Resumes HTML and Graphical text issues **Activity five:** Develop your web portfolio site storyboards

continued on following page

Table 3. continued

Six	Graphic Design for Print Developing print and web graphics Developing print pages and web screens Developing chapters, headings, and navigation Digital Artifact Production (MS Office and Photoshop) Using Adobe Photoshop and Fireworks **Activity six:** Develop your book design using a sketch pad and then create the publication framework using Word, or Indesign -MIDTERM REVIEW
Seven	*MIDTERM EXAM Web Page Design Developing web graphics Developing web screens Developing navigation Digital Artifact Production (MS Office) Using Adobe Photoshop **Activity seven:** Develop your web portfolio site screens using a sketch pad and then using Adobe Photoshop or Fireworks. -Write a journal entry on your experiences
Eight	Web Authoring Using Macromedia Dreamweaver with Fireworks web page functionality issues web page development demonstrations and tutorials **Activity eight:** -Using sliced graphics from Photoshop or Fireworks -Add links, page properties, and other functionality in Dreamweaver
Nine	Web Authoring (continued) Using Dreamweaver to create web pages Creating rollovers Creating pop up windows **Activity nine:** -Add rollovers, pop up windows, and other web functionality -Create a web resume -Place artifacts and descriptions in pages and pop up windows
Ten	Work week/Catch up/One on One meetings **Activity ten:** Complete web authoring or print layout
Eleven	Multimedia Authoring Using Macromedia Flash Motion Graphics Animation Text effects **Activity eleven:** -Create a Flash based text animation using fades -Complete web content of HTML pages

continued on following page

Table 3. continued

Twelve	Publishing and Finishing the Portfolio Book Creating an interactive PDF of the book Finishing techniques **Activity twelve:** -Print and finish the portfolio book comp -Proof and submit final comp for review -Write a journal entry on your experiences on the print project GET FINAL APPROVAL -Print and assemble the final portfolio book(s)
Thirteen	Uploading the web portfolio using Macromedia Dreamweaver and FTP Testing the web portfolio Checking download time Checking links and popups Testing Usability Final Exam Review **Activity thirteen:** -Test your web portfolio for technical functionality -Test your web portfolio for usability using the provided evaluation instrument -Create a test report on usability and functionality from two classmates -Make edits and fixes -Write a journal entry on your experiences and post it to the portfolio
Fourteen	Presentation skills How to use the portfolio in an interview (job, client, grad school app) **Activity fourteen:** -Pseudo role plays
Fifteen	*Final Portfolio Presentations and Assessments

continued on following page

students having some form of web portfolio on the internet and accessible through the Syracuse University website.

As web portfolios were found, authorship, content, and context were recorded. The following is a breakdown of region, setting, total enrollment and percentages for web portfolios together with enrollment data for the nine schools where web portfolios were found.

The study focused on gaining insight into the existence or non existence of student web portfolios available via the websites of New York Colleges and Universities. As the web portfolio situation unfolded during this study, the data revealed some interesting facts that might yield additional parameters for investigating web portfolio activities within academia. Two facts seemed to evolve from this study. The first fact was few web portfolios are readily available for viewing at the websites of colleges and universities in New York State. Also, when web portfolios were found, there were few in relationship to the total enrollments of the institutions. The largest web portfolio stake is claimed by Syracuse University who present web portfolios for over 26% of the total enrolled student population.

The second fact found was that many academic websites posted documents regarding the virtues and involvement of web portfolios, yet these institution's websites showed no tangible implementation of web portfolios by students. To elaborate on this point, when searching colleges and university websites using the site search engines, much data was found on various case sites that discussed web portfolio programs and technology grants received by institutions to create comprehensive or pilot web portfolio programs. Although evidence of this literature was found in the form of faculty newsletters, academic memos, and research proposals, there were few or no web portfolios found at the institution's website.

This was observed in more than a dozen case sites examined. At Adelphi University, for example, a web article was found that was written by education faculty describing the use of an IT based system for web portfolios in use by the Special Education Department. Despite the presence of the article, there was no evidence of web portfolios. The point here is that there seems to be data pointing to web portfolios at many institutions, but little or no action research to implement a viable, usable, embraceable solution.

The most active web portfolio program appeared to be Syracuse University. This conclusion was based on the proportion of users to enrollment. In the case of Syracuse University, the institution provides web space to all students for the creation and development of a web portfolio. It also provides instructional technology resources such as a dedicated system for web space users, basic tutorials on uploading files, and frequently asked questions. The support and creative freedom provided by the institution might be part of the reason why so many students are taking advantage of the university web space.

The most pedagogically integrated web portfolio program existed at LaGuardia community college. Formal examples of web portfolios are showcased along with digital tools and processes. These items are outlined in a dedicated web portfolio section of the college website that resides only one click away from the college home page. Although organized and well funded, the LaGuardia e-portfolio project only showcases 17 web portfolios, a mere .13% of enrollment. This is substantial because the college has initiated a strict policy of web portfolio development within the curricula. However, few web portfolios actually exist for viewing at the college website.

Another area that further research might target is web portfolio usage within technology and non-technology related disciplines and web portfolio use by faculty members to see whether a relationship exists between the existence of teacher web portfolios, disciplines, and the development of student web portfolios. Are students more likely to create a web portfolio if they have a professor facilitating or mentoring the process? Several trends in this direction were observed when analyzing the data. Some sites showed student projects were a large part of the portfolio population within educational technology departments. SUNY Cortland had a web portfolio population that existed of students from several concurrent sessions of a course in the use of educational technology.

Literature review has shown that the promise of web portfolios is apparent and usage is growing among academic institutions throughout the world. Variations of web portfolio usage are seen in building information literacy skills, branding teachers, applying PowerPoint in the web portfolio process, and performing assessment. The value of these domains are evident. However, the present analysis of web portfolio activities at New York colleges and universities has demonstrated small numbers of web portfolios in relationship to overall enrollment. I set out to answer the question: What is the level of web portfolio usage (as measured by the presence of a web portfolio) and activity (as measured by content and artifact incorporation) by students within the 294 colleges and universities of New York State? The answer is the level of usage is extremely low in relationship to enrollments. This fact leads me to view the web portfolio phenomenon as a research topic that might be studied using exploratory methods. Future research might help indicate why there is such a lack of web portfolios available for viewing on the academic websites of colleges and universities in New York.

REFERENCES

Callison, D. (2007). Portfolio Revisted with Digital Considerations. *School Library media Activities Monthly. Vol.* 23 (6): 43-47.

D'Ambrosio, J. (2003) *E-Teaching: Creating Web Sites and Student Web Portfolios Using Microsoft PowerPoint.* Columbus: Linworth.

DiMarco, J. (2007). Web Portfolio Design for Teachers and Professors. *Proceedings of the 2007 Information resources management Association International Conference.* IGI Global: New York.

DiMarco, J. (2006). *Web Portfolio Design and Applications.* Hershey: Idea Group.

Goldsby, D., Fazal, M. (2001). Now that your students have created web-based digital portfolios, How do you evaluate them? *Journal of Technology and Teacher Education*, Vol. 9, (4): 607-616. http://proquest.umi.com

Greenberg, G. (2004). The digital convergence: extending the portfolio mode [electronic version]. *Educase 39*(4): 28 37.

Keller, B, Stuve, M (2006). Teacher as Brand: Pursuing Professional Identities in a Digital Domain. In: Sharon Y. Tettegah and Richard C. Hunter, Editor(s), Advances in Educational AdministrationJAI. *Technology and Education: Issues in Administration, Policy, and Applications in K12 Schools.* Vol.8: 57-70.

Listing of colleges and universities managed by NYS Dept. of Higher Education. http://www.nymentor.edu [retrieved college websites 2/1/2006 through 4/2/2006]

Moonen, J. Tulner, H. (2003). *E-learning and electronic portfolio: Some New Insights.* Universiteit van Twente, The Netherlands. Retrieved Sept 5, 2006, from http://www.connict.nl/pdf/moonen-tulner-portfolio.pdf.

Pecheone, R.L., Pigg, M.J., Chung, R.R., & Souviney, R.J. (2005). Performance assessment and electronic portfolios:Their effect on teacher learning and education. *The Clearing House, 78,* 164–176.

Sanders, M. (2000). Web-based portfolios for technology education: a personal case study. *The Journal of Technology Studies,* 11. Retrieved Sept 5, 2006, from http://scholar.lib.vt.edu/ejournals/JOTS/Winter-Spring-2000/pdf/sanders.pdf.

Sharma, S (2007). From Chaos to Clarity: Using the Research Portfolio to Teach and Assess Information Literacy Skills. *The Journal of Academic Librarianship.* 33 (1). Retrieved Dec 5, 2006, from http://www.sciencedirect.com/science/article/B6W50-4MK0HTH-4/2/de-225b2c2d9329a19623c1fe1365d0db.

Stansberry, S and Kymes, A. (2007). Transformative learning through "Teaching With Technology" electronic portfolios. *The Journal of Adolescent & Adult Literacy. Vol.* 50 (1):488-497.

State-wide portfolio project for the state of Minnesota. (http://www.efoliominnesota.com) [retrieved 3/10/2006]

Web Portfolio Commercial Site. http://www.portfoliovillage.com

Chapter XVI
Competent Web Dialogues: Text-Based Linking of Thoughts

Marianne Döös
Stockholm University, Sweden

Eva R Fåhræus
Stockholm University, Sweden

Karin Alvemark
Dalarna University, Sweden

Lena Wilhelmson
Stockholm University, Sweden

ABSTRACT

Conducting a dialogue on the Web is a matter of linking thoughts in digital conversations. Dialogue differs from discussion by not being aimed at beating or convincing other participants in the conversation. The present chapter highlights group dialogues as conversations in which people learn with and from each other. Learning dialogues have the potential of developing the learners' capacities for critical thinking and complex problem solving. The model of dialogue competence is suggested in order to improve the linking of thoughts in web dialogues. The chapter concludes with considerations when developing dialogue-based communication forms for learning purposes and contributes to teachers' demand for more support in pedagogic and educational issues.

INTRODUCTION

Digital technology has led us into new forms of conversation, and in some digital exchanges the similarity is so great that it has become natural to say that we are talking, even though we are actually writing. People have very different images of what communicating on the web and at a distance implies. There are great variations of experience and competence. For example, there are generation differences and different kinds of software as well as different degrees of digital skill. The lack of common images, i.e. similar understanding of what digital conversations can be, is an important basic precondition which will have to be taken into consideration for many more years to come.

Web-based learning is arranged in university educations world wide, in the US (Finkelstein, 2006), in Turkey (Yuzer, 2007), Sweden (Fåhræus, 2003), Italy (Francescato et al, 2007) and China (Ng, 2007), to mention a few. Many are also the contributions aiming at improving and making sense of these learning situations, and in that endeavour several aspects come to the fore. Some researchers investigate broad collaborative aspects (Francescato et al., 2007; Fåhræus, 2003) and student participation (Hrastinski, 2006a). Others problematise more specific issues like e-learners experiences of time and the connection between participants' time management and their use of certain metaphors (Allan, 2007), or eye-contact as an example of the important non-verbal communication (Finkelstein, 2007; Yuzer, 2007).

Conducting a dialogue on the web is a matter of linking thoughts in digital conversations (Fåhræus & Döös, 2007). Contacts and conversations via the computer are growing in volume. With technology development people are successively changing their understanding of what this kind of conversation is, and of its possibilities and difficulties. When conversing with others via the computer we are moving in the border zone between solitude and company, being simultaneously present and non-present. In order to achieve learning qualities in these digital conversations, the interlocutors need to wise up on the digital interchange of thoughts as a communication form.

The present chapter is intended to contribute in that direction, essentially by borrowing knowledge concerning group dialogues and dialogue competence in real face-to-face situations (f2f), i.e. where a group of people sit in the same room talking to one another. This knowledge is applied to written web-learning conversations, where mainly electronic conferencing systems (e.g. FirstClass) are used. Thus, the focus of attention is on the learning dialogue in connection with the text-based, asynchronous meeting, and in particular on digital group conversations in connection with courses, training programmes and university education. Often, though not always, these conversations are conducted with an element of examination or with a stipulation of presence and activity. Above all, though, they are meant to be good learning conversations, i.e. conversations in which people learn with and from each other. The chapter concludes with considerations when developing dialogue-based communication forms for learning purposes and thus contributes to teachers' demand for more support in pedagogic and educational issues, than in support of technical kind (Sällström, 2005). Briefly, the possibilities for synchronous digital dialogues brought about by technological development are also touched upon.

The term "dialogue" as used in this chapter refers to a special kind of learning conversation that will be presented below, one having both a structure and a learning purpose (cf. Wilhelmson, 2006). Such a dialogue does not come easily, even when people see and hear each other. A group dialogue requires practised skills, but its learning potential, i.e. the possibility it affords, in the company of others, of increasing one's understanding and pondering one's experience, makes it worth the trouble (Isaacs, 1999). When here applying knowledge from f2f dialogues this is done with the ambition to highlight also the potential for learning. As Tham and Werner (2005) conclude in their evaluating review of e-learning in higher education the "concern should not be just with whether online learning is conducted successfully using the available technology, but also whether the institutions did what they set out to, i.e., educating students" (p. 24). As shown by Wilhelmson (2002; 2006) learning dialogues do have the potential of developing the learners' capacities for critical thinking and complex problem solving.

Two Digital Conversations by Way of Illustration

The participants in a distance course are set the task of conversing in writing with each other, by computer, about gender aspects of IT use. They are each to have read an article or a book chapter on the subject and are now to give their views on it and converse with each other in groups of about ten people. This conversation is expected to continue for a fortnight in a conference system of the First Class type. In one group the conversation makes rather slow progress. When the participants have described the content of the article they have read, communication comes to a standstill. The course leader intervenes to ask whether anyone has had personal experience of gender discrimination in connection with IT use, whereupon a number of participants describe experience of the kind and others reply with affirmative comments on some of these contributions. Nothing much more happens after this.

In another group the communication gets off to a somewhat hesitant start but soon gets quite lively. In connection with describing the article, one participant relates personal experience of the subject. Another participant relates a similar experience while a third objects that the reason must have been something other than gender discrimination. A lively conversation ensues as to what the reasons might have been. There are still some people who haven't voiced an opinion, and so the person who was first to speak asks the "silent ones" to give their opinions. The conversation flows on and now gets quite heated. Someone writes that she feels insulted by the doubt cast on her story. The person who queried what she wrote apologises and clarifies his meaning. The misunderstanding is cleared up and the conversation continues on a frank and friendly basis.

What did the participants in the two groups learn from all this? Those in the first group presumably learned a certain amount from the articles they read and perhaps too from the summaries provided by other group members. No one, presumably, got much out of the web talk. At worst, the participants may have come to the conclusion that electronic conversations are worthless. The other group probably learned quite a lot, both facts about gender issues and various possible perspectives on IT and gender. Perhaps they even changed their opinions about certain forms of IT use and saw greater opportunities for their own part. We shall be returning to the examples later on in this chapter.

Factors Influencing Digital Group Conversations

There are a large number of factors influencing the way in which a group discussion on the web develops and how much the participants get out of it. The nature and wording of the task, the choice of subject, the size and composition of the group and the participants' private situation and attitude to the task make a difference to motivation and activity. This in turn makes a difference to what and how much people learn and how they allow themselves to be influenced. Another important factor is the way in which the dialogue is conducted and how conscious the participants are of the role they themselves are playing in it. The electronic dialogue in text, like the f2f variety, can be improved by the participants developing their dialogue competence (Wilhelmson & Döös, 2002). Ability to link thoughts together in digital conversations makes a substantial addition to learning, especially in the case of distance learning. As a participant in digital conversations one needs, then, to cultivate the ability to conduct learning conversations and also think about creating good preconditions both for oneself and for others.

The person planning and conducting a course has great possibilities of influencing how instructive the digital conversations become and the extent to which they attain qualities concerned with linking thoughts between and within the individuals taking part. This is primarily a question, not of intervening in the actual conversation but rather of course planning, and of course leaders and teachers understanding why, and by what means, a digital conversation proves successful. In this way they can contribute towards positive opportunities and conditions in the form of tasks, group composition and requirements.

When arranging dialogue group conversations f2f, one can speak of the importance of reserving "a bubble of time" affording scope and tranquillity for reflection (Wilhelmson & Döös, 2002). A counterpart in dialogue on the web could be concerned with each participant assuming personal responsibility for allotting time to be present in the task of writing and reading, i.e. the web conversation's two counterparts speaking and listening. This implies a degree of carefulness which differs from the impulses characterising a great deal of computer communication – availing oneself of the possibility inherent in choosing for oneself, not only the point in time but a period of time when web talk takes place with peace and quiet for study. Just as the members (Dixon, 1994) of an organisation (meaning, in everyday speech, associates/employees) have to shoulder their responsibility for the learning required for performance of the task in hand, so the members of the conversation have to assume personal responsibility for rigging up good preconditions for getting something out of the digital conversations which are a part of the course they are taking. This assumption of responsibility, however, is connected with the nature of the task which, as a participant, one sees ahead of one. Articulating this to oneself and to the others can help to prevent problems later on, e.g. regarding work inputs and digital attendance, problems rooted in participants' differing perceptions of the task. For example, is it something you undertake because you are bursting to learn, or are you attending the course for some other purpose and aiming for it to take up as little of your time as possible?

LEARNING PROCESSES

Dialogue for the purpose of learning is based on different people having different ways of looking at things because they have led different lives and acquired personal experience. In dialogue, the fact of us thinking and feeling more or less differently is both a precondition and an obstacle. We shall now turn to consider learning processes of relevance in the sense of helping us to understand how a competent web dialogue can be conducted.

Learning in Action

Experience-based learning takes place in concrete contexts, while the learning individual is occupied with quite different things from learning, namely the performance of tasks or duties, conversing, solving problems and understanding things (Döös, 2007; Kolb, 1984). Experiential learning, takes place within tasks; in other words, you busy yourself with something and you learn into the bargain. The individual person's learning proceeds by short steps, with the greater part of learning consisting, not of new things but of confirmations and more of the same. One experience is added to another and everyday knowledge is constructed and consolidated continuously. Against this background differences and deviations appear. Only when one knows what something is like normally, is it possible to be surprised by deviations (Döös, 1997). Learning through conversations with others can be described in similar terms (Bjerlöv, 1999).

When learning takes place via the web, Fåhræus (2003) has identified three parallel learning processes: learning to communicate and converse electronically, learning to learn together (the collaborative aspect) and learning the matter which the course is concerned with (the content). These three learning processes proceed in parallel and, at best, support one another. As teacher and course leader one should bear all three processes in mind and remember that they need time and attention in order to come about and acquire quality.

Collective Learning

Reflection is a vital ingredient of learning, with the result that dialogue, conversation, communication etc. has attracted a great deal of attention in recent decades, on research as well as practice. When people learn together, collaboratively (Fåhræus, 2003), in an interactive and communicative process, then in favourable circumstances we can also find collective learning taking place (Dixon, 1994; Ohlsson, Granberg, & Stedt, 2004). If so, we get synergistic effects, with 1+1 making more than two, i.e. the participants change their conceptions in a different way from what each of them could do on his or her own. Instead of just each individual learning by belonging to a group, learning results in the participants' conceptions during the conversation becoming collective, i.e. a great deal more similar, almost communal.

In Swedish work life education research, collective learning has been made theoretically comprehensible through studies of f2f meeting conversations of different kinds. It has been described as something that happens in teams or other clearly delimited, (most) often formalised and small groups in which it is established and known which people are included and what their functions are.

Collective learning, has partly different principles and sequences in different specific environments, in different contexts (Döös & Wilhelmson, 2005). The engendering process results in changed understanding and similar preparedness for action. In more distributed contexts in working life, it has been possible to identify the importance of a common arena for action (op. cit.) in that product development in the telecommunications and data communication industry has led to collective learning through a host of paired contacts by telephone, Internet, email, electronic subscription and meeting points. The common arena for action consisted of coordination in the task, i.e. joint development of the technology. Accordingly, the results of one's own actions and other people's were gathered together in technical artefacts.

WHAT IS DIALOGUE COMPETENCE?

Dialogue is something more and different than ordinary conversation. It differs from discussion by not being aimed at beating or convincing other participants in the conversation. Bohm (in Senge, 1990) describes discussion as a subject being tossed between the participants: the subject is analysed and arguments for and against presented. Each participant's purpose "is to win, i.e. to gain support from the others for his own opinion" (p. 221). The purpose of the dialogue, in contrast, is described as that of "extending the boundaries, reaching further than the individual person can do unaided". In a dialogue one is not out to prevail over others, "in a well-conducted dialogue everybody wins." (p. 221).

Dialogue conversation is difficult and demands a competence which many people need to practise for. It contains (see fig. 1) four different parts: speaking, listening, critical self-reflection and critically review other people's standpoints (Wilhelmson, 2002). In learning dialogues people can meet by making different perspectives visible, conversing on a subject and helping each other to relate it to a wider context, gaining an overview and seeing one's own part of the overall picture.

Group conversations can be powerful learning situations providing an opportunity for the communal creation of meaning. Arranging for preconditions entails the creation of conversation topics which make the conversation meaningful and necessary in relation to what is to be learned on this particular occasion. It is recommended that the topic of conversation be of such a kind that there are no right or wrong answers, and that if possible it should concern something of which the participants have personal experience or knowledge. Dialogical quality of conversation (Wilhelmson, 2006) helps to make conversations in themselves instructive in a qualitatively good sense.

The point of departure for learning dialogues is for each individual taking part in the conversation to have specific experiences of his or her own which are different from other people's. Wilhelmson (2006) likens the conversation topic to a statue, arguing that the statue is an imaginary picture of the conversation topic that the participants in the conversation build up together when, from their various perspectives (each individual person's conceptions/experiences); they make different aspects of one and the same subject to reality visible. Together the people conversing figuratively walk round the statue and in their conversation highlight as many aspects as possible of the subject which they are together endeavouring to understand.

Speaking and listening are the core of dialogue competence. This is more complicated than it sounds at first. In order to learn from and with each other in the course of conversation and perhaps build new shared knowledge, every participant must be capable of pleading for his own standpoint in the conversation but must also be open to other people's arguments. One has to be prepared to query one's own standpoints and to critically review other people's.

Figure 1. Ingredients of dialogue competence. Source: Wilhelmson, 2002.

	Self	Others
Closeness	Speak	Listen
Distance	Critical self-reflection	Critical review of other people's standpoints

A dialogical group conversation has both integrating and differentiating qualities (Wilhelmson, 2002). Contributing as a participant towards integration means building on what others say, gathering the threads of conversation into a fabric. Not instantly dismissing other people's differing views but instead trying to get inside their way of thinking. Contributing towards differentiation means problematising and questioning on the basis of one's own experience and knowledge, contributing one's own viewpoints and experiences, with personal integrity. The friction arising out of difference provides an opportunity of getting to the bottom of things and scrutinising one's own and other people's conceptions.

DIFFERENT FORMS OF INSTRUCTIVE E-MEETINGS

There are many different kinds of e-meetings. Here, as has already been made clear, we are mainly concerned with electronic meetings conducted within the framework of a course in order for the participants to learn something. Meetings communicated electronically are primarily concerned with text communication, possibly supplemented by image or video transmission in which the participants can see each other. An e-meeting can be conducted in chat form, with all participants active simultaneously, but often an asynchronous (non-simultaneous) meeting is opted for and allowed to last longer. Even the relatively synchronous chat, however, is more asynchronous than f2f conversations, in which the speaker can be heard in mid-sentence, can have eye-contact and read off the other participants' facial expressions and can make the rest of what he or she says take a different course from what it was about to do.

Degrees of Liberty and Structure

Just as a dialogical f2f group conversation can be facilitated by adhering to a certain structure, governing must be considered a natural part of the teaching context. Pedagogy is about processes of influence.

To arrange digital conversations for others' learning requires skilful awareness of the need for governing, i.e. awareness of the appropriate adjustment between freedom and structure.

Research has shown that computer-based training needs a certain amount of guidance in order to get a rewarding discussion started at a distance (Wännman, 2002). It is not the intention here to define a certain degree of suitable control but rather to consider how it can be done. Basically, governing guidance can be exercised in two ways (Döös & Ohlsson, 1999). Firstly by intervening, through the medium of conversation, in the individual person's world of thoughts and ideas, directly in people's thoughts and meaning contexts. Use is then made of the changed possibilities of the conversation itself, e.g. through a direct question from the course leader to the participants. The other path to guidance involves conditioning surroundings and outward arrangements in such a way as to change the preconditions, e.g. in this case through the choice of subject or procedure.

There are ways of imposing a structure on the conversation without necessarily controlling it too closely. One can present a vignette which serves as a starting point and stimulus for thoughts and ideas which the participants want to share with each other, the plan being for this to arouse curiosity and trigger conversational activity. The conversation which then ensues can take quite a different turn from what the leader had envisaged, but if it is to be a free conversation one should refrain from intervening and directing it. Another way of providing a common kick-off point is by asking the participants to see one and the same film or read an article, as in the introductory examples. They can also engage in brainstorming, role play or a debate. (Fåhræus, 2000).

The next step towards more structure may be to indicate a certain sequence. All participants, for example, are to begin by describing personal experience of the chosen topic before embarking on more two-way conversation. To further structure the procedure, they may be required to read and summarise an article and then ask each other questions which have to be answered within a certain time. This structure can be made highly detailed, and computer support can be used for shaping it up. There can be different electronic conferences or folders in which the participants have to post their article summaries, or question boxes can be used with linking answers.

An extra clear structure is obtained if the system is made to verify that the participants complete all their assignments according to the procedure and remind them of what they have to do next. There are systems which create groups by pairing participants off when they have reached a certain stage in their studies (see for example http://www.theducation.se/english/index.asp). This can provide teachers and course management with support for the administration and monitoring of large groups. Systems can also direct students' questions in digital conferences, in such a way that they are first processed and thus potentially answered by the student group. This can have the effect of increasing communicative activity as well as lightening the teacher's workload and saving him or her from becoming a bottleneck.

Figure 2. Degrees of governing guidance.

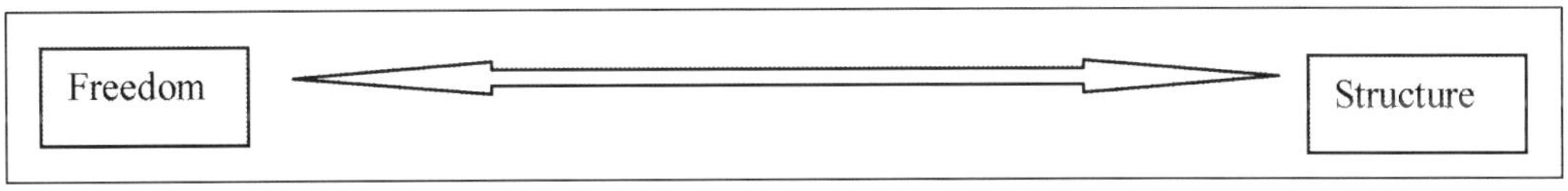

Synchronous or Asynchronous Text Meetings?

The meeting assumes a completely different character if conducted with all participants present and active at their computers simultaneously, compared with the situation of its lasting for a fortnight and the participants deciding for themselves when to access, read and post contributions. The synchronous text meeting can get quite hectic: you have to be pretty nimble-fingered to get a word in edgeways. Without a predefined procedure for the meeting, things often seem pretty chaotic.

In f2f conversations, some moderators use the expedient of a talking stick or ball which has to be replaced in the middle before being picked up by the next speaker. Part of the purpose of this arrangement is to slow down the tempo. The time lag in the asynchronous conversation rules out the need for any such dodges. Basically it gives us more time to work out what we want to say, reread what we or other people have written previously and amend what we write before sending it off. Each individual is enabled to contribute with his or her piece, without being interrupted by those who have "the gift of the gab". Instead, another sort of time shortage is liable to occur, owing to competition from everyday commitments, reserving a bubble of time isn't always easy.

An asynchronous meeting is sometimes found to be unenthusiastic and slow-moving. One variant which can pep things up and help to overcome the difficulty of reserving time is for the group to fix certain times when everyone will be on line at once, an on-line seminar, or at least a day or two when everyone participates more frequently in a dialogue. This is something midway between the synchronous and the asynchronous, an attempt to combine the advantages of both.

Choosing Different Forms

It is natural, in a learning process during a course, to vary and utilise different forms at different stages. Perhaps you begin by getting a group to formulate their own questions in a subject on a brainstorming basis. What are the problems? What do we need to know? Or do? For arriving at a decision, e.g. on a working plan, the synchronous meeting is the best alternative. Afterwards perhaps the participants will search individually for information on the web, updating each other on important discoveries as they go along. Then a communal asynchronous dialogue ensues on the most interesting or controversial points, ending perhaps with a joint verbal debriefing to a larger audience.

On a course it may be a good idea to ask the participants to reflect on the form of dialogue in itself. One initial task could be to link up thoughts on one's notions and experiences of conducting a learning conversation and of doing so in texting digitally. This is one way of getting the participants, right from the outset, to create their own understanding and insight concerning what is expected to take place. And what can take place, so long as they themselves make something of it. For example, the group could be issued with three pictures to begin with. For example: linking thoughts, web talk and alternation. Such digital starting conversations can be returned to, reread and talked about again later on during the course, in order to reflect on their application in practice, what has been learned and what experiences have been gained. And what can be done differently for the remainder of the course.

THE DIALOGUE POTENTIAL OF THE MEDIATED CONVERSATION

Conversation at a distance *versus* f2f presents both similarities and dissimilarities. Among other differences, whereas communication in a group conversation on the web is text-formulated, delayed and enduring, f2f communication is oral, immediate and evanescent. Dialogue, as already remarked, is a concept originally referring to a particular variant of oral f2f conversations.

Can one really speak, then, of conducting a dialogue when people cannot see each other, cannot read off the small but often distinct signs of the interlocutors being on the same wavelength or objecting to what one is saying? In the electronic meeting, a lot of the things people communicate to each other in the physical meeting through vision and hearing are lacking. To make up for this in the electronic meeting, extra care has to be taken not to offend each other and to verify that the person we are conversing with has understood us properly. Habitual distance speakers often supplement the written word with emoticons[1] such as ☺, "laugh" or <J> (for "joking"). Emoticons serve to clarify the participants' feelings and the meaning behind their words, which in turn can contribute towards involvement and help to underpin the dialogue. Emoticons can also have a disarming effect or provide a touch of humour and friendliness.

The time lag is one characteristic of the asynchronous electronic meeting which can both impede and facilitate dialogue. Perhaps we are sitting at the computer, describing a harrowing experience we have been through or a new idea that has come to us. We would then prefer to know instantly what the others think of it. Instead, hours or days may pass before anyone picks up the thread of one's contribution. This can be frustrating, indeed frightening, and make us lose interest in the conversation. On the other hand a modicum of delay can be a good thing. There are advantages to allowing oneself time for reflection between the lines. One way of reducing the disadvantages of the time lag is for all participants to try to answer each other's contributions as promptly as possible, sometimes perhaps just to show that they have read what was posted and want to give it more thought before pitching in with their own viewpoints. That way I am spared the feeling of writing in a vacuum; someone out there does "hear" me. (Fåhræus, 2000).

Once the e-meeting conversation has got underway, the participants sometimes get very active, writing long contributions and many of them. This may complicate matters, with the participants finding it hard work and perhaps also tedious to read everything. There is a risk of one or two participants dominating the proceedings and crowding out others, just as in a f2f meeting. If the participants themselves do not notice this and propose a different speaking order, the course leader may have to intervene. Often just making everyone aware of what has happened and suggesting they keep it a bit shorter will help. Or again, one can identify different topics of conversation and assign them to different chat spaces.

In an electronic conversation, whatever is said/written is perpetuated. Perhaps you happened to write something unusually silly, offensive or ill-considered. And once sent away it can't be retracted, it can be read and reread over and over again. What has been written can offend somebody seriously. That kind of damage is extra difficult to repair from a distance. But the permanence of the text also gives us instruments with which to proceed from other people's contributions, deepening the arguments. It all adds up to a different dialogue from the face-to-face variety, offering still greater opportunity for reflection, which may suit some people better than others.

In order for the thoughts about dialogue competence to be applicable to e-meetings, concepts like "speaking" and "listening" have to be reinterpreted. After all, people are not really talking but writing to each other. The speech component means presenting viewpoints, arguments and known facts in writing. Participants state who they are and what their background is. This applies not only to the content – the subject matter of the dialogue – but also to matters concerning the communication and learning process. Instead of just shaking their heads at a contrary opinion, they have to write about it. Not being visible in purely physical terms, participants have to be made visible by what they write.

The listening component in the dialogue competence matrix is concerned with reading what others have written and showing curiosity to know more, to learn from the others' contributions. In f2f meetings, just looking at a participant may be sufficient inducement for them to speak. Here we have to write a message and send it off, as an ingredient in active listening.

Dialogue competence includes critical self-reflection. In an e-meeting this is achieved by writing in a way which shows openness to changing habits of mind and points of view (Mezirow, 2000). Participants should also tell the others when they have changed their minds how they have received and understood other people's experiences and knowledge. Showing agreement with and suggesting other ways of understanding and thinking, as the case may be. As well as query other persons' input data, requesting more facts and arguments.

Dialogue-competent behaviour on the asynchronous web consists of reading and writing with openness and scrutiny, as shown in Figure 3. Talking becomes writing, offering a deepening of the arguments and an opportunity for reflection. You make yourself visible through writing. Active listening becomes reading and showing curiosity by sending a message in return where the original message is commented and further developed from your own perspective. Critical self-reflection is done through showing openness to changing your understanding by describing your learning process when writing. And also how and when you are influenced by what others say in their writing. You tell when and how your mind has changed. Critically scrutinising is done through describing your interpretation, show agreement or suggest other ways of understanding, requesting more facts and arguments. Thus supporting a deepening and critically reflecting joint learning process.

Figure 3. Ingredients of dialogue competence in the text-based asynchronous meeting. Development from Wilhelmson (2002).

	Self	*Others*
Closeness	Writing	Reading - writing
Distance	Describe openness	Describe scrutiny

DEVELOPING DIALOGUE-BASED COMMUNICATION FOR LEARNING PURPOSES: SOME CONSIDERATIONS

The knowledge of Fåhræus (2003) triple helix of learning processes among distance-education students and the dialogue competence model of Wilhelmson (2002) in f2f meetings has been elaborated upon in this chapter. Here follows in summarized form some points to be considered by teachers, students and technical support.

Considerations for Teachers and Students

Three learning processes were identified by Fåhræus (2003) and a fourth is here added: learning to become dialogue competent, which might also be seen as a deepening of the communication ingredient in learning to communicate electronically. All four represent processes that teachers as well as learners (students) benefit from being aware of and reflecting. Such meta-learning is regarded as productive and enhancing learning. Thus, teachers as well as students benefit from:

- Being aware of the four parallel learning processes:
    - learning the content in the course (subject)
    - learning to learn together (collaboration)
    - learning to communicate electronically (technology)
    - learning to reflect on the dialogue in itself (dialogue competence)
- Being aware of the beneficial balance between the integrating qualities and the

differentiating qualities of learning group dialogues.

Considerations for the Teacher

- As a teacher you have the task to create opportunities for understanding and insight concerning what is expected to take place among your students. Different perceptions of the task of digital conversation among the students may be articulated in the beginning. Problems later on regarding work inputs and digital attendance can thus be prevented.
- As a teacher you govern and influence the learning process, you are the guide to a rewarding web communication. This can be done in two main ways:
  - Intervening directly in student's thoughts and meaning context, e.g. direct questions to the students.
  - Arranging conditioning surroundings, the preconditions for e.g. choice of subject or procedure.
- Impose a structure but do not control.
  - Create a starting point for sharing thoughts by arousing curiosity and trigger conversation activity. Common kick-off points could be a vignette, a film, an article, brainstorming, role-play or a debate.
  - Indicate a certain sequence to create structure: describing a personal experience; read and summarize an article; ask each other questions which have to be answered within a certain time; return to the starting point conversation: what has been learned?
  - Intervene when domination occurs.
- Synchronous meetings can get hectic without a predefined procedure.
- Now and then during a course it is convenient to reflect on meeting forms and tone of conversation.
- Asynchronous meetings can be slowed down in tempo and become boring, a solution can be to answer promptly and to reserve time, e.g. through on-line seminars lasting a day or two when everyone visits the website frequently. It can be tedious to read a lot, a solution can be to identify different topics and assign them to different chat spaces.

Considerations for the Student

- Allot time to be present in the task of writing and reading, you need peace and quiet for study.
- Assume responsibility for the creation of good preconditions.
- Search information individually in between the meetings.
- Take extra care not to offend each other, the lack of body language easily creates misunderstandings. Verify properly your understanding, e.g. through emoticons. Once sent away can't be retracted and damage is difficult to repair from a distance.

Considerations for the Technical Support

Computer support for shaping up a detailed structure is needed, e.g.: different electronic conferences/folders where to post summaries; question boxes with linking answers; verification from the system when assignments are completed; reminders of what to do next; creation of groups by pairing participants off; support for administration and monitoring large groups; possibility to direct students questions in digital conferences. The aim is to increase student activity and decrease the teacher's workload.

CONCLUDING REMARKS

The electronic meeting can be regarded as an inferior variant of face-to-face conversation, but this is an oversimplification and an outmoded way of looking at things. Instead, possibilities are opening up for different forms of communication. These have advantages and disadvantages of their own, knowledge of which increases with use. The present chapter draws attention to what is different in relation to ordinary conversation and, more specifically, it highlights group dialogues as learning conversations.

An electronic meeting is here seen as an offer, an arena which I visit of my own volition. There I can find the thoughts contributed by my fellow-students and can decide for myself what to read and what to reply to. I can choose what to show of myself, but at the same time I should realise that my choice, my arena behaviour, will affect the other participants and the group's interaction, and with them the benefit to myself from our coming together. It is hard to divest oneself entirely of the comparison with the f2f meeting and not to see the latter as the ideal worth pursuing. This makes it appropriate to recall the manifold disruptions occurring in the course of ordinary conversation, the fact of some things being easier to talk about on the phone and the fact of many lessons and meetings being deadly dull and basically just a way of killing time. But the parallel to conversations in which people have, face to face, developed their capacity for engaging in a dialogue (Wilhelmson, 2006) is a useful one; knowledge from this field spurs each and every one to ponder the nature of its digital counterpart.

Returning to the examples with which this chapter began, and considering what happened in the two groups, we may note that the participants in the first group showed little of themselves or where they stood. When, at the course moderator's request, they presented personal experiences, these did not meet with any critical appraisal. Assenting remarks can be encouraging to hear, but similarities alone will not deepen the conversation, and the dialogue fails to materialise. In the other group both differentiating and integrating conversational qualities (Wilhelmson, 2002) were present. The misunderstanding which occurred threatened to detract from confidence within the group, but the open dialogue which had already been treated enabled the misunderstanding to be cleared up. Arguments concerning the nature of gender discrimination and its causes gave the participants new perspectives on their own experiences. In this instance the participants can be said to have demonstrated their dialogue competence in the electronic meeting.

It may help to pause every now and again during a course to reflect on meeting forms and the tone of conversation. The concepts presented in this chapter can be a good help. One cannot expect everyone to possess dialogue competence without having been given an opportunity for practice and reflection. It may, for example, be appropriate to have course tasks underpinning the participants' own awareness of what it takes for conversations to attain dialogical quality. This is a matter of understanding the basic ingredients of the f2f dialogue and of jointly considering, in the digital conversation, how it can be applied to e-meetings.

Talking in writing must be termed essentially alien to our human nature, and yet it is common enough. Digital exchange in the learning context involves linking thoughts and for each individual also changing one's thoughts, not just exchanging words. Verbal exchanges by computer are associated, not infrequently, with hasty action and impulsive responses, sudden statements and brief questions. On the contrary, pensiveness, exchange of thoughts and individual learning are associated with paper, books and armchairs. Perhaps technology today offers a digital context which can be likened to the common arena of action which is important for the occurrence of collective learning (Döös & Wilhelmson, 2005). Computer conferences have become a new way of meeting, and people now have the possibility of also being dialogue competent in digital conversations.

Before closing we will briefly comment upon the progressive technical development that has resulted in the learning management systems mainly focused in this chapter being supplemented by instant messaging systems and net-based meetings (e.g. Parnes' Marratech and Microsoft's NetMeeting). This is making communication both more synchronous and also enabling participants to see each other on the computer screen. This development has brought about the possibility of combining the written word with digital f2f meetings. Sound, voice and video are being used. Using downloaded software, a web camera and headset, the students themselves, with or without teachers, can communicate in real time and both hear and see one another. On the screen they can see all the participants and a whiteboard on which they can all write spontaneously or insert previously processed texts and images. In this way the digital dialogue can take on an additional dimension of presence, thanks to the possibility of supplementing asynchronous conversations with digital f2f meetings.

Experience of distance education programmes using this technology to supplement other learning platforms (course websites and suchlike), digital teaching material, e-mail, chat groups etc., indicates several positive benefits. In the virtual rooms the students have an opportunity of getting together to discuss reading matter, questions and experiences, in a way resembling f2f meetings. They can write their questions and reflections on the whiteboard which is also shown on the screen, and an opportunity is provided for direct reaction and feedback. This kind of digital conversation can to a great extent be likened to f2f meetings, even though the technology as yet cannot fully convey gestures and body language, partly because for the most part the web camera only captures the participant's face and also because the image lags a little behind the sound. Partly because the web camera transmission quality is not good enough to allow for seeing details in facial expressions and gestures. Nevertheless people's intonation, stressing of words and pauses can be heard, and the presence of others felt. Hrastinski (2006a; 2006b) concludes that students seem to become more involved and active when using instant messaging and other forms of synchronous communication as a supplement to the asynchronous text-based. It is assumed that this is to a major part the effect of the synchronous aspect.

Like ordinary meetings, asynchronous as well as synchronous net-based f2f meetings both require and provide training in dialogue competence. Going round in order to include everyone in the conversation often becomes a natural way of organising the digital f2f conversations, and in order for these to convey the benefit desired, those taking part must be able to keep their distance, be open-minded and build on other people's thoughts. The structure of the conversations imparts clear limits to speaking/writing and listening/reading, which can be supportive of the students' reflection process and, consequently, their learning.

A variety of evaluations are required in order to learn more about how to improve the learning qualities of our contemporary communication possibilities. This chapter contributes with suggesting the model of dialogue competence to be used in order to improve the linking of thoughts in web dialogues. It concludes by suggesting the ingredients of dialogue competence to be part of that evaluative undertaking.

REFERENCES

Allan, B. (2007). Time to learn?: E-learners' experiences of time in virtual learning communities. *Management Learning, 38*(5), 557-572.

Bjerlöv, M. (1999). *Learning in work based discourse.* Doctoral dissertation. Arbete och Hälsa 1999:1, Solna: National Institute for Working Life. (In Swedish, English summary)

Dixon, N. (1994). *The organizational learning cycle. How we can learn collectively.* London: McGraw-Hill.

Döös, M. (1997). *The qualifying experience.* Doctoral dissertation. Arbete och Hälsa 1997:10, Solna: National Institute for Working Life. (In Swedish, English summary)

Döös, M. (2007). Organizational learning: Competence-bearing relations and breakdowns of workplace relatonics. In L. Farell & T. Fenwick (Eds.), *World Year Book of Education 2007. Educating the global workforce* (pp. 141-153). London: Routledge.

Döös, M., & Ohlsson, J. (1999). Relating theory construction to practice development – Some contextual didactic reflections. In J. Ohlsson & M. Döös (Eds.), *Pedagogic interventions as conditions for learning* (pp. 5-13). Stockholm: Department of Education, Stockholm University.

Döös, M., & Wilhelmson, L. (2005). Collective learning - on the importance of interaction in action and a common action arena. *Journal of Swedish Educational Research, 10*(3/4), 209-226. (In Swedish; English summary)

Finkelstein, J. E. (2006). *Learning in real time: Synchronous teaching and learning online.* San Fransisco: Jossey-Bass.

Finkelstein, J. E. (2007). *Back to school: Virtual eye contact.* October 7th, 2007. http://www.learninginrealtime.com/minute/.

Francescato, D., Mebane, M., Porcelli, R., Attanasio, C., & Pulin, M. (2007). Developing professional skills and social capital through computer supported collaborative learning in university contexts. *Int. J. Human-Computer Studies* (65), 140-152.

Fåhræus, E. R. (2000). *Growing knowledge - How to support collaborative learning e-discussions in forum systems.* No. 00-005. Dept. of Computer and Systems Sciences, Stockholm University/KTH. http://www.dsv.su.se/~evafaahr/lic/index.html

Fåhræus, E. R. (2003). *A triple helix of learning processes - How to cultivate learning, communication and collaboration among distance-education students.* Doctoral dissertation. Dept. of Computer and Systems Sciences, Stockholm University/KTH, Report No. 03-015. ftp://ftp.dsv.su.se/users/jpalme/eva-f-doktorsavhandling.pdf

Fåhræus, E. R., & Döös, M. (2007). Competent web dialogue: Thoughts linked in digital conversations. *The International Journal of Information and Communications Technology Education, 3*(3), 14-24.

Hrastinski, S. (2006a). Introducing an informal synchronous medium in a distance learning course: How is participation affected? *The Internet and Higher Education* (9), 117-131.

Hrastinski, S. (2006b). The relationship between adopting a synchronous medium and participation in online group work: An explorative study. *Interactive Learning Environments 14*(2), 137-152.

Isaacs, W. (1999). *Dialogue: The art of thinking together.* New York: Doubleday.

Kolb, D. (1984). *Experiential learning. Experience as the source of learning and development.* Englewood Cliffs, NJ: Prentice-Hall.

Mezirow, J. (2000). Learning to think like an adult. Core concepts of Transformation Theory. In Mezirow J (ed.) *Learning as transformation. Critical perspectives on a theory in progress* (pp. 3-33). San Fransisco: Jossey-Bass.

Ng, K. C. (2007). Replacing face-to-face tutorials by synchronous online technologies: Challenges and pedagogical implications. *The International Review of Research in Open and Distance Learning, 8*(1).

Ohlsson, J. Granberg, O., & Stedt, L. (2004). Collective learning - the bridge between the individual and the learning organization? In *Workplace learning - From the learner's perspective.* LearningLab Denmark. (Conference paper).

Senge, P. M. (1990). *The fifth discipline: The art and practice of the learning organization*. New York: Doubleday/Currency.

Sällström, A. (2005). *Best practice net-based meetings - a pilot study*. Report 11:2005. Härnösand: Nätuniversitetet. (In Swedish)

Tham, C. M., & Werner, J. M. (2005). Designing and evaluating e-learning in higher education: A review and recommendations. *Journal of Leadership & Organizational Studies, 11*(2), 15-25.

Wilhelmson, L. (2002). On the theory of transformative learning. In A. Bron & M. Scheemann (Eds.), *Social science theories in adult education research* (pp. 180-210). Münster: LIT Verlag.

Wilhelmson, L. (2006). Dialogue meetings as non-formal adult education in a municipal context. *Journal of Transformative Education, 4*(3), 243-256.

Wilhelmson, L., & Döös, M. (2002). *Dialogue competence for development in working life*. Stockholm: National Institute for Working Life. (In Swedish)

Wännman, G. T. (2002). *Women creating knowledge on the web*. Doctoral dissertation, Umeå: Umeå University, Department of Education. (In Swedish)

Yuzer, T. V. (2007). Generating virtual eye contacts through online synchronous communications in virtual classroom applications. *Turkish Online Journal of Distance Education, 8*(1).

ENDNOTE

1 emotion + icon

Chapter XVII
Employing Interactive Technologies for Education and Learning:
Learning-Oriented Applications of Blogs, Wikis, Podcasts, and More

Jeffrey Hsu
Fairleigh Dickinson University, USA

ABSTRACT

A number of new communications technologies have emerged in recent years which were originally used primarily for personal and recreational purposes. The emphasis of these is on social networking and communications. However, these "conversational, constructivist Web 2.0 learning tools", coupled with the power and reach of the Internet, have been identified and employed effectively for both educational learning and knowledge-oriented applications. In particular, the technologies given attention in this paper include Instant Messaging (IM), weblogs (blogs), wikis, and podcasts. A discussion of these technologies and their uses, underlying educational and cognitive psychology theories, and also applications for education and the management of knowledge, are examined in detail. The implications for education, as well as areas for future research are also explored.

INTRODUCTION

In any current discussion of educational methods and techniques, the influence of technology is ever present. While the impact and perseverance of mainstream methods have still maintained their hold on many in the educational realm, it is without a doubt that technology is now an important component of the teaching and learning in the 21st century. The concept of "teaching with technology" is now an accepted part of the educational literature (Goffe and Sosin, 2005).

While for many years the mediums employed for education have remained fairly constant and traditional, including tried and true methods such as the blackboard and chalk, whiteboards, flip-charts, and overhead projectors, the employment of computing technologies has resulted in the usage use of PowerPoint, e-mail, and web-based course portals/enhancements such as Blackboard and WebCT. These have remained in widespread use in education for a wide variety of courses and programs.

In connection with this, there have been numerous studies done, and papers written, about the use of technology in the classroom, together with work on the related areas of e-learning, web-based learning, and online learning. The usage of computing technologies in education has been examined widely, and there is a sizable body of work on web and online learning, including the studies by Ahn et al. (2005), Liu and Chen (2005), and Beck et al. (2004) and numerous others.

In particular, some of the relatively newer technologies have been recognized as being particularly useful in the classroom, and have been engaged in innovative ways. These technologies of particular interest , are referred to as "conversational technologies," which allow for more effective creation and sharing of information (Wagner, 2004; KPMG, 2003). Another term often used to describe these technologies is the concept of "constructivist learning tools," which encourage, and are focused on, users creating, or constructing, their own content (Seitzinger, 2006).

The interest in employing these kinds of technologies stems not only from the unique pedagogical benefits gained, but also from the basic need to stay in tune with the focus and strengths of today's students. Prensky (2001) suggests that the students being taught today are "no longer the people our educational system was designed to teach" and that while the students of today can be termed "digital natives", many educators could be regarded as newcomers and relative novices, perhaps termed as "digital immigrants." Yet another way to look at this is to view earlier educational approaches as "print-based" while those of the current environment can be called "digitally-based, secondly-oral" (Ferris and Wilder, 2006).

Another key benefit of these kinds of technologies is that they can help to support what is called "collective cognition." This concerns thought processes and insights which are the product of the combined efforts of two or more people, and which did not arise from the ideas of one individual alone (Stahl, 2006; Giere, 2002). A way of viewing this is that this kind of work is designed to solve problems which are "beyond the capabilities of any individual member (Lund and Smordal, 2006; Hutchins, 1995). This concept is related to "communities of practice" (Lave and Wenger, 1991), and also a more unstructured, but insight and activity oriented, "collectivities of practice" (Lindkvist, 2005)

Another term which has become widely used to describe these kinds of technologies, which as its distinguishing characteristic emphasizes sharing, collaboration, and participation, is Web 2.0. This class of technologies emphasizes participation rather than static information presentation, and encourages ease in social networking. Users can get involved in contributing and commenting on the material presented, instead of passively reading and receiving information (O'Reilly, 2005). The scope of Web 2.0 include the technologies focused on in this paper, (blogs, wikis, and podcasts), as well as some of the popular online networking sites such as Flickr (picture sharing), MySpace, del.icio.us, and YouTube (Millard and Ross, 2006; Huang and Behara, 2007). A newer component of Web 2.0 is the web based service known as a mashup, which is a website application which integrates information from various sources to create an "integrated" resource on a topic, such as Craigslist or Google Maps (Dearstyne, 2007).

The influence of these come not only from their advances in technology, but also from wide use in our society and that of the college student population in general. According to the Pew Internet and American Life Project survey, 86% of college students have used the Internet, 80% feel that its use has helped their educational experience, and overall, students in general are far less likely to be satisfied with pedagogical experiences which do not use the Internet (Levin and Arafeh, 2002).

The purpose of this paper is to examine these technologies, and explore both the evolution of their use from personal applications to that of educational tools, and also to examine the key educational applications for which these are being used. Relevant research and applications are examined and analyzed. The future of these technologies for educational and professional use, together with viable research areas, is examined as well.

CONVERSATIONAL TECHNOLOGIES AND CONSTRUCTIVIST LEARNING TOOLS

The notion of conversational technologies is not a new one, as it encompasses many types of systems which have been widely used for some time, including e-mail, video conferencing, and discussion forums.

The term "conversational technology" is derived from the work of Locke et al. (2000) relating to conversational exchanges and his Cluetrain Manifesto. One of the key concepts here is that "markets are conversations" and that knowledge is created and shared using question and answer dialog. Specific theses which relate to this form of "conversational knowledge management" suggest that aggregation and abstraction of information hels to create new information. Other characteristics of conversational knowledge management include the fact that it is fast, stored in different locations, and does not require sophisticated technologies in order to be accomplished (Wagner, 2004).

Conversational technologies encompass a wide range of systems and software, many of which are familiar, including e-mail, instant messaging, web pages, discussion forums, video and audio content/streaming, wikis, and weblogs. While there are specific aspects of more mature technologies which are used for teaching and learning , the ones which will be given greatest attention in this paper are the issues, impacts, and applications relating to instant messaging (IM), weblogs (or blogs), wikis, and podcasts. These are technologies which are relatively newer, have a growing base of users, and are starting to become recognized as viable tools for education.

The term "constructivist learning tool" has also become associated with these, particularly blogs and wikis, in that they have the key characteristic of allowing users to develop and maintain their own content. Some of the characteristics of constructivist learning include engagement, active learning, collaboration, real-world basis, and reflection as a part of the learning process (Seitzinger, 2006).

It should be noted that maximizing the benefits of these technologies and tools requires that they be used in courses and exercises where class collaboration and communication are encouraged or required, rather than those with an emphasis on lectures and the presentation of factual information. More specifically, in courses where there is substantial group work, or projects where a collaborative document or product is created, the use of these would be especially helpful and useful. Both hybrid and full distance learning courses would also be situations where these could be used effectively.

TEACHING AND LEARNING: NEW TRENDS

Before discussing further about conversational and constructivist technologies, it would be useful to briefly review learning styles and phases. These are certainly critical to how students learn, and also what kinds of technologies would be used for certain kinds of activities, courses, and students. Fleming and Mills (1992a, 1992b) discuss various types of learners, including visual, auditory, writing/reading, and kinestic (or tactile). These are related to a preference for graphs and diagrams; for spoken lectures and information; for written words and documents; and for hands on, experiential, and simulations; respectively.

From another perspective, Kolb (1984) discussed several phases which comprise learning, which includes listening and observations, concrete experiences, generating ideas, and making decisions using active experiences. While the emphasis of conversational and constructivist technologies is on experiential and interactive facilities to support courses, these technologies are flexible enough in that they can also support the various types of learners and phases of learning mentioned.

Conversational and constructivist technologies are certainly here to stay, as evidenced by their extensive role in our society. It would therefore be useful to examine their applicability in the educational realm. While usage based mainly on popularity or student preference can be one factor to consider, there are also theoretical and conceptual bases for employing these kinds of technologies in the classroom.

Earlier paradigms of teaching emphasized print-based materials for instruction, which included printed textbooks, paper-based instructional materials, and written tutorials, all of which are grounded in the notion that the teacher, lecture, and instructional materials form not only the basis, but also the authority in the educational process. The transmission of material from the teacher (lecture) and/or textbook to the student (called the "print model") is still the central basis of most teaching, even if they are supplemented with other methods including discussion and other forms of student interaction/participation. (Ferris and Wilder, 2006). The use of e-mail, PowerPoint, and course management systems/portals could be regarded as supporting the print model, by transmitting information in electronic form.

However, the advent of digital and conversational technologies has brought forth the new concept of secondary orality (Ong, 1982). This concept emphasizes that teaching and learning should go beyond printed materials towards a greater emphasis on supporting group work, fostering student communities, and encouraging student participation. The concept encourages a greater sense of interaction with and "ownership" of knowledge, emphasizing self awareness and expression, and effectively using electronic tools to support collaboration (Gronbeck, Farrell and Soukup, 1991).

The use of conversational technologies can have a positive impact, because it attempts not only to improve upon the print approach, but also to employ secondary-oral techniques. In other words, while a student can still be presented with material in different formats using the print model, the introduction of secondary-oral methods can be used to improve the overall learning experience. Using the latter, there is the opportunity to work collaboratively, explore, analyze, engage in discussion, and otherwise "learn" in new and innovative ways (Wallace, 2005; Ferris and Wilder, 2006).

In relation to these theories and methods, there have been models developed which attempt to capture the relation between learning activities and the use of technology. Bonk and Zhang (2006) developed a R2D2 (Read, Reflect, Display, Do) model which links general learning activity categories with more specific learning activities; this is then mapped to the use of specific technologies and methods suitable for that learning application.

Included in the discussion is an examination of how each quadrant of the R2D2 model can assist in helping a specific type of learner. For instance, verbal and auditory learners (Read) would be able to acquire information more readily through the use of podcasts, online documents, online discussion boards, and chat, emphasizing both spoken and written words as its main medium.

Reflective and observational learners (Reflect) would be better suited to online discussion forums, the use of blogs, and other means where thinking, critical analysis, and observation are paramount. The analysis of writing, literature, or issues would be useful here. Visual learners (Display) would benefit more from technologies and methods which emphasize the use of videos, animations, online whiteboards, and other graphical and visual-oriented displays. The final focus is on tactile/kinestic learners (Doing) which emphasizes interactive projects, case simulations, online surveys, and online gallery creation. The goal is to emphasize hands-on, experiential, and group interaction based model of learning (Bonk and Zhang, 2006).

Huang and Behara (2007) discuss the advantages of using Web 2.0 technologies to help support experiential learning. The wide use, acceptance, availability, and popularity of these technologies and sites contribute to their success as educational tools. One positive benefit of these is the ability for students to have access to a greater set of learning content, instead of obtaining it strictly from a text or the instructor. In addition, the very nature of technologies such as wikis and blogs, for example, encourage experiential learning, interaction, and collaborative learning. Finally, the use of Web 2.0 can help to foster greater intergration of the knowledge gained. In many cases, while the type of assignment given may not appear to be any different, the effective deployment of the technologies in conjunction with the assignments will help to bring about the greater levels of interactivity and collaboration.

MESSAGING TECHNOLOGIES: INSTANT MESSAGING (IM)

It's unlikely that there would be many college students who are unfamiliar with the use of Instant Messenger (IM). Allowing for interactive and real-time synchronous communications with instant response, instant messenger is truly conversational in that it allows for "chat" and communications between both individuals and groups. The major instant messaging systems in use include AOL (AIM), MSN Messenger, Yahoo! Messenger, and ICQ. Both this and related messaging-oriented technologies, which include Twitter and SMS, may be viable for educational use. Twitter is a newer technology which supports both social messaging and updating, as well as microblogging. Messages sent via Twitter are known as "tweats" and can be accessed through the web, cell phone text messages, and through RSS feeds. The other major method, SMS (Short Messaging Service) has been in place for some time, and is generally referred to as "text messaging" or "texting." (Stephens, 2007).

Instant Messaging (IM) is a means for users to "chat" and communicate in real-time. While originally the domain of personal users, over time the unique benefits and effectiveness of this medium was realized, and IM started to become accepted as a form of communications, particularly informal communications in businesses (particularly high-tech firms), and now has been studied and tested as an educational tool (Kinzie, Whitaker, and Hofer, 2005).

The important features of IM include both its synchronous nature and its ability to support both chat and phone-like interaction. Aside from its real-time interaction allowing for rapid communications to occur, there is also no need to enter an interaction "space" as with chat rooms. Instead, the main usage of IM is in one-on-one communication, which can be more formally termed as a dyadic "call" model, closely resembling phone call interaction. It should be noted that even though much of the communications is done between 2 individuals, there are some systems which support multiparty instant messaging. Some of the salient features of IM include both the ability for users to see user details as to current status (online, idle, away, out to lunch), and also a user's changes in status (active, logged out, etc.). Lists of users can be displayed on the screen, so that contact can be made when desired. If a "chat" is initiated, a special window comes up and the interaction can commence, provided that both parties are online and willing to proceed.

The real-time nature of IM has resulted in the techology being used for reasons aside from personal "chat" . In business, IM has become in some industries an accepted form of communication. A number of studies have concluded that instant messaging is ideal for informal interaction. There are claims of benefits including less formality required, being able to get an answer quickly, and not requiring a formal message structure as with many e-mails. In particular, the use of IM has been shown to be helpful in cases where collaborative coordination and problem solving is involved. The benefits of social bonding and interaction, which is a component contributing to the success of more complex collaboration situations, is also enhanced by using instant messenger technology (Nardi et al., 2000).

An important difference between IM and e-mail is the tendency for instant messenger interaction to be more casual and informal than e-mails, which helps to bring about a more "friendly" communication atmosphere. This may in part be due to a reduction in the formalities which are typically involved when using e-mail or the phone. In particular, IM has been considered more suitable for such tasks as scheduling meetings, asking or answering quick questions, and for other kinds of tasks which are brief, require a prompt response, or are less formal. It is perceived to be far simpler to IM someone to ask a quick question , for example, or to confirm a meeting or lunch, rather than to e-mail or call (Nardi et al., 2000).

It is also of interest that IM communications tend to be more flexible in terms of its uses (everything from task-related questions to a new joke), and can allow for greater expressiveness in terms of emotion, humor, and personality (Nardi et. al., 2000). Another interesting aspect is what Nardi et. al. (2000) refers to as "outeraction," which focuses on the processes associated with IM. These include conversational availability, communications zones, intermittent conversations, awareness, and conversational progress/media switching. IM is useful in certain communications situations, since it tends to be less disruptive and interrupting, while at the same time a user's availability is more clearly known (scanning buddy list status, for example). It is also a convenient means for setting up more formal interactions, such as arranging a conference call (media switching). Intermittent, dispersed communications can be conducted over a longer period of time, which includes interruptions. Other benefits include the knowledge that others are "there" and available even if not currently in chat mode, however there is always the opportunity to make contact, whether through IM or a different form of communications.

While some educators may scoff at and even express criticism at the thought of Instant Messaging as a viable educational tool, others believe there is potential in the medium. In terms of educational uses for IM, they are being explored and tested. Clearly, IM allows students not only to collaborate more effectively on homework assignments and projects, but also helps to maintain a closer social network between students, which could have a positive impact on learning. In addition, if IM is carefully targeted and focused towards the material or lecture topic in hand, the use of IM may actually help and stimulate deeper and more active learning.

Having mentioned possible benefits, it has also been hypothesized that the distraction of working on various other tasks in parallel with IM, known as "distracted attention, " may have a negative impact on learning (Hembrooke and Gay, 2003).

Active learning (Grabinger, 1996) and dual (verbal and visual) processing (Clark and Paivio, 1991) are at work here. It could be said that using IM to encourage greater discussion and reflection on the course contents would be likened to the use of discussion boards, however since IM is a real-time technology, the interaction is conducted during the lecture or class, not afterwards. Some studies have reported positive effects and student satisfaction from IM being used to discuss course subjects real-time (Guernsey, 2003).

A study by Kinzie, Whitaker, and Hofer (2005) examined the use of IM during classroom lectures, and found that while the general idea of using IM on-line discussions was positively received, the actual process and experience of using IM to conduct discussions during class lecture sessions was found to be less than a positive experience by both teachers and students. It was suggested that the difficulties of multitasking and dividing one's attention between the lecture and instructor, while doing the IM discussion, contributed to the lack of satisfaction with the process.

Burke (2004) used Instant Messaging as a medium for creating course diaries in 3 different mathematics courses. IM was chosen since it was thought to be popular, widely used by students, and considered more "fun," so there was some hope that this popularity would transfer over to greater and more enthusiastic usage by students. In fact, the choice was made to use IM over a seemingly more suitable choice, blogs. A bot was created which would retrieve IM diary entries from student and store them in a PostgreSQL database, and there was also a program set up to display diary entries from each student, viewable by both the student and the instructor. The main finding of the study was that the IM media was not ideally suited for all kinds of courses, especially those which involved creating longer portions of text, or involved diagramming. Error detection and recovery was also not that well developed, and also there was a need for better editing tools.

In summary, while Instant Messenger can be appropriate for various applications, in particular for information communications in a business setting, the results from educational studies appear to be mixed, with both positive and negative effects noted. While there seem to be advantages to real-time communications between students, between students and instructors, and also between groups working on a project, it appears that there are problems and limitations if the technology is used in a classroom setting. The challenges of focusing on a class lecture, together with maintaining a conversation online, seem to be a problem which has not yet been resolved. In addition, while instructors can often establish closer relationships with students using IM, there is also the problem of unreasonable student expectations of continuous teacher access, which may not be present if IM was not available as an option. In connection with this, using IM for student help can result in a greater time commitment, since sessions can become lengthy with many questions and responses being sent back and forth. In terms of related technologies, SMS text messaging has been available for some time, but has not been worked with much as an educational tool. The new technology, Twitter, may have potential applications, but is relatively new compared to the others.

BLOGS (WEBLOGS)

Blogs, or weblogs, started as a means for expressive individuals to post online diaries of themselves. Complete with text and photos, these logs were essentially an individual's online narrative or diary, with events, stories, and opinions. While its original use was for personal expression, recently its effectiveness as a tool for education has been discovered, including its use as an extension of "learning logs" which are created online (Barger,

1997). One of the earliest blogs, as we know and use them today, was Dave Winer's Scripting News , which was put online in 1997. While the use of weblogs can be considered generally new, the concepts of keeping a "log" or "learning log" is not.

The concept of "learning logs" have been in use before the advent of the weblog. The concept behind this is to enable someone to document his or her learning, and also to do some critical reflection (Fulwiler, 1987) and self-analysis. The use of a learning log or journal is related to action research learning strategies (Cherry, 1998) and attempts to link previous knowledge and new information learned. Blogs are a natural extension of learning logs/journals in that they are electronic and can be made available ("published") more easily (Armstong, Berry, and Lamshed, 2004).

The use of electronic weblogs as educational tools offer the benefits of increased information sharing, simplified publication of information, and improved instructor monitoring and review (Flatley, 2005; Wagner, 2003). The use of blogs has been expanding, as Perseus Development reported that there were some 10 million blogs in 2004, and the number is ever increasing (Nussbaum, 2004). The growth in this area is expected to increase in the years to come.

Blogs can be defined more formally as being "frequently updated websites consisting of dated entries arranged in reverse chronological order" (Walker, 2005), and can take several forms, including the personal diary/journal, knowledge-based logs, and filter blogs. The first, an electronic, online diary of one's life events and opinions, is probably the most common. The online diary/journal blog is one which, being on the Internet, is public, as opposed to the traditional (typically paper) diaries which are generally kept private. It should come as no surprise that there are many different online diary/journal blogs which are currently online, where one can find out details, often much more than one might want to know, about someone's life and thoughts. Personal blogs form the majority of the blogs which are currently online and available, which makes up roughly 70% of all the blogs in existence (Herring et al, 2003).

The second type (knowledge-based) captures knowledge, and places it online in various formats. The third type (filter) attempts to select, rate, or comment on information contained in other sites (Herring et al., 2004).

There are software packages which are designed to help users create blogs, including Blogger, Xanga, Blurty, MovableType, and others. While the basic features of most blog software emphasize the creation of blog pages, some of the more sophisticated ones offer the capability to track readership, see who followed what links, add photos, and set up more advanced structures. When online, blogs can range from being totally public (listed in the blog service directory), to being "unlisted" but still open, to being highly restricted (password-protected).

Blogs are also interesting and unique in that they are not merely online versions of paper diaries and journals. Rather, as a communications medium under the control of the main writer (author), it is reflective of the fact that an audience is always "watching and listening." What is put on a blog is not merely a one-sided set of thoughts and reporting of events; there can also be responses to feedback, and reactions from the "viewing audience." Therefore, blogging is considered a highly social activity, rather than a personal one. In fact, recent work has indicated that the momentum for creating, and also updating a blog, came about as a result of encouragement from friends and the viewing audience (Nardi et al., 2004). In addition, blogging can have negative repercussions when posted information is perceived to be confidential, proprietary, or improper. In some cases, employees posting what was considered by their employers as "confidential" information caused problems.

Blogs do not, in general, exhibit a free-flow of information between the blogger and the outside audience. While feedback is often requested, received, and desired by the blogger, the level and quantity of feedback from readers is generally limited compared with the output from the blog writer. In addition, while blogs may have sections where hyperlinks are mentioned, the number of hyperlinks in blog pages is frequently not very large (Schiano et al., 2004).

More formally, weblogs can be considered to be objects motivating human behavior, which is related to activity theory. Activity theory states that there are objects which have motives which respond to a human need or desire, and that they manifest a person's desire to accomplish that motive (Leontiev, 1978; Vygotsky, 1986). The objects which connect bloggers to their own social networks include letting people know what is happening in their lives, voicing opinions, asking for feedback, and "letting off steam" about certain challenges or difficulties currently being experienced, to name a few (Nardi et al., 2004). Blogs have been categorized by Krisnamurthy (2002) as being categorized into four different types, along the dimensions of individual vs. community, and personal vs. topical. A blog can therefore range from being very individual and personal, all the way to being open to the community, however very focused on a particular topic.

The acceptance of blogs for educational purposes is gaining interest, with one university, the University of Maryland, attempting to implement blogging software campus-wide (Higgins, Reeves, and Byrd, 2004).

In addition, the educational uses of blogs take advantage of their ability to encourage expression and the development of online relationships. Blogs allow for learning and interaction to be more knowledge-centered, especially if the assignments are structured in the format of encouraging feedback and input from the instructor and outside experts. Blogs also allow students to gain a better understanding of a subject's knowledge domain (Glogoff, 2005). As an example of this type of blog-enhanced class structure, students might be provided with a weblog from which to complete certain course assignments. After researching the indicated subject, the student would present the new information by "publishing" it to the weblog. The weblog would constitute the student's assignment, which would then be subject to review and critique by not only the instructor, but also by other students in the class. Supplementing this could be discussion boards, where threads would be devoted to discussions on the weblogs created by the students. This kind of assignment and interaction would be especially useful for both hybrid and fully online distance learning courses (Glogoff, 2005).

There are other benefits of weblogs. These could be expressed using the learning theories and concepts of guided discovery, directive learning, receptive learning, and social/community-centered instruction.

Guided discovery allows for the exploration and study of a certain topic, which is then followed by assignments which emphasize the synthesis of information. In effect, a student can be asked to research an area and "construct knowledge" using the weblog as a medium. Part of the assignment goes beyond merely explaining or presenting the material, and asks for the application of the concept using a real-world situation. The ability for students to post and make comments about other students' blogs provides an atmosphere of interactivity and collaboration. One of the advantages of using blogs together with guided discovery is that it encourages the use of cognitive scaffolding, where students would approach learning (together with the use of blogs and interaction) by repeatedly seeking information, reflecting and thinking about what has been learned, then going back and obtaining more information, so as to build upon and dig deeper into the subject area. This can result in a more active and productive form of learning (Betts and Glogoff, 2004; Glogoff, 2005).

Directive learning, where responses from students are followed by prompt feedback from instructors, can also be supported using blogs. In this case, students would not only use a blog to submit assignments, but also to review instructor feedback. In additional to feedback, there would be opportunities for the instructor to ask additional questions; in effect, to encourage further exploration and "drilling down" into the subject (Betts and Glogoff, 2004; Glogoff, 2005).

Receptive learning is where instructional modules are presented which focus on certain broader areas, from which certain sub-areas within these are highlighted for a student to research and report on. Generally, these areas are contained within a designated theoretical context (Betts and Glogoff, 2004; Glogoff, 2005).

Social/community-centered instruction is a logical component of educational work using blogs, and in particular the use of peer and social interaction as a part of the learning process. The use of blogs functions as an easily accessible medium for students to present their findings (and to be read by others) and also to integrate not only the information presented, but also related links, and references to other resources. This form of interaction helps to encourage further exploration by students. A blog-based discussion can then be continued by conducting peer reviews of other students' blogs, which may include commentary, critique, the posing of questions, and opening up the way for further inquiry. The ability to share and benefit from the findings of other students (and to explore further) is another important outcome. The theories of community practice (Snyder, 2002), social cognition (Vygotsky, 1978) and communities of inquiry (Lipman, 1991) provide support for the blog-related techniques mentioned above.

Ducate and Lomicka (2005) discuss their experiences in using weblogs to support foreign language classes. Weblogs help the foreign language student to learn during the process of reading, and then creating blog entries. Students can learn by reading blogs which are written in the new, target language; including learning new vocabulary, checking out links and further information on words, and learning associated cultural information. The reading and absorption of blogs on the culture associated with the target language, including literature and lifestyles, all would contribute to the learning process.

Another approach would be to have students maintain blogs written in their new, target language, and then the goal would be to seek commentary and critique on these blogs by others in the class. Yet another innovative method might be to share blogs with other classes studying the same language, and for students to read and comment on each other's postings. In the case where students travel to a country where the target language is spoken, then the compilation of travel blogs would be a useful learning tools well (Ducate and Lumicka, 2005).

Wang and Fang (2005) looked at whether the use of blogs enouraged or enhanced cooperative learning in a English rhetoric/writing class taught in Taiwan. The main premise was that blogs can encourage students to spend more time working within a class "community," and can benefit from a greater sharing of contributions and inputs. In general, cooperative learning benefits can be divided into three different types: formal, informal, and base groups . Formal cooperative learning is where the instructor explicitly provides course materials and assignments to a group , and then observes the students' learning processes When the instructor provides information more generally (such as detailing how to use a blog for course assignments), and then lets the group work out their own methods for handling an assignment , that is known as informal cooperative learning. When a learning-oriented group is maintained for an extended period of time, such as throughout a semester, then this form of cooperative learning is known as a cooperative base group (Johnson and Johnson, 1998; Johnson, Johnson and Holubec, 1991). The study, run over the course of a semester, found that the use of blogs contributed not only to cooperative learning in general, but also to autonomous learning. Autonomous learning is focused on how much students take responsibility for their own learning, and also develop self-confidence in the task or skill (Wenden, 1991). The use of blog technologies was also found to help improve information processing, learning self-evaluation, and effective time management (Wang and Fang, 2005).

Martindale and Wiley (2005) also used blogs in their courses, and looked at two cases of the impact of this technology on teaching and learning. Martindale taught a doctoral-level course on Instructional Design and Technology. In it, students were introduced to blogs and used them throughout the course, which overall tended to promote higher levels of quality in their course work. Blogs were used to post ideas and abstracts of their projects, and also to place links for relevant research papers and web-based resources. The end result was a course "knowledge base" which represented the cumulative output of the students in the course. The blogs were also used for article critiques, which were an integral part of each weekly class. Students were given access to the blogs of other students and were able to offer feedback.

Wiley taught two different courses, one on online learning research, and the other on online interaction culture. Both included the use of blogs as a supporting technology; the first employing a general course blog where information about the course, student assignments, and class updates and student/instructor interaction exchanges were posted on an ongoing basis. In the second, blogs were used to discuss experiences using different online communications technologies, causing students to become highly engaged, resulting in passionate discussions and detailed commentaries posted to the blogs, far exceeding the level and depth of feedback that was expected (Martindale and Wiley, 2005).

In summary, blogs can be useful for educational purposes, particularly where there is the need to encourage and stimulate critical thinking and reflection on a work, concept, or idea. The submission or publication or a document or text as a blog can then lead others in a class to review and comment, setting the stage for greater analysis and study. In particular, blogs are well suited to writing courses, where a text can be analyzed and critique, or for a situation where a peer review is desired. The use of blogs is also ideal for the situation where someone keeps an online journal of one's learning, or wants to display her/her work to an audience. The blog approach is also considered useful for group study of a certain focused problem or case, such as those used in graduate courses and seminars.

WIKIS

Yet another technology, known as the wiki, has emerged, which allows for improved collaboration compared with weblogs. While the major emphasis of weblogs is the creation of a set of pages and documents primarily by a single individual, the strength of a wiki is the ability for numerous interested readers and users to express ideas online, edit someone else's work, send and receive ideas, and post links to related resources and sites. As a result, wikis go a step further and allow for greater collaboration and interactivity (Chawner and Gorman, 2002). Wikis have been found to have value for educational purposes, and their use have begun to be integrated into a number of university courses (Kinzie, 2005).

The term "wiki" comes from the Hawaiian word "wikiwiki" which means "fast." (Leuf and Cunningham, 2001). The technology is computer-based, and can be generally described as a knowledge sharing and creation system which has as its basis a set of web pages, which can be created and updated on an iterative and collaborative basis, and is in many ways a form of groupware. A wiki is designed to run on the Internet and World Wide Web, uses the HTTP protocol, and resemble traditional web sites in terms of its basic underlying structure. Some of the benefits of wikis include the ability to easily create pages (using a simplified form of HTML or basic HTML), and the ability for a document to be authored collaboratively and collectively. The first wiki was created to support software development, and ones which followed were for applications as diverse as travel, online encyclopedias, and education applications.

Another characteristic, and one which distinguishes the wiki, is that there is no clear dividing line between reader and contributor. A user in effect can be both. In addition, a wiki structure allows for the creation of a "knowledge base," which, inherent in its structure, is a chronological history of how the information developed and evolved. The latter is enabled by the wiki's built in versioning capabilities, where changes in content are tracked and recorded (Mindel and Verma, 2006). In particular, simplicity is the key to wikis, and wiki pages have been designed to be easy to create, and in general simpler than the process of creating standard web pages. Wikis are therefore especially useful for team projects and other collaborative educational exercises where a group interacts to create a certain result or deliverable. The exchange of and contribution of information, support for many to many interaction, and maintaining autonomy of users are found in wikis (Cheung et al., 2005).

Problems which can occur due to the open nature of wikis include the possibility of inaccurate information being posted, together with the ever-present potential for vandalism. One way to help maintain the integrity of wikis is to have a solid user community, the members of which will monitor the wiki and ensure that its content is accurate and complete (Mindel and Verma, 2006).

Other issues specific to wikis include the problems which occur when two or more users attempt to modify a section at the same time, for which the solutions include locking a certain portion of the content from changes, or making notifications to all parties involved whenever there is a conflict. The wiki structure also allows content sections to be set to be open to everyone (open wiki), to be readable by everyone but created or modified by some (fishbowl wiki), or or where both reading and creation/modification is restricted to certain users only. The effective use of these options, together with designation of various roles, rules, and guidelines within the user community, can help to alleviate the potential problems of structural inconsistency, poorly written entries, and even total chaos and disorganization (Mindel and Verma, 2006).

Based on the information presented so far, it should be obvious that wikis can be regarded as a type of groupware, which supports collaborative learning and asynchronous work across distances. An important difference between traditional classroom instruction and doing an exercise using a wiki is that the process is less instructor-centered. Simply put, the instructor is more of a facilitator or coach rather than the deliverer of information.

Current wiki systems include those designed for content management (DocuWiki), agile system development (Trac), project group work (Twiki), as well as more general systems such as MediaWiki and various others.

One of the better known examples of a large-scale wiki in operation is www.wikipedia.org, which is an online encyclopedia, with entries authored and edited by different persons worldwide, and in several different languages as well. In essence, it is an online information resource which is authored by interested and knowledgeable persons from around the world.

Wagner (2004) developed a set of design principles which relate to wikis. These are the principles of open, incremental, organic, mundane, universal, overt, unified, precise, tolerant, observable, and convergent wikis. *Open* means that anyone can edit a wiki, creating an "open source" environment for the sharing of knowledge. *Incremental* means that new pages can be added, even if they do not yet exist. *Organic* means that the information can be continuously evolving, as changes and edits are made. Wikis are *mundane* because they involve the use of a simplified set of essential commands. The design of wikis is also *universal*, meaning that writing and editing is a "combined" activity, formatting is related to input (*overt*), page names are not context specific (*unified*), and pages are generally named with some precision *(precise*). Wikis should be *tolerant* of error, activity should be *observable* by all, and duplications are undesirable and should be deleted (*convergent*).

There are a number of software programs which enable the effective creation of wiki pages, including TikiWiki, TWiki, and Pmwiki. These allow for the effective creation, modification/editing, and management of wikis, including creating pages, creating links, formatting, and feature modules (discussion forums, photo pages, download areas, etc.) (Chawner and Lewis, 2006)

Wikis are set up to allow for easy collaboration, and more specifically, editing. Rather than passively reading a passage of text or related information (which may include graphics, multimedia, hyperlinks, etc.), a reader of a wiki can also take on the role of a writer, making changes to the text (re-organizing, editing, re-writing, and marking up) at will. In essence, the document is open to changes by a "collaborative community," which allows for the secondary-oral model in education to be applied.

One reservation on the part of educators to embrace wikis is the fact that wikis are designed to allow for open and free access and editing by all members of a "community. " As a result, if improperly managed, a wiki's content can become an unreliable, inaccurate, or biased source of information due to its largely unmonitored format. There is also the issue of having little or no "quality control," resulting in a wiki not being trusted by its readers and users. An example of this was the controversy over the accuracy and reliability of Wikipedia in the case of John Seigenthaler, in which the subject alleged that false and incorrect statements were posted in his biography (Seigenthaler, 2005). However, other studies have attempted to prove that Wikipedia was overall, an accurate information source. One article reported that after an analysis of a set of Wikipedia's science articles by the British journal Nature, they were judged to be as reliable as the Encyclopedia Britannica (BBC News, 2005).

Wikis are useful for education in that they help to promote student participation and also a sense of group community and purpose in learning. Indeed, an important element of this is the relaxed sense of control over the content, allowing students to have a greater role in managing its focus and direction.

Wikis are not all the same, and there is significant diversity between various forms and implementations of wiki systems. In fact, it could be debated what features truly characterize a "true" wiki." The features inherent in most include the ability for users to both read and edit information, without the need for security or access restrictions. The emphasis is on simplicity, and the informal, "never finished" nature of wikis, which may constitute the contributions of multiple authors, is another key characteristic. While the emphasis of many wikis is on simplicity and a lack of access restrictions, that does not mean that all wikis work this way. In reality, there can be a continuum of features from simple to complex. At the complex end of the scale can be capabilities for security/access restrictions, complex organizational structures, and for integrated with content management systems (Lamb, 2004).

Now that the strengths and weaknesses of wikis has been established, it would be useful to examine the educational applications of wikis. In general, the most suitable applications are those that take advantage of the wiki's free, open structure. As such, the use of wikis as discussion/bulletin boards, brainstorming tools, and online sketchpads are appropriate. Meeting planning is another viable application area, in that the organizer can start with a preliminary agenda, from which the other participants can then add their own additions or make modifications or comments.

An important application area for wikis has been identified in knowledge management (KM). The use of wikis for knowledge management may allow for an improvement over existing systems and technologies. Currently, with existing KM systems, there do exist a number of bottlenecks relating to knowledge acquisition, namely acquisition latency, narrow bandwidths, knowledge inaccuracy, and "maintenance traps." Basically, these knowledge acquisition bottlenecks result from a time lag between when the knowledge is created, and then distributed. In addition, there are the problems of limited channels of knowledge input, possibilities of erroneous information being received, and also the difficulties of maintaining the knowledge base as it grows larger (Wagner, 2006; Land, 2002; Waterman, 1986).

The use of wikis to elicit a "bazaar" approach to knowledge management, rather than a "cathedral" approach, is proposed as a better alternative. These terms are derived from software development, whether the "bazaar" allows for more continuous and open access to code (or information), as opposed to the "cathedral" approach where access is only provided on certain (release) dates to certain persons. The difference between the "cathedral" (closed) sources of knowledge acquisition management and "bazaar" (open) could be illustrated by the difference between encyclopedias which are created by a single firm, such as Encarta or the Encyclopedia Brittanica, and those which obtain information from readers and users, such as the well-known Wikipedia.

The emphasis therefore is on teamwork, continuous review and testing, and the development of conversational sharing. (Wagner, 2006). Inherent in the workings of wikis is support for an open, collaborative environment, where many people can contribute to the development of knowledge instead of being limited to a set of "experts." It appears that conversational knowledge acquisition and management are appropriate for wikis (Cheung et al., 2005). As for educational applications and KM, a study by Raman et al. (2005) examined the use of a wiki to help encourage and support collaborative activities in a knowledge management course. More specifically, using wikis in the course helped to encourage openness, and better sharing and updating of knowledge bases. Many-to-many communication is supported, and the persistence of the created pages formed the basis of a knowledge repository. In short, the impact of easy page creation, improved updating and editing, together with effective maintenance of knowledge histories were seen as positives (Raman et al, 2006; Bergin, 2002).

Activities in the KM course included group article review assignments, answering questions about sharing knowledge and uses of the wiki technology, and also creating a wiki-based knowledge management system. Students were asked to create, update, refine, and then maintain a class knowledge management system. In terms of these experiences, while the use of the wiki technology was generally viewed positively, feedback received indicated that, since the goals of using the wiki were not made clear, using one was perceived to be counter-productive. More specific guidance on goals and objectives, a clearer system structure, and advance training were suggested as ways to make the wiki a more effective educational tool. The availability of too many features made the task of doing the course activities more difficult, since much time was spent learning the various features rather than focusing on the task at hand. A simpler, less feature-rich version, was therefore preferred (Raman et al., 2005).

Another popular application of wikis in the classroom is in supporting writing courses. The use of this technology can help to foster the impression that writing is "fun," and that there can be a shared and collaborative side to writing, revising, and commenting on written work. In other words, the technology can benefit not only the writing and editing process, but also in bringing about an awareness that writing is being done for a specific audience.

An example of the use of wikis in English is the Romantic Audience Program at Bowdoin College, where students used a wiki to discuss and examine Romantic literature, focusing on poetry, poets, and related topics. The technology was used to elicit discussion and analyses by the group, encourage elaboration on areas where further study or insight was sought, and to seek out linkages to additional sources and commentary. Another project is Auburn University's Wikifish, which was created by one school within the university, where questions are posed, and opinions and comments by all are encouraged.

Difficulties encountered in using wikis for education include the difficulty of tracking the new pages and contributions made by students, since modification can be made to another student's text without specifically identifying the author or editor. As a result, it can be difficult to monitor, and properly attribute, what contributions were made by whom, on a particular page. A proposed solution to this was an instructor's use of categories and topics, requiring that students link to and add, rather than simply modify, the contributions of other students. Another issue was how much of a balance in terms of the tradeoff between total freedom, and total control, was ideal. Since a clear benefit of a wiki is the emphasis on free expression and on spontaneous inputs, reducing this may have a negative effect on open interaction and student contributions (Lamb, 2004).

An interesting application of the use of wikis in the classroom was the work by Bruns and Humphreys (2005), where wikis were used in a New Media Technologies course to collaboratively develop entries for an online wiki encyclopedia called the M/*Cyclopedia of New Media* , an actual live wiki resource made available online. The effort to develop entries involved over 150 students spanning six classes in the program. Feedback indicated that while students had little difficulty with using the wiki system, obstacles came about more with the writing and format of the encyclopedia entries.

Many kinds of educational exercises which are group oriented could be supported by a wiki. As mentioned, wikis are suitable for writing summary reports, doing peer reviews, examining case studies, or doing many other kinds of class-oriented group work. Examples of those which have been in use for educational purposes include knowledge management exercises (Raman et al., 2005), technology analyses, literature reviews, and journal creation (Higdon, 2006).

An important component of delivering an effective wiki exercise is to "stage" the wiki effectively, which means to set up and prepare, either as blank pages (no content or structure), presenting some kind of rudimentary outline, or by providing some "starter" content and structure. The establishment of who can read, write, and modify the wiki is another component of this, and this needs to be established before the exercise is assigned. Earlier, the concept of structure was mentioned. This could include an outline layout or structure of how the final product is to be formed, and there could be designated whether the final result is a single document, or a set of hyperlinked pages. The establishment of a "wiki etiquette policy" is also important, so that some guidelines are given as to content additions, modifications, and deletions. (Mindel and Verma, 2006).

A study of the use of a wiki for education at San Francisco State University spanned from early 2005 to mid 2006, employing ten faculty from four different colleges within the university. The use of wikis in courses ranged from business analyses, to literature review creation, to online simulation game interaction, project summary creation, or simply as a means to promote student interaction. The wiki staging and assignment/collaboration structure varied by course and purpose, but a number of conclusions were garnered from an analysis of the outcomes (Mindel and Verma, 2006).

It was found that acceptance of wikis in a class setting varied. Some students were unfamiliar with the technology, and were not enthusiastic about learning how to use it. Some preferred to use existing technologies such as e-mail and were reluctant to use the wiki unless it was specifically required. However, there also tended to be greater acceptance of wiki use in classes when training was provided (Mindel and Verma, 2006).

The results also suggested that students were more likely to add new content (aggregate) rather than to modify the work of others (collaborate). As a result, the wiki tended to receive many new entries, but students were reluctant to change the postings of others. However, the interaction would be far better if students were encouraged to do more collaboration rather than just adding new material. (Raman et al, 2005).

Overall, students found the wiki to be a useful tool given its many advantages for group collaborative work. However, given its lack of familiarity to some students, and also the need to learn the wiki system, the instructor would be advised to mandate its use, providing some form of training, and then modify the exercise to take advantage of the wiki's strengths. It may be helpful to suggest a structure to be followed, display a sample "finished" wiki, and provide guidelines for creation, modification, and other collaboration tasks (Mindel and Verma, 2006).

Critical tasks which should be given attention in terms of wikis for educational purposes include focusing on the creation and design of the wiki-based learning activity, together with providing greater support and guidance in terms of actually using the wiki. Lund and Smordal (2007) argue that while many instructors are well equipped to design learning activities, some fall short in terms of providing support and guidance to students in using the wiki most effectively to complete the assignment. The ways to help foster the learning process using the unique features of wikis using creation, reflection, and discussion could be more effectively managed by instructors.

Another study examined the use of wiki technology in German-speaking communities in Switzerland (Honegger, 2005), while Desilets et al. (2005) looked at how usable wikis are, and found that the main problems encountered related to the management of hyperlinks. Wikis were also examined from the perspective of structure: how did the use of certain templates affect the appearance and layout of wiki pages? The results suggested that they are useful in certain circumstances and could be helpful overall to end users (Hakke et al, 2005).

In summary, wikis are best suited to course and activities where there is a document, text, or other project to be worked on jointly by a class or group. In a sense, it is a tool for collaboration, and a form of groupware. The compilation of a class or group report or project, the creation of a knowledge base, or brainstorming sessions appear to be viable applications. The free and open structure, however, can fall prey to disorganization and a degradation in quality, and so it is important to have safeguards and procedures in place to ensure an effective result.

PODCASTS

Podcasts are considered to be the "most in thing" by undergraduate college students, according to a poll by CNN (CNN, 2006). This would not seem far-fatched, given the prevalence of iPods and other mobile devices, MP3s, and the popularity of "portable" audio and video. Podcast files can be obtained from any Internet-enabled computer loaded with podcatching software (such as Juice or iTunes), then listened to or watched on either the mobile device or from the computer. Both the convenience and the ubiquity of the devices and files to hear both audio and video broadcasts (vodcasts) has led to the concept of "radio on demand." Using podcasts for educational purposes has spawned a new term, that of "coursecasting." (Bongey, Cizaldo, and Kalnbach, 2006). There are numerous sources on the specifics of capturing lectures and presentations, producing them in MP3 form, and case studies on experiences working with these technologies (Chandra, 2007; Wolf, Linckels, and Meinel, 2007; Aldrich, Bell, and Batzel, 2006).

Now an inevitable question – what is a podcast? While the term "pod" and "podcast" at first mention might evoke visions of "Invasion of the Body Snatchers," for most tech people in the know, the reference to Pod is almost certainly a reference to Apple's popular and ubiquitous iPod.

However, podcasts are in actuality not what their name might imply them to be. A "podcast," a combination of "iPod" and "broadcast," neither refers to a technology specifically requiring an iPod, nor broadcasts information to users. Instead, podcasts are multimedia files (typically audio or video), which are downloaded to users on a subscription basis. Because of the potential confusion due to the use of the word "pod," some have called for the letters to mean "personal option digital" or "personal on demand, " rather than iPod.

Podcasts can be played back on any device or system which can play digital audio (typically .MP3) or video files, and are not broadcast to a large audience, in the way that television, radio, or spam e-mails are sent. Instead, they are sent to users who have specifically subscribed to a podcast service, and as such files are automatically downloaded to the user's computer when they are ready and available. In addition, podcast files are generally not streamed (as video is streamed), but rather are downloaded for later playback (Lim, 2005; Lum, 2006). Podcasts are delivered to subscribers through the use of RSS or RFD XML format media feeds, rather than more traditional forms of downloading (Descy, 2005).

Podcasts are considered to be viable educational tool for several reasons. First, because of the popularity and wide use of devices such as the iPod and similar units, it would seem like a good medium from which to distribute educational materials. Secondly, the ease with which information can be retrieved and accessed make this a good choice for students, who are using these devices on a regular basis for music and should have few technical difficulties or learning curves (Lum, 2006).

There are multiple facets to podcasts. First, there is the consumption perspective, where someone downloads the podcast and then listens or views it. This involves subscribing to a service (or enrolling in a course), finding the relevant file and then downloading and playing it. Alternately, someone can actually create podcasts; an instructor can produce lessons for students, or students can produce an assignment in the form of a podcast file (Lum, 2006).

Education is one area where the power of the podcast has been used in various ways. At Duke University, orientation material was distributed as podcasts, loaded onto iPod units, and given to students in its 2004 incoming freshman class. The units were provided not only for orientation purposes, but also for use in playing podcast lectures when the students take certain courses at the university. At Mansfield University, students were sent podcasts about various student life issues, and at Arizona State University, President Michael Crow used podcasts to deliver messages to the university community. (Lum, 2006). Purdue University has made available podcasts of various courses to students, and Apple has opened what is known as the "iTunes University." (Apple, 2006; Read, 2005).

While there appear to be sound reasons for using podcasts, there is also a theoretical basis behind the use of podcasting. This is based on cultural-historical activity theory (Engestrom, 2002), and is based on the fact that podcasting can be considered a tool that can be used to help learners to better interact with or understand a task and its environment. Vygotsky (1978) argues that the effectiveness of podcasts rests in its linkage between social/cultural influences present in the environment, and the cognitive development of the learner. Expressed another way, the concept is that since so many student have access to iPods and MP3 players, it would make sense to explore the viability of using such a device for learning and educational purposes.

Lim (2005) discussed experiences in using podcasts for teaching geography. Because of the nature of the subject, it was found that video would have a greater impact and result than audio. Students were asked to submit assignments to be created and submitted as podcasts. Overall, this helped to bring about satisfaction and interest in terms of the subject. Ractham and Zhang (2006) looked at the potential for podcasts in a classroom setting. While there are the obvious benefits in terms of being able to distributing class materials, there also are benefits in terms of contributing to social networking in the class and continuing a flow of academic knowledge. Class discussions, announcements of research activities and conferences, and also campus activities could be distributed as podcasts. Review materials could be distributed effectively on an automatic basis to interested students. In addition, the ability for students to create podcasts to be distributed to others would also be a new means of submitting assignments or expressing creativity.

Based on the experiences of Bongey, Cizardlo, and Kalnbach (2006), podcasts were employed as supplemental material in a biology course. Class lectures were recorded as MP3 files and made available to students as a way to makeup missed lectures, and also for reviewing previous lectures for greater understanding as needed. The results revealed that podcasts were a boon to students since the ability to listen to and review lectures anytime, anywhere was very helpful and welcome. The availability and usage of these podcasts is easily spread in a "viral", word of mouth manner (as in viral marketing) which transcended the boundaries of the college, to other schools and overseas. While there were some concerns whether the availability of podcasts would reduce class attendance, it was found in this study that this was, overall, not the case. Students preferred live lectures over recorded podcasts, but also liked to have podcasts as a supplemental backup resource.

Wolff (2006) provided podcast lectures as a supplement to a statistics course on the PhD level. There was significant usage of these files, and were positively received as a useful supplement to having PowerPoint lectures made available online.

Podcasts, unlike IM, blogs, and wikis, offer a greater emphasis on providing engaging auditory and visual course materials to students, rather than on collaboration and group work. While not generally a substitute for traditional lectures and knowledge presentation, they offer the benefits of easily accessible, and "digestible" course material. Whether it be excerpts from class lectures, highlights from a guest speaker, or an oral test review, the use of podcasts provides a means by which students can obtain and easily receive course-related information. In addition, it also provides a means by which students can express and present one's work, which can then be "published" and distributed in podcast format.

DISCUSSION AND CONCLUSION

The face of education, whether online, hybrid, or classroom, is constantly changing, and it is important for educators to stay abreast of the many opportunities and possibilities which are available.

In this paper, several technologies, generally termed as conversational (and Web 2.0) technologies due to their interactive and collaborative nature, were discussed in detail, together with their capabilities, benefits, and educational applications. Relevant research and case studies as they relate to classroom and educational applications were discussed.

In general, the tools discussed here fall into the class known as "conversational technologies" or "constructivist learning tools." As such, they emphasize student interaction, group learning, collaboration, rather than the more traditional classroom lecture mode. In light of this, they are more suited to educational courses or environments where the emphasis is on student communications, where students have access to technology, and where creative output and thinking is encouraged.

In the situation where a course is primarily lecture-based, or is mainly concerned with the delivery of factual or conceptual information, these tools may have limited applicability. The one application which may be helpful in this case may be for interaction to be extended outside of the classroom, through the use of instant messenger; or for supplemental materials to be distributed as podcasts. For instance, a group can use a wiki to keep a record of the group's work and contributions during the times outside of class. Students in the group can communicate informally using instant messaging. Important lecture concepts can be reviewed by listening to podcasts.

Since each of the tools has its own characteristics and suitable applications, it would be up to the educator to select those which are most appropriate to one's course and activity.

Instant messenger (IM), which is commonly used by students for personal use, has found its way not only into the business community, but also into the classroom, because of its strengths in terms of informal communications which are conducted real-time. There are some mixed results regarding the use of IM for use for education; benefits are claimed by some but there are limitations as well. The use of IM would best be employed in situations where group work, student communication, and real-time discussion would be helpful. However, it should be used cautiously, since it can be distracting, and students may end up carrying on personal, rather than course-related discussions.

Both blogs and wikis have been hailed as useful tools for education, and the specific advantages and disadvantages of each are noted and discussed. Blogs tend to be more one-sided, with an author presenting his/her information, with generally limited input from the readers and public. However the use of the technology has been used effectively to promote information sharing, and to support writing courses. The use of blogs to support online learning journals, class article/peer reviews, creating online portfolios and galleries, and for solving a specific problem or case would be advantageous. It would also appear to be a good medium for graduate research seminars, where papers and studies are analyzed and critiqued.

Wikis, which allow freedom in creation, and in editing and enhancement by others, are especially useful in collaborative situations where the input of an entire group is desired instead of a single person's contribution. In the classroom, wiki support for collaborative writing and group activities, where students contribute to the creation of a common result, would be useful. In general, any kind of learning activity which involves the collection and maintenance of knowledge or information, may benefit from the use of wiki technology.

The use of podcasts, which may include audio or video, is growing in popularity, and is being used for delivery of information to subscribers on an automatic basis. Educational podcasts, both for the delivery of audio and video-based knowledge to students, and also as a new medium for the creation and presentation of assignments, appear to have potential. While podcasts are useful as a means for publishing and distributing files of multimedia-based class materials, there also exists the potential to create new podcast content, both for educators and as a course activity. The interest in educational podcasts have been reflected by many instructors producing podcasts of lectures delivered in class, or for additional supplemental material to be distributed as audio or video files to students. This has found to be popular with students, and did not, in the studies reviewed, contribute to a drop in student class attendance.

The availability of podcasts on a subject was found to be transmitted quickly through word-of-mouth.

Clearly, the use of these new conversational technologies is allowing for the enhancement, and continued evolution of, new and innovative forms of support for teaching and learning.

FUTURE RESEARCH AREAS

Certainly, there are many benefits to the use of conversational technologies, constructivist tools, and Web 2.0 services for educational use. However, more research needs to be done, not only in terms of identifying additional types of applications and uses, but also in terms of how to more effectively identify and apply new approaches to learning with the aid of these kinds of technologies.

Some of the broader research issues which can be examined include measuring both learning, and the perceived quality of education, depending on the specific technology or tool employed. Are there measurable benefits to using a certain technology in terms of the material learned, better class performance, or more subjective factors? It would also be useful to determine, particularly when using blogs and wikis, the neutrality or bias in the entries, and how much these contribute to (or detract from) the material submitted for a course assignment. It would also be useful to study how different technologies, used for the same assignment, would vary in terms of usability, usefulness, and outcomes.

Other research areas are more technology specific. It was mentioned earlier that wikis can be a useful tool in knowledge management. The application of wikis to effective knowledge management deserves further attention, both in terms of developing or adapting the wiki structure and features to knowledge management uses, and also for identifying various kinds of user interfaces and functionality which would improve usability . The establishment of wiki-supported communities of practice is one area where the tool could possibly be useful.

Podcasting also has many areas which are ripe for further investigation. There are issues which can be explored in the areas of knowledge management, collaboration, and the adoption of podcasts. Some of the specific topics of interest include the management and sharing of knowledge using podcasts, examining whether their use actually improves learning, studying their effects on collaboration and networking, and what are the factors (or features) which would help to promote its use. As an example, if an instructor provides podcasts to students, does the availability of lectures to review increase learning and grade performance and/or retention? How heavily used are they, and if the use of these is optional, in what ways do students employ them in their studies?

Some of the newer technologies, including Twitter, a form of instant messaging, and mashups, which are integrated web resource sites, could benefit from research to determine their usefulness for various educational applications.

There also has been work on the psychological aspects of distance learning and online courses (Dickey, 2004), and a study of learners' reactions to using IM, blogs, and wikis, for example, would yield insights into its appropriateness for its further use in education. Does the use of these technologies contribute to the satisfaction of students, or is more classroom face-to-face still better? Through further research, it may be possible to determine which technologies work best in which situations, and for what kinds of exercises and assignments. Aside from satisfaction, rapport, performance, retention, and perception of learning would also outcomes to study.

The realm of these new technologies is certain ripe with a host of opportunities for both interesting and meaningful research studies.

REFERENCES

Ahn, J., Han, K. , and Han, B. (2005). Web-based education: characteristics, problems, and some solutions. *International Journal of Innovation and Learning 2005, 2*(3), 274 - 282.

Aldrich, D.,Bell, B. and Batzel, T. (2006). Automated podcasting solution expands the boundaries of the classroom, Proceedings of SIGUCCS'06, Nov 5-8, 2006, Edmonton.

Apple (2006). Today's Top Podcasts—Education Genre, www.apple.com, Sep 28, 2006.

Armstrong, L. Berry, M. and Lamshed, R. (2004). Blogs as electronic learning journals. *e-Journal of Instructional Science and Technology,* 7(1). Retrieved from http://www.usq.edu.au/electpub/e-jist/docs/Vol7_No1/CurrentPractice/Blogs.htm

Barger, J. (1997). Weblog resources FAQ, Robot Wisdom Werblog. Retrieved from http://www.robotwisdom.com/weblogs

BBC News (2005). Wikipedia survives research test. BBC News, Dec 15, 2005.

Beck, P., Kung, M., Park, Y., and Yang, S. (2004). 'E-learning architecture: challenges and mapping of individuals in an internet-based pedagogical interface,' *International Journal of Innovation and Learning 2004, 1*(3) 279-292.

Bergin, J. (2002) Teaching on the Wiki Web. *ACM SIGCSE Bulletin*, 1.

Betts, J. D. and Glogoff, S. (2004). Instruction models for using weblogs in e-learning. Proceedings of Syllabus 2004, July, San Francisco.

Blood, R. (2004). How Blogging Software Reshapes the Online Community. *Communications of the ACM, 47*(12), 53-33.

Bongey, S. Cizadlo, G. and Kalnbach, L. (2006). Explorations in course-casting. *Campus-Wide Information Systems, 23*(5), 350-367.

Bonk, C. and Zhang, K. (2006). Intrpducing the R2D2 Model: Online learning for the diverse learners of this world. *Distance Education, 27*(2), 249-264.

Bruns, A. and Humphreys, S. (2005). Wikis in Teaching and Assessment: The M/Cyclopedia Project. Proceedings WikiSim '05, Oct. 16-18, San Diego CA.

Burke, M. (2004). Instant Messaging and Course Diaries. *CCSC Northeastern Conference proceedings, May 2004, 19*(5), 207-274.

CNN (2006). Apple surpasses beer on college campuses. Retrieved June 7, 2006, from www.cnn.com

Chandra, S. (2007). Lecture Video Capture for the Masses.

Chawner, B. and Gorman, GE (2002).Wikis are wicked. *New Library World, 103*(1182/1183), 483.

Chawner, B. and Lewis, P. (2006). Wiki wiki webs: new ways to communicate in a web environment. *Information Technology and Libraries, March, 25*(1), 33-43.

Cherry, N. (1998). *Action Research: A Pathway to Action, Knowledge, and Learning.* RMIT Publishing.

Cheung, K., Lee, F., Ip, R. and Wagner, C. (2005). The development of successful online communities. *International Journal of the Computer, the Internet, and Management, 13*, 1.

Clark, J.M., & Paivio, A. (1991). Dual coding theory and education. *Educational Psychology Review, 3*, 149-210.

Cold, S. (2005). Using RSS to Enhance Student Research. *ACM SIGITE Newsletter, 3*, 1.

Copley, J. (2007). Audio and video podcasts of lectures for campus-based students. *Innovations in Education and Teaching International, 44*(4), 387-399.

Dearstyne, B. (2007). Blogs, Mashups, & Wikis, Oh My! *Information Management Journal, 41*(4), 25-33.

Dearstyne, B. (2005). Blogs: The New Information Revolution. *Information Management Journal, Sep/Oct*, 38-44.

Descy, D. (2005). Podcasting: Online media Delivery…with a Twist. *TechTrends, Sep/Oct, 49*(5), 4-6.

Desilets, A., Paquet, S. and Vinson, N. (2005). Are wikis usable? *Proceedings WikiSim'05, Oct.* 16-18, San Diego.

Dickey, M. (2004). The impact of web-logs on student perceptions of isolation and alienation in a web-based distance learning environment. *Open Learning, 19*(3), 279-291.

Doering, Beach, and O'Brien (2007). Infusing Multimodal Tools and Digital Literacies into an English Education Program, *English Education, 40*, 1.

Ducate, L. and Lomicka , L. (2005). Exploring the Blogosphere: Use of Web Logs in the Foreign Language Classroom. *Foreign Language Annals, Fall, 38*(3), 410-421.

Efimova, L. and Fiedler, S. (2004). Learning Webs: Learning in Weblog Networks. In Kommers, Isaias and Nunes (eds.) Proceedings of the IADIS Intl. Conference on Web Based Communities 2004. Lisbon: IADIS Press.

Engstrom, M. E. and Jewitt, D. (2005). Collaborative Learning the Wiki Way. *TechTrends, Nov/Dec 2005, 49*(6), 12-15.

Engestrom, Y. (2002). Cultural Historical Activity Theory. Retrieved from http://www.helsinki.fi/~jengestr/activity/6a.htm

Ferris, S.P. and Wilder, H. (2006). Uses and Potentials of Wikis in the Classroom. *Innovate, 1*(5).

Fisher, M. and Baird, D. (2005). Online learning design that fosters support, self-regulation, and retention. *Campus-Wide Information Systems, 22*(2), 88-107.

Flatley, M. (2005). Blogging for Enhanced Teaching and Learning. *Business Communication Quarterly, 68*, 2005.

Fleming, N. and Mills, C. (1992a). Not another inventory, mrater a catalyst for reflection. *To Improve the Academy, 11*, 137-149.

Fleming, N. and Mills, C. (1992b). VARK a guide to learning styles. Retrieved from http://www.vark-learn.com/English/index.asp

Fuchs-Kittowski, F. and Kohler, A. (2005). Wiki Communities in the Context of Work Processes. Proceedings WikiSim'05, Oct. 16-18, San Diego CA.

Fuchs-Kittowski, F. and Kohler, A. (2002). Knowledge Creating Communities in the Context of Work Processes. *SIGGROUP Bulletin, 23*, 3.

Fulwiler, T. (1987). *The Journal Book*. Portsmouth, NH: Boynton/Cook.

Giere, R. (2002). Distributed Cognition in Epistemic Cultures. *Philosophy of Science, 69*(4), 637-644.

Glogoff, S. (2005). Instructional Blogging: Promoting Interactivity, student-centered learning, and peer input. *Innovate, 1*, (5).

Goffe, W.and Sosin, K. (2005). Teaching with Technology: May you Live in Interesting Times. *Journal of Economic Education, 36*, (3).

Grabinger, R. S. (1996). Rich environments for active learning. In D.H. Jonassen (ed.) *Handbook of Research for Educational Communications and Technology*. New York: Macmillan, 665-692.

Gronbeck, B. Farrell, T., and Soukup, P. (1991). *Media Consciousness and Culture*, Newbury Mark CA: Sage.

Guernsey, L. (2003). In the Lecture Hall, a Geek Chorus. *The New York Times*, July 24.

Haake, A., Lukosch, S. and Schummer, T. (2005). Wiki Templates. Proceedings WikiSym'05, Oct 16-18, San Diego, 41-51.

Hargis, J. and Wilson, B. (2005). Fishing for Learning with a Podcast Net. Instruction and Research Technology, Univ. of North Florida.

Hembrooke, H. and Gay, G. (2003). The laptop and the lecture. *Journal of Computing in Higher Education, 15*(1).

Hemphill, L. and Yew, J. (2007). Twiki and WetPaint. *Proceedings of GROUP'07*, Nov 4-7, Sanibel Island FL, 273-276.

Herring, S., Scheidt, L., Bonus, S. and Wright, E (2004). Bridging the Gap: A Genre Analysis of Weblogs. Proceedings of HICSS 2004, Big Island, Hawaii.

Higdon, J. (2006). Pedagogies of Wikis. EDUCAUSE NLII Meeting, San Diego, Educause.

Higgins, C. Reeves, L. and Byrd, E. (2004). Interactive Online Journaling. Proceedings SIGUCCS'04, Oct 10-13, 2004, Baltimore MD.

Honegger, B. (2005). Wikis—a rapidly growing phenomenon in the German-Speaking School Community. Proceedings WikiSim'05, Oct. 16-18, San Diego CA.

Hsu, K. and Tanner, J. (2006). An assessment of classroom technology usage in business. *International Journal of Innovation and Learning 3*(5), 488-498.

Huang, C. D., and Behara, R.S. (2007). Outcome Drivem Experiential Learning with Web 2.0. *Journal of Information Systems Education, 18*, (3), 329-335.

Hutchins, E. (1995). *Cognition in the Wild*. Cambridge MA: MIT Press.

Johnson, D.W. and Johnson, R. T. (1998). *Cooperation and Competition*. Edina MN:Interaction.

Johnson, D.W., Johnson, R.T. and Holubec, E. (1991). *Cooperation in the Classroom* (3/e) Edina MN: Interaction.

Kinzie, S. (2005). Colleges Land Lines nearing an end. The Washington Post, Feb 12.

Kinzie, M., Whitaker,S. and Hofer, M. (2005). Instructional Uses of Instant Messaging During Classroom Lectures. *Educational Technology and Society, 8*(2), 150-160.

Kolb, D. (1984). Experiential Learning. Englewood Cliffs NJ: Prentice-Hall.

KPMG (2003). Insights from KPMG's European Knowledge Management Survey 2002/2003. KPMG Knowledge Advisory Servicees, Amsterdam.

Krisnamurthy, S. (2002). The Multidimensionality of Blog Conversations. Internet Research 3.0, Maastricht, Netherlands.

Lamb, B. (2004). Wide Open Spaces Wikis ready or Not. *Educause, Sep/Oct*, 36-48.

Lave, J. and Wenger, E. (1991). *Situated Learning, Legitimate Peripheral Participation*. New York: Cambridge University Press.

Leontiev, A. (1978). Activity, Consciousness, and Personality. Englewood Cliffs, NJ: Prentice Hall.

Leuf, B. and Cunningham, W. (2001). *The Wiki Way*. Reading, MA: AW Professional.

Levin, D. and Arafeh, S. (2002). The Digital Disconnect. Pew Charitable Trust, 2004.

Leuf, B. and Cunningham, W. (1999). The WIKI Way: Quick Collaboration on the Web. Reading, MA: Addison Wesley.

Lim, K. (2005). Now Hear This—Exploring Podcasting as a Tool in Geography Education. Nanyang Technological University.

Lindkvist, L. (2005). Knowledge Communities and Knowledge Collectivities. *Journal of Management Studies, 42*(6), 1189-1210.

Lipman, Matthew (1991). *Thinking in Education.* New York: Cambridge University Press.

Liu, C. and Chen, S. (2005). Determinants of knowledge sharing of e-learners. *International Journal of Innovation and Learning 2005, 2*(4), 434-445.

Locke, C. et al. (2000). The Cluetrain Mainfesto: The End of Business as Usual, Cambridge, MA: Perseus.

Louridas, P. (2006). Using wikis in software development. IEEE Software, Mar/Apr, 88-91.

Lum, L. (2006). The Power of Podcasting. *Diverse issues in Higher education, 23*(2), 32.

Lund, A. and Smordal, O. (2006). Is There Space for a Teacher in a Wiki? Proceedings of WikiSym'06, August 21-23, Odense Denmark.

Martindale, T. and Wiley, D. (2005). Using Weblogs in Scholarship and Teaching *TechTrends, 49*, 2.

Millard, D. and Ross, M. (2006). Web 2.0: Hypertext by any other name? Proceedings of HT'06, August 22-25, Odense Denmark, 27-30.

Mindel, J. and Verma, S. (2006). Wikis for teaching and learning. *Communications of the AIS, 18,* 1.

Mock, K. (2002). The Use of Internet Tools to Supplement Communication in the Classroom. *CCSC Northeastern Conference proceedings, 17*(2), 14-21.

Nardi, B. and Bradner, E. (2000). Interaction and Outeraction. Proceedings of CSCW'00, Dec. 2-6, 2000, Philadelphia PA.

Nardi, B., Schiabo, D.J. , Gumbrecht, M. and Swartz, L (2004). Why We Blog. *Communications of the ACM, 47*(12), 41-46.

Nardi, B., Schiano, D.J. and Sumbrecht, M. (2004). Blogging as Social activity. Proceedings of CSCW 2004, Chicago, Nov. 6-10, 222-231.

Nichols, R. and Anderer (2007). Using Video Podcasts Outside the Classroom in *Proceedings of SIGUCCS'07,* Oct 7-10, 2007, Orlando FL, 277-279.

Nussbaum, E. (2004). My so-called blog. *The New York Times*, Jan 11.

Oliverm K. Leveraging Web 2.0 in the redesign of a Graduate-Level Technology Integration Course. *TechTrends, 51*(5), 55-61.

O'Neill, M. (2005). Automated Use of a Wiki for Collaborative Lecture Notes. Proceedings SIGCSE'05, Feb 23-27, 2005, 267-271.

Ong, W. (1982). *Orality and Literacy.* New York: Routledge.

O'Reilly, T. (2005). What is Web 2.0. Retrieved September, 2005, from http://www.oreilly.com

Prensky, M (2001). Digital Natives, Digital Immigrants. *On the Horizon NCB University Press, 9*(5).

Quible. (2005). Blogs: A Natural in Business Communication Courses.

Ractham, P. and Zhang, X. (2006). Podcasting in Academia. Proceedings SIGMIS-CPR'06, April 13-15.

Raman, M, Ryan, T. and Olfman, L. (2005). Designing Knowledge Management Systems for Teaching and Learning with Wiki Technology. *Journal of Information Systems Education, 16*(3), 311-320.

Read, B . (2005). Lectures on the Go. *The Chronicle of Higher Education, 52*(10), A39.

Schaino, D., Nardi, B. Gumbrecht, M. and Swartz, L. (2004). Blogging by the Rest of Us. Proceedings CHI 2004, 1143-1146.

Seigenthaler, J. (2005). A false Wikipedia biography. *USA Today,* Nov 29.

Seitzinger, J. (2006). Be constructive: blogs, podcasts, and wikis as constructivist learning tools. *Learning Solutions,* July.

Seng, J. and Lin, S. (2004). A mobility and knowledge-centric e-learning application design method. *International Journal of Innovation and Learning 2004, 1*(3) 293-311.

Stahl, G. (2006). *Group Cognition.* Cambridge MA: MIT Press.

Stephens, M. (2007). Messaging in a 2.0 World. *Library Technology Reports, 43*(5), 62-66.

Vygotsky, L. (1986). Thought and Cognition, Cambridge MA: MIT Press.

Wagner, C. (2006). Breaking the Knowledge Acquisition Bottleneck through Conversational Knowledge Management. *Information Management Resources Journal, 19*(1), 70-83.

Wagner, C. (2004). Wiki: A Technology for Conversational Knowledge Management and Group Collaboration. *Communications of the AIS, 13,* 265-289.

Walker, J. (2005). Weblog. Definition from the Routledge Encyclopedia of Narrative Theory.

Wallace, M. (2005). Notes towards a literacy for the digital age. Retrieved from http://uclaccc.ucla.edu/articles/article-digitalage.htm

Wang, J. and Fang, Y. (2005). Benefits of cooperative learning in Weblog networks. Proceedings APAMALL2005 and ROCMELIA2005, December 9-11, 2005, Kun Shan University.

Wenden, A. (1991). *Learning Strategies for Learner Autonomy.* New York: Prentice-Hall

Wickramasinghe, N. and Lichtenstein, S. (2006). Supporting knowledge creation with e-mail. *International Journal of Innovation and Learning, 3*(4), 416-426.

Wikipedia (2006). Mobile Computing, definition. Retrieved from http://en.wikipedia.org/wiki/Mobile_computing, accessed 8/25/2006

Wolf, K. Linkels, S. and Meinel, C. (2007). Teleteaching Anywhere Solution Kit (tele-TASK) Goes Mobile, In Proceedings of SIGUCCS'07, October 7-10, 2007, 366-371.

Wolff, T. (2006). Podcasting Made Simple. In Proceedings of SIGUCCS'06, Edmonton Alberta Canada, Nov 5-8 2006, 413-417.

Chapter XVIII
Assessing Online Discussion Forum Participation

Matthew Shaul
Kennesaw State University, USA

ABSTRACT

As a socially constructive learning tool, discussion forums remain central to online education. They have continued to evolve in functionality, acquiring ever-increasing usability features. However, development has lagged in providing instructors the means to assess student work in forums. The author submits an overview of his software program that provides instructors with the means to evaluate forum work quickly, easily, and repeatedly. The software accomplishes this by accessing the forums' underlying database, searching for manifest and latent data, and calculating data associated with an array of metrics. This is a Web-based tool built on Open Source and standards-based languages, providing opportunities to port the program to numerous Learning Management Systems. It is the intention of this author to provide this tool, when completed, for such use as a free, Open Source tool. Interested parties may e-mail the author for progress updates. Currently, however, further work on the project must await the completion of another project, the author's dissertation.

INTRODUCTION

Learning management systems (LMS) continue receiving expanded toolsets and quickly assimilating new Web-technologies to provide users an increasingly interactive, richer experience. Chat, streaming media, "blogs," "video-casting," and "podcasting" found their way into online educational settings soon after being generally accepted on the Internet. Yet, discussion forums, an old (in Internet time) technology, seemingly remain the core from which many instructors build online classes. These technological descendants from long-ago bulletin boards and listservs, one of the earliest tools integrated into online education, remain central to the design and success of many distance education courses.

More so than the newer technologies, discussion forums approximate a replacement for the give-and-take of the brick-and-mortar experience, mimicking many-to-many discussions found in traditional classrooms. In addition, the recognizable conversational structure reflected visually in the tree-like output, simplicity and flexibility of the tool likely contribute to its continued success and acceptance, granting users an immediate sense of familiarity. The importance of such comforting effects cannot be discounted, especially in a field still relatively new.

Discussion Questions

However, despite the history and wide, though not full, acceptance of the importance and use of forums, lack of awareness on how best to use them persists. Note that this unawareness does not pertain to the implementation of forums, or designing them to encourage adoption. In fact, Markel (2001) notes that forums have developed beyond simple, plain text message boxes, incorporating emoticons, HTML formatting, images, and hyperlinks to provide a more enticing tool to draw students into their use. Yet, while these features encourage participation, there is no clear way for instructors trying to devise effective forum evaluation schemes.

This article, therefore, examines forum technology assessment. Given the importance of assessment in learning, it is apparent that such a widely used distance leaning tool must provide instructors with sound options for evaluating student work. Moreover, effective assessment options, with associated feedback, provide the added benefit of encouraging an increase in student postings, thus adding to the forums' potency. Yeh (2005) notes that student participation increases as instructors place an importance on posting by assigning grades to forum use. This is unsurprising, as one would expect graded assignments to garner more attention from students than non-graded activities. Swan (2001) finds this true as students calculate reward versus effort when determining whether to participate in forums. Forums with a larger percentage of influence on grades receive more use. However, while most LMS do provide instructors some means of forum assessment, current tools remain either overly limited or too time consuming to use.

Forum Types

Note, different forum types exist, and not all contain posts needing assessment. The first might be termed "social" forums. These forums furnish students with an informal area to discuss class- or non-class-related matters. Often, instructors state they will not view these forums' contents, thus creating a space in which students are free to speak openly, criticizing or praising the instructor, course, curriculum, or school without concern the comments will influence grading. Instructors often refer to these forums as "water coolers" or "student lounges." While these forums may provide students social benefit, instructors almost never assess them (Nelson et al., 2005).

A second type of forum might be labeled "general discussion." Like the social forums, these tend toward a free flowing, less structured style. However, unlike the social forums, these pertain to the course material and are less informal. Instructors may select broad topics or simply ask students to post any course related questions or material. Whereas social forums resemble hallway discussions among students, general discussion forums mimic an open question discussion in the classroom. Like their classroom counterparts, online general discussion students might receive grades based upon participation, insight, argument, initiative, and other factors.

The last forum type considered here is the "topic driven" forum. These forums are the most structured in terms of content and correspond to classroom assignments in which the instructor picks topics and expects students to come prepared to debate and defend stances. Similar to general discussion forums, instructors may wish to assess topic driven forums as they would in the classroom—on participation, style, scholarship, argument, insight, and other subjective and objective factors.

Quantitative and Objective Forum Assessment

As stated previously, forums currently do possess methods to assess student work. The LMS and the instructor provide quantitative and objective means of evaluation. The concern, however, is whether these methods offer instructors the tools needed to accurately and meaningfully measure student work.

One of the most basic methods, in fact, exists in most LMS's, and is simply a count of each students' postings, allowing the instructor to assess based upon predetermined levels matched to grades. For instance, a minimum of five posts per week may warrant an "A," four a "B," and so on. The advantages of this method appear plentiful. For one, students easily understand the measurement and are clear on expectations, which Dennen (2005) notes may promote participation. In addition, assessing is easy for instructors, as the LMS likely provides reports listing a count total for each student. Instructors need do no more than run weekly reports and award grades accordingly.

Disadvantages, however, exist as well. Forum threads often contain a number of insignificant posts consisting of little more than "me too" or "well said." Since the LMS in this case only considers counts, these posts weigh equally with well-written, researched posts, a situation many students and instructors would find unfair.

Additionally, relying solely on counts encourages post submissions, but not necessarily forum participation, if the expectation is that students will also read their classmates' offerings. As a result, the forum may devolve into a writing exercise with each student posting detached, unrelated essays rather than interrelated posts, building an interconnected discourse (Dennen, 2005). Consequently, the forum in this case remains underused as an interactive, communicative learning tool (Dunlap, 2005) and becomes little more than a channel to submit electronic topical papers.

Instructors often address this shortcoming with a seemingly sensible solution: requiring students to comment or respond to a few of their peers' posts each week, in addition to submitting original posts. The requirements remain simple for the students to follow and easy for instructors tally. The new rule's intent, of course, is to forge threads from the posts, and subsequently, discussions from the threads, by mandating a level of interactivity. Although the idea seems reasonable and does produce at the least the instructor-assigned degree of interaction, the threads may consist of little more than this minimum. Students post to expectations (Dennen, 2005) and without more guidance than simple quantitative requirements will post the minimum type of reply necessary. Moreover, the response or comment posts can lack a depth matching the original post, and a repetitive pattern quickly ensues in which responders follow up initial posts with inconsequential replies, adding little to the aggregate knowledge (Ivankova & Stick, 2005). Thus, the forums attain some interaction, but the requirement change may not bring about the desired higher-level discourse or debate.

The next step, then, is to raise the degree of discussion while maintaining an easy means to measure student work. Some instructors attempt this by requiring all posts, including the response postings, to have citations from peer-reviewed publications. This seeks to infuse a degree of scholarship, thus raising the discussion level. At minimum, the hope is that the new rule forces students to research and form their responses around that research.

This remains a simple quantitative and objective measure. Instructors may alter the number of citations needed, or the sources allowed, but the measure remains merely a type of count. There is the beginning of significant change, however. Note, though in essence simple, such a measure is beyond the abilities of typical LMS because they lack the capability to differentiate a citation from any other string of text. Thus, they cannot count or note citations, and the burden of tallying this assessment, then, moves for the first time from the LMS to the instructor.

Qualitative and Subjective Forum Assessment

Part of the allure of quantitative and objective assessment of forum work is the simplicity for the students and the instructors. For the instructors, this translates into speed and timesaving, as quickly accessed reports reveal each student's standing; grading is very straightforward. Yet, perhaps depending upon the student's age and grade level, relying solely upon quantitative measures may not offer the clearest evaluation of performance. In the case of tallying postings, the measure is ultimately an electronic attendance sheet, tracking whether students checked in and "participated." However, at the undergraduate level and higher, instructors likely prefer grading on criteria that are more substantial. Absent in most pure counting methods, subjective measures such as writing style, initiative, strength of argument, and originality offer a more robust grading rubric.

In fact, many such rubrics exist. For example, Edelstein and Edwards (2002) devised a forum assessment rubric called "Assessing Effectiveness of Student Participation in Online Discussions." This rubric consists of five categories: Promptness and Initiative; Delivery of Post; Relevance of Post; Expression within the Post; and Contribution to the Learning Community. The instructor is to consider each student's work as a whole, evaluating each post on a scale of one to four, with one the lowest, in each of these categories. Edelstein and Edwards provide explanations for each rating of each category to aid the assessor. For instance, a one rating in the "Relevance of Post" category translates to "Posts topics that do not relate to the discussion content; makes short or irrelevant remarks."

Such qualitative and subjective measures afford the instructor far more leeway in assessment than mere counts. Whereas a "me too" post is awarded standard points in a count assessment, such a post would draw the lowest score in the just mentioned Edelstein and Edwards category. These subjective ratings also provide the instructors the means to use their judgment, allowing them to weigh factors differently, perhaps for instance, heavily rewarding creativity in thinking or writing style, while affording less weight to the inclusion of citations.

Additionally, by moving to richer assessment rubrics, instructors lead students to submit richer messages, which elicit richer peer responses and lay the foundation of scholarly discussion. These subjective assessments attempt to measure higher-level learning, including analysis, synthesis, and evaluation and direct students to post accordingly. Bhagyavati, Kurkovsky, and Whitehead (2005) note that students adjust their posts to meet these expectations, and the forums' quality wholly moves upward. As stated earlier, students adjust their work to meet expectations.

Unfortunately, the instructor time expended is a considerable drawback of detailed rubric scoring and subjective qualitative assessment. Consider the time needed to work through a five-category rubric in a class of 20 students in which postings could easily total between 500 and 1000. Unsurprisingly, instructor fatigue becomes a concern as forum management evolves into a significant time- and effort-intensive activity (Dunlap, 2005).

Another drawback, often overlooked, is the return to manual effort for the assessor such rubrics require. This is unfortunate considering most LMS's run atop powerful computers very capable of intense calculations. Moreover, relational databases contain the actual forum data, availing a trove of information to SQL querying. Forgoing such computational resources is inefficient and counterintuitive considering online courses exist only through the use of such advanced technologies.

Proposed Solution

The need for a forum assessment aid seems clear. The requirements are also apparent—develop a tool to assist instructors in assessing forums using measurements beyond the simple quantitative counts. The tool should incorporate some degree of qualitative or subjective measure and should utilize the power of the host computer. Certainly, the tool should provide a simple, usable interface to encourage adoption.

Current Packages

Ideally, forum software would offer an integrated tool to assist in assessing forum messages; unfortunately, this is presently not well advanced. However, Wu and Chen (2005) have developed software that attempts automating assessment of student forum work. Their software is similar to Qualrus, mentioned by Gilbert (2005), in that it parses written submissions and grades them based on the instructor's preprogrammed criteria. Whereas Qualrus and other similar essay-grading software purportedly evaluate style, grammar, structure, quality, and argument, Wu and Chen's software appears to measure fewer writing criteria. Instead, the software algorithmically determines knowledge density, or message quality, using instructor-specified keywords. The software then accesses the forum's database, and using the message field's length and participant-sorted message counts, assigns values for student effort and participation.

Wu and Chen's method attempts less to assess message content than essay grading software such as Qualrus, and examines the database for supplemental information; this may be preferable for forum evaluation. If the instructor uses forums for socially constructive learning, for instance, much of the "construction" exists not individually in each message, but in the forum structure, in the interfaces between and among messages, captured latently in the database. Any assessment must consider the forum as a whole, messages inter-related in a web (Schellens & Valcke, 2004). To appraise each message separately and solely limits the instructor to an assessment of independent "mini essays," each presumably unaffected by others' postings.

Developing a Solution

Although Wu and Chen (2005) propose gleaning modest amounts of data from among the tables, this author believes the tables and the relations between the tables hold enormous amounts of valuable information. As an example, Dringus and Ellis (2005) believe mining the database has significant potential to reveal information hidden, for instance, in timestamps and sequence numbers. These numbers, referred to here as "manifest" information, lie in fields readily available to SQL queries. However, properly manipulating this manifest information potentially reveals additional "latent" information that is also useful in assessing the forums' contents.

For instance, message timestamps reveal relative temporal information that may show evidence of student initiative by indicating first postings or responses. Additionally, a post with many responses, evidenced by subsequent sequence numbers and parent-child pointers found in the database, may show a post's effectiveness. Certainly, for instance, it is arguable in a socially constructive environment that a student post, eliciting numerous responses and thus acting as the impetus for peer involvement, warrants a positive assessment to some degree. Yet, a methodology focused wholly on the individual message, and not tuned to consider the message "tree," misses this information entirely.

The solution, therefore, appears to require first the formation of meaningful measures, followed by a search of the database fields and relationships for latent and manifest data supporting the measures, and finally the development of the algorithm and SQL to pull the data from the database. The author has proceeded through these steps, designed such a program, and has developed a simple Web-based interface, allowing instructors to quickly and repeatedly use the tool. The hope is that the simplicity will compel instructors to use the tool often to provide students frequent feedback on forum performance.

Following is an overview of the measures used by this author, some insight as to the reason each is included, and a brief explanation of how the program attains the data:

- **Initiative:** Being first to provide an opinion is not easy, so the system rewards students in two situations: one if the student starts a thread, and two, if the student is the first to respond to an instructor post. The system captures this in the posts timestamps.
- **Effectiveness-Depth:** In a socially constructive environment, one measure of effectiveness is the amount of involvement elicited. Thus, the system calculates the number of responses and sub-forums spawned. Students are rewarded for being able to draw others into a discussion.
- **Effectiveness-Breadth:** This is similar to the previous measure in that it rewards students for educing classmate responses. However, a "deep" thread (many responses) may involve only two or three students, perhaps still valuable but less of an indication of the enthusiasm for thread than the number of responses indicate. On the other hand, this measure calculates and rewards effectiveness by tallying unique responders, thus revealing the scope of the enthusiasm.
- **Value:** In the current system, students can anonymously rate each other's posts on a one-to-five scale representing "Not Valuable" to "Very Valuable." The system then measures value by averaging the peer rating a student's posts receive. This is a very important measure since each student defines value for each post differently. What may seem to many students a simple, low-value posting may clarify a point and offer high value to other students. This metric accounts for the possible variances.
- **Timeliness:** This measure is best used when the instructor does not impose a posting deadline on threads. Rather, students continue threads for as long as there is interest. In this way, students may revisit older threads as they learn more, or as they come upon new information. However, differences exist between legitimate, interesting late posts and messages submitted well after a thread is exhausted. The system recognizes this by calculating timeliness as the standard deviation of a thread's posting time, and assumes interesting late posts will draw responses and move the standard deviation toward itself. Merely late posts will not alter the standard deviation and will not receive credit.
- **Participation:** Post count is not a good measure of participation. For instance, one student may log in and post several times a week over the extent of a course. Conversely, a peer may log in at the last minute and post an equal number of messages. Clearly, they participated at different levels. To reward consistent participation, this system determines whether each student's Average Time Between Posts (ATBP) is within the standard deviation of the class's ATBP. Students whose ATBP is outside do not receive credit.

- **Scholarship:** Instructors may expect posts to contain certain keywords, phrases, or names. Additionally, they may require citations. The system searches each post for words from an instructor-determined list and scans for citations. Posts receive credit for containing either keywords or citations. The citation search is not flawless, as the system uses regular expression patterns to match what are likely citations.
- **Style:** Perhaps misnamed, the system does not examine the prose for writing style in this metric, but instead performs a word count. Students receive points for posts above a specified count but below another count. The attempt is not to reward short, unsubstantial posts, or long, rambling posts. Therefore, more precisely, the metric attempts to encourage succinct, concise writing.
- **Instructor Points:** While the previous metrics seek to cull needed data directly from the database, certainly some subjective measures cannot be calculated from the tables' fields. Thus, the system provides the instructor an opportunity to add or subtract points from each student's assessment. Therefore, for instance, the instructor may reward a student who has consistently put forth original arguments, or who has carried discussions to a higher level. Likewise, an instructor may subtract points from a student who has done well, but has consistently used poor grammar or spelling.

These metrics are varied enough to provide a flexible array of point opportunities to students. Some students may be comfortable, or in a fortunate position to attain points from the Initiative measure. Others may better at writing concisely or with a style the instructor appreciates, and positioned to acquire Style points. Still others may have the time to post frequently and receive points for Participation. All can pursue Value or Scholarship points.

The author's system seeks to add further flexibility for the instructor as well. Rather than considering each measure equally, the tool allows instructors to weight each metric to their preference. For example, one instructor may find the Effectiveness-Breadth measure most compelling in assessing forum work. Therefore, the instructor may weight this as 40% of the overall assessment calculation. On the other hand, another instructor may weight this at 10%, and weight Instructor Points much higher. By allowing varying weights, the system incorporates another degree of instructor subjectivity into the assessment.

An instructor configures the desired weights on a simple Web page that lists each measurement with an accompanying explanation for the measure. Each is followed by a dropdown box listing numbers from zero to 100, and the instructor weights each measure so the total of all selected weights, added together, equals 100. This screen also has a textbox input for keywords, used in text searches, in the event the scholarship measure is chosen. Naturally, instructors weight measures they feel important for assessment highly and those they consider less important lower or zero.

If the instructor selects Instructor Points as a measure, the next screen displays a class roster with a dropdown box with numbers from negative 100 to positive 100 associated with each student name. Here, the instructor assigns positive or negative points to each student. Note, these are the actual points; the weight of these Instructor Points was configured on the first screen.

The final screen, whether or not Instructor Points is used, displays each student's calculated rating. It is important to note the system scores students as a percentile rank of all points awarded and not from a finite allotment of points. Thus, first the system determines total points awarded for everyone, and then ranks each student based upon the student's earned points. In this way, students are not aiming to amass a specific number of points for associated grades (i.e., 100 points for a C, 200 for a B, 300 for an A), but instead realize they must maintain pace with classmates through participation by accumulating as many points as possible. As classmates participate, the pool of awarded points grows, compelling students to continue to post, less their awarded allotment shrinks as a percentage.

CONCLUSION

Assessing student work in discussion forums remains difficult for busy instructors, especially if one wishes to use measures beyond simple tallies. However, the author's software solution offers some hope by providing an easy to use, flexible solution. The tool is Web-based, written in Open Source and standards-based languages that should provide the basis for easy portability. Interested parties can e-mail the author for progress updates. Unfortunately, for the moment other project requirements demand a—hopefully brief—respite from the project. When finished, though, it will be freely available as Open Source.

Because of use of the LMS's underlying hardware and software, the tool performs its calculations quickly. Additionally, because of the simple interface, the tool encourages instructors to run assessment reports often, thus enabling the instructor to provide continual feedback to students. Consequently, the forums rise to higher levels of discussion and debate and become true socially constructive learning environments as students learn to post, read, and respond accordingly.

REFERENCES

Bhagyavati, Kurkovsky, S., & Whitehead, C. C. (2005). Using asynchronous discussions to enhance student participation in CS courses. In *Proceedings of the 36th SIGCSE Technical Symposium on Computer Science Education* (pp. 111-115). Retrieved December 22, 2005, from the ACM database.

Dennen, V. P. (2005). From message posting to learning dialogues: Factors affecting learner participation in asynchronous discussion. *Distance Education, 26*(1). Retrieved December 23, 2005, from the ProQuest database.

Dringus, L., & Ellis, T. (2005). Using data mining as a strategy for assessing asynchronous discussion forums. *Computers & Education, 45*(1), 141-160. Retrieved December 8, 2005, from the Science Direct database.

Dunlap, J. C. (2005). Workload reduction in online course: Getting some shuteye. *Performance Improvement, 44*(5). Retrieved December 8, 2005, from the ProQuest database.

Edelstein, S., & Edwards, J. (2002). If you build it they will come: Building learning communities through threaded discussions. *ELearn, 2002*(4). Retrieved December 12, 2005, from the ACM database.

Gilbert, A. (2005). *Teachers leave grading up to the computer.* CNET News.com. Retrieved December 30, 2005, from http://news.com.com/Teachers+leave+grading+up+to+the+computer/2100-1032_3-5659366.html?part=rss&tag=5659366&subj=news

Ivankova, N. V., & Stick, S. L. (2005). Collegiality and community: Building as a means for sustaining persistence in the computer-mediated asynchronous learning environment. *Online Journal of Distance Learning Administration, 8*(3). Retrieved December 15, 2005, from http://www.westga.edu/%7Edistance/ojdla/fall83/ivankova83.htm

Markel, S. L. (2001). Technology and education online discussion forums: It's in the response. *Online Journal of Distance Learning Administration, 4*(2). Retrieved February 23, 2006, from http://www.westga.edu/~distance/ojdla/summer2001/markel42.pdf

Nelson, M., Bhagyavati, Miles, G., Settle, A., Shaffer, D., Watts, J., et al. (2005). Online teaching practices (both best and worst). Panel discussion. *Journal of Computing Sciences in Colleges, 21*(2). Retrieved December 14, 2005, from the ACM database.

Schellens, T., & Valcke, M. (2004). Fostering knowledge construction in university students through asynchronous discussion groups. *Computers & Education, 46*(4). Retrieved June 20, 2006, from the ScienceDirect database.

Swan, K. (2001). Virtual interaction: Design factors affecting student satisfaction and perceived learning in asynchronous online courses. *Distance Education, 22*(2), 306-331. Retrieved February 18, 2006, from the ProQuest database.

Wu, Y. B., & Chen, X. (2005). E-learning assessment through textual analysis of class discussions. In *Proceedings of the Fifth IEEE International Conference on Advanced Learning Technologies* (pp. 388-390). Retrieved January 8, 2006, from the IEEE database.

Yeh, H. (2005). The use of instructor's feedback and grading in enhancing students' participation in asynchronous online discussion. In *Proceedings of the Fifth IEEE International Conference on Advanced Learning Technologies*. Retrieved December 14, 2005, from the IEEE database.

This work was previously published in International Journal of Information and Communication Technology Education, Vol. 3, Issue 3, edited by L. Tomei, pp. 39-46, copyright 2007 by IGI Publishing (an imprint of IGI Global).

Section III.b
Synchronous Tools

Chapter XIX
Synchronous Hybrid E-Learning:
Empirical Comparison with Asynchronous and Traditional Classrooms

Solomon Negash
Kennesaw State University, USA

Michelle Emerson
Kennesaw State University, USA

John Vandegrieft
Blackstone & Cullen, Inc., USA

ABSTRACT

An empirical analysis was conducted to compare synchronous hybrid e-Learning environment with traditional classrooms. Empirical study with 165 students from eight colleges at a large public university was used. The results show (1) contrary to prior research students taking unfamiliar subjects online, in synchronous format, were satisfied; (2) no statistical difference was found in student satisfaction between synchronous online and traditional face-to-face formats; and (3) overall satisfaction, measured by intent to us the same format again, found no statistical difference between the two formats.

INTRODUCTION

Advances in technology have increased the popularity of virtual learning environments (VLEs) in both the educational arena and corporate world (Alavi, Marakas, & Yoo, 2002; Dagada & Jakovljevic, 2004). VLEs are defined as "computer-based environments that are relatively open systems which allow interactions and encounters with other participants and providing access to a wide range of resources" (Piccoli, et al., 2001, p. 402; Wilson, 1996). Advances in information technology (IT) continually expand the capabilities of VLEs (Seng & Al-Hawamdeh, 2001).

VLEs can be characterized by six dimensions which distinguish them from traditional classrooms and computer aided instruction: time, place, space, technology, interaction, and control (Piccoli, Ahmad, & Ives, 2001). VLEs are comprised of different formats including synchronous and asynchronous. This study delineates between synchronous and asynchronous learning to better understand student satisfaction when using online learning. The Piccoli, et al study implicitly refers to asynchronous instruction delivery format when defining the six dimensions. For this reason we refer to VLE in the Piccoli, et al. study as asynchronous. Research still remains to uncover the effectiveness of these environments and whether the differences alter student learning outcomes (Alavi & Leidner, 2001; Alavi *et al.*, 2002; Hodges, 2005; Seng & Al-Hawamdeh, 2001)

Prior research indicates that students are less satisfied when using asynchronous e-Learning format for unfamiliar topics like databases and more satisfied using asynchronous format for more familiar topics like word processing (Piccoli, et al., 2001). This study compares synchronous e-Learning format with traditional formats to understand the research question: Are students satisfied with unfamiliar topics when using synchronous e-Learning format?

This research presents the findings of an empirical study conducted to compare a VLE using synchronous hybrid eLearning environment with a traditional classroom setting. Synchronous hybrid e-Learning environment is one where portions of the interaction among the participants takes place in real-time, albeit virtually, and the remaining portion is taught in a traditional (face-to-face) classroom format. The next section presents the research background followed by hypothesis definition, research design, results, discussion, limitations of the study, future research direction, and conclusion.

BACKGROUND

In this section we discuss the research background on hybrid e-Learning followed by how the six VLE dimensions differ when using synchronous VLE. We conclude this section with a discussion on the differences between synchronous and asynchronous VLEs.

Hybrid E-Learning

Online learning is an ongoing focus of researchers, overall there is a need to gain a deeper understanding into the effectiveness of online learning (Alavi & Leidner, 2001; Alavi et al., 2002). To address this issue recent research has focused on understanding the effectiveness of different pedagogical approaches in different content areas; as a result, there have been a number of studies examining hybrid approaches. A hybrid approach involves providing content in a variety of formats with a mixture of online and in-class instruction. Synchronous hybrid e-Learning combines virtual and face-to-face learning. The ratio of virtual and face-to-face class sessions vary greatly; some include face-to-face portion only the first day of class, others equally divided the semester between virtual and face-to-face.

Current research provides a mixed response on the subject of advantages and disadvantages of using a hybrid approach to teaching. Comparisons of learning outcomes for courses taught in hybrid, virtual, and traditional classrooms is needed; furthermore, research in this area highlights the importance of self regulation (ability to control actions and decisions) and learning environment control (Hodges, 2005; Piccoli et al., 2001).

Webb, Gill, & Poe (2005) examined the differences between pure versus hybrid approaches to teaching using the case method and found that students' online discussions may enhance learning in case methods when taught using a hybrid approach. In a comparison of traditional and technology-assisted instruction methods in eight sections of a business communications class, where live versus hybrid formats were compared, an improvement in writing skills was found in students who participated in the hybrid course, particularly for whom English is a second language (Sauers & Walker, 2004). McCray (2000) found courses which combine online learning with the traditional classroom can help students become more engaged in rich classroom interactions by appealing to different learning styles through variety in content delivery.

Piccoli et al. (2001) examined differences in learning outcomes for students training in basic information technology skills in a traditional classroom with those in asynchronous classroom. No major differences were found in the performance of students in the two environments.

Brown & Liedholm (2002) compared the outcomes of three different formats (traditional classrooms, hybrid, and asynchronous) for a principles of microeconomics course and found that the students in the asynchronous course did not perform as well as the students in the traditional classroom and that differences between students in the traditional and hybrid sections, versus those in the asynchronous section, were shown to increase with the complexity of the subject matter. The researchers also concluded that ultimately there is some form of penalty for selecting a course that is completely asynchronous.

Synchronous View of VLE Dimensions

Piccoli et al. defined the six learning environment dimensions and how asynchronous e-Learning apply as shown in Table 1. This study provides how these dimensions apply in synchronous formats.

Time Dimension

In comparison to traditional classrooms "when instruction is delivered asynchronously in [asynchronous format], participants retain control as to when they engage in the learning experience. Learners determine the time and pace of instruction" (Piccoli et al., 2001, p. 404), the time constraints for participants in Asynchronous formats are therefore removed (Piccoli et al., 2001). In synchronous formats participants have to be present, albeit virtually, at the time of instruction delivery. Participants in synchronous formats therefore do not have control over when they can engage in the learning experience. Learners' ability to be able to control their engagement in the learning experience is often cited as an advantage in asynchronous learning, this option is not available in synchronous formats. In the current study recording of each class session was

Table 1. Classification of dimensions of learning environments

Dimension	Definition
Time	The time of instruction. Asynchronous formats free participant from time constraints.
Place	The physical location of instruction. Asynchronous formats free participant from geographical constraints.
Space	The collection of materials and resources available to the learner. Asynchronous formats provide a wide array of resources.
Technology	The collection of tools used to deliver learning material and to facilitate communication among participants.
Interaction	The degree of contact and educational exchange among learners and between learners and instructors.
Control	The extent to which the learner can control the instructional presentation. Control is a continuum enabling the design of varying degree of learner control (Newkirk, 1973).

made available to students giving them the ability to access the archived session at any time. Consequently, the "anytime" benefit of asynchronous formats was achieved.

Place

Participation can be done from anyplace. This frees participants from geographical constraints. In this case, synchronous and asynchronous e-Learning are identical.

Space

The collections of materials and resources available to the learner are comparable in asynchronous and synchronous formats. The difference however is on the level of access, while both synchronous and asynchronous formats are online in synchronous formats instruction is being delivered live so learner have to differ much of their access to the collections of material and resources until after the instruction.

Technology

The technologies available in both synchronous and asynchronous formats are similar. The main difference here is synchronous formats use technologies that support real-time delivery in addition to all the technologies used in asynchronous formats.

Interaction

Asynchronous interaction is time delayed. Synchronous interaction on the other hand has real time interaction between learner and instructor and among learners.

Control Dimension

Control in an asynchronous format is defined as: "The extent to which the learner can control the instructional presentation." (Piccoli, et al., 2001, p. 404). When compared to traditional classrooms, asynchronous formats allow participants to control the pace and sequence of material during instruction (Piccoli et al., 2001). In synchronous formats participants' control over the pace and sequence of instruction material is somewhat limited. In the current study for example, while participants were able to move around the instruction material in the sequence and pace they chose, they were re-directed to the instructor-led page each time the instructor changed the page. In the archived session however, participants had control over the pace and sequence just like in the asynchronous format.

Differences between Synchronous and Asynchronous VLEs

Piccoli, et al. (2001) state that participants of asynchronous VLEs have difficulty managing the high degree of control, feel overburdened by the shift of responsibility and control to the learner, feel isolated, experience anxiety and encounter difficulty in time management. These challenges have not yet been found to be evident in synchronous VLEs.

Difficulty Managing the High Degree of Control

Asynchronous VLEs avail a high degree of control for participants. In traditional classrooms the instructor provides direction and structure. In asynchronous VLEs participants are challenged when managing the high degree of control in the absence of the instructor direction and structure. Synchronous VLEs on the other hand, enable students to participate in a familiar strategy consisting of instructor direction and structure, in addition to having the benefit of archived sessions for further elaboration.

Overburdened by the Shift of Responsibility and Control

In asynchronous VLEs the learner is responsible for finding solutions when the concept is not clear. In synchronous VLE on the other hand, learners can ask questions at the time of instruction delivery.

Feeling Isolated

Learners in asynchronous VLEs go through instruction material on their own and learn at their own pace. On the other hand, video and audio connections in synchronous VLEs allow learners to see and hear the instructor and fellow learners during instruction delivery. Thus synchronous VLEs overcome the challenge of feeling isolated.

Experiencing Anxiety

In synchronous VLEs learners who experience anxiety at the time of instruction delivery are able to get immediate assistance from the instructor through audio and video communication. Getting assistance at the time of instruction, however, is not available to learners in asynchronous VLEs. In asynchronous VLEs learners may request for help at the time of instruction but typical response from the instructor is delayed. Many instructors in asynchronous VLEs include a statement in their syllabus stating their policy for electronic response as 24 or 48 hours response time.

Difficulty in Time Management

Learners in asynchronous VLEs have to manage their time, whereas students in synchronous VLEs use a familiar fixed-time format with "anyplace" connection.

HYPOTHESES

Piccoli et al., (2001) found that the level of learner satisfaction in a VLE for unfamiliar topics like Microsoft Access dropped when compared to familiar topics like Microsoft Word and Microsoft Excel. Brown and Liedholm (2002) also found that when exam questions required more complex applications of basic concepts learners in asynchronous VLEs performed worse than those in a traditional classroom. Online learners in unfamiliar courses are therefore expected to be less satisfied; this leads to the following hypotheses:

H1: Synchronous learners in unfamiliar courses will report lower levels of satisfaction than traditional learners.

Time flexibility and learner control are found to be benefits of asynchronous VLEs (Piccoli et al., 2001), however synchronous VLEs fix the time of delivery, eliminating the time flexibility advantage. In asynchronous VLEs, the learner has a greater degree of control during the time of instruction; learner control in synchronous VLEs is reduced. The lack of time flexibility and loss of control in synchronous VLEs may discourage learners from using synchronous formats. Therefore we hypothesize:

H2: Synchronous learners are less satisfied with the course than learners that used traditional formats.

Synchronous learner's feeling of isolation and experience of anxiety are removed due to the real-time learner-instructor interaction at the time of instruction delivery. The real-time interaction is similar to traditional classrooms and we expect students overall satisfaction in synchronous VLEs to be similar to traditional classrooms. The following is therefore hypothesized:

H3: Synchronous learners will report similar level of overall satisfaction as learners in traditional classrooms.

RESEARCH DESIGN

The VLE framework (Piccoli et al., 2001) shown in Figure 1 was used as the theoretical background for this study.

The University Setting and the Courses

The setting for the study was a large public four-year AACSB-accredited university. The university is comprised of eight colleges including arts, business, education, health and human services, humanities and social sciences, science and mathematics, undergraduate college, and graduate college. A total of 165 students from all colleges participated in this study. The study was conducted during the 2006-2007 academic year. Students enrolled in 15 different courses across the eight colleges.

The Learning Environment

All courses were supported by a learning management system from WebCT.[1] All course material including lecture notes and assignments were posted on WebCT. Discussion forums and email communication for the three classes were conducted through WebCT.

Synchronous and traditional face-to-face classroom formats were used as learning environments. The synchronous formats used video and

Figure 1. Dimensions and Antecedents of VLE Effectiveness (adopted from PIccoli et al., 2001)

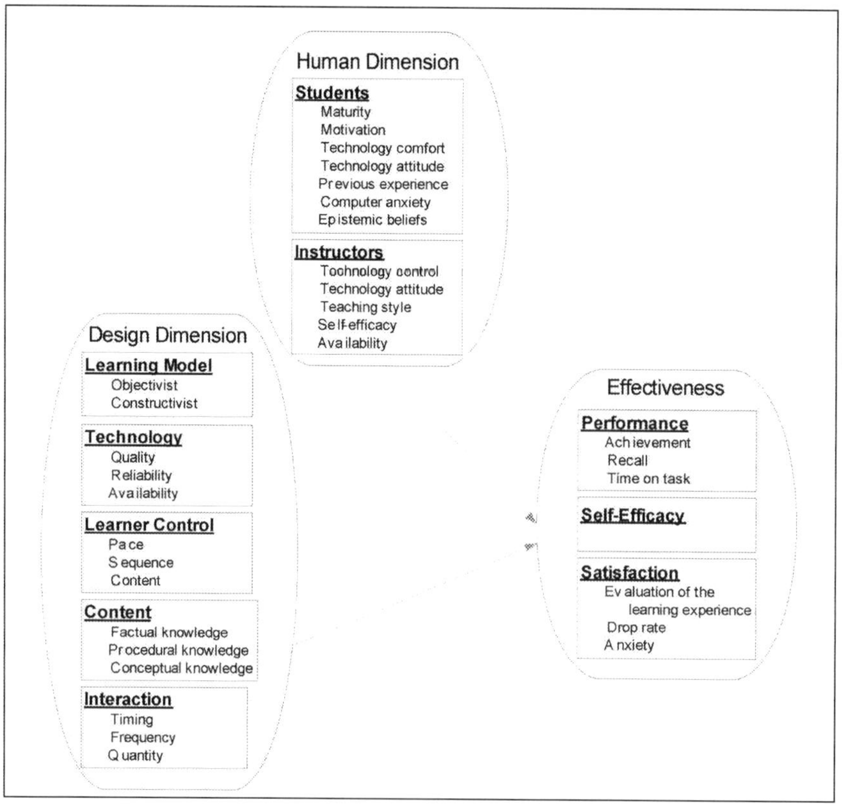

audio for instruction delivery, learners connected to the synchronous class from their home. In the traditional format PowerPoint slides were posted on WebCT, learners access the course material on their own; learners communicated with their professors via email and discussion board.

The Synchronous Technology

The technology used for the synchronous VLE is called Marratech. Marratech was used in conjunction with WebCT. The Marratech system has video, audio, chat, whiteboard, web-browsing, recording and playback features. The user-interface has five panes, see Figure 2. The left pane, approximately ¾ of the screen width is used for whiteboard or web-browsing. The right pane, approximately ¼ of the screen width has four vertically stacked up windows: video or talking head pane, list of participants' pane, display pane for chat messages and a text box for typing chat messages.

Whiteboard and Web-Browsing Pane

Participants can alternate between the whiteboard and web-browsing screens. Lecture notes, lecture slides and annotations are displayed on the white-board. The lead person (typically the instructor, but often participants were given the privilege to lead) has the capability to highlight or write additional notes in real-time. Participants can browse around to different pages of the whiteboard independent of the instructor. All participants however, are guided back to the instructor page each time the instructor changes to a new page. The web-browsing screen is used for exploring the Web; the lead person can take everyone on a guided tour of the web. Participants can browse around on their own on the Web independent of the instructor, but again, they are re-directed to the same page as the instructor each time the instructor changes a page.

Figure 2. Marratech user interface

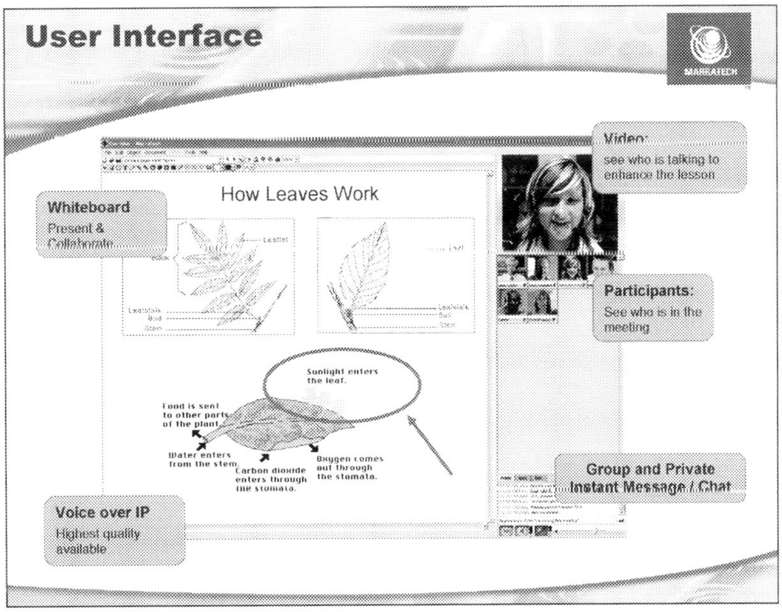

Video or Talking Head Pane

The video screen displays the active speaker. For this study, video was used at all times by the instructor. Although video for students was optional, some students used a webcam. The video screen is typically 20-inch x 18-inch (50 cm x 45 cm), however it can be expanded to 40-inch x 32-inch (100 cm x 80 cm). Real-time video display is projected and video can be enabled or disabled by clicking on the video ON/OFF button. The video delay in this study was negligible (less than 1 second).

Audio

The audio uses voice over internet protocol (VoIP) format. Participants only require a speaker to listen to the class interaction, however if a participant wants to be heard, they need to have a microphone connected to their computer. Audio can be enabled or disabled by clicking on the audio ON/OFF button and participants can also adjust the audio volume.

List of Participants' Pane

The user name for each participant is displayed in the participant list screen. For participants using a video connection, a thumbnail picture and username is displayed.

Chat Display Pane

All chat messages during a session are tracked in the chat window. Each chat message is tagged with a time stamp and sender's user name. Participants have the ability to scroll back and forth to view chat messages.

Chat Entry Pane

A text box is used for typing chat messages.

Public vs. Private Mode for Video Audio, and Chat

In the public mode participants who are logged in can see and hear everyone's communication. In private mode only selected individuals can see, hear and chat. Private mode privileges must be set by the instructor. When a student wants to ask a private question, the instructor can pull them to the private mode for discussion. Private mode can also be used among students. Private mode is available for video, audio and chat communications.

Recording and Playback

Recording can be set at anytime during a session. All interactions in a Marratech session including the whiteboard, web-browsing, video, audio and chat are recorded; recordings can be archived and played back at anytime. If students miss class or want to review concepts they can play back the recording. With permission, all participants can record a session. When a participant sends out a request to record a session, it is displayed in the participant list window; the instructor can disable the recording at will.

RESULTS

Descriptive Statistics

A total of 165 students completed the survey. Descriptive statistics by learning format is given in Table 2.

Respondent breakdown by courses enrolled is given in Table 3.

The distribution of the age ranges is shown in Table 4.

The gender mix of survey participants was 50.4% (83) male and 49.1% (81) female, one person (0.6%) did not provide a response to this question. Gender distribution is shown in Table 5.

Nearly all respondents indicated that they had computer (98.8%) and internet access (99.4%) from home. Computer experience for participants was reported as 42.4% professional users; 51.5% frequent users; 4.8% occasional users; and 1.2% somewhat experienced. In a five point Likert scale, one being do not enjoy computers at all and 5 enjoy computers very much respondent selected 0%, 1.2%, 18.2%, 24.2%, and 56.4% for scales 1, 2, 3, 4, and 5, respectively.

Some respondents (31%) indicated that they expected the courses to be somewhat or very difficult, while 8% indicated that they expected the course to be easy or very easy.

Hypothesis 1: Overall Learning Format Satisfaction

The first hypothesis compared unfamiliar courses in synchronous learning format with traditional learning format. Ninety percent of our respondents using synchronous format expected the course they are taking to be difficult (somewhat difficult to very difficult). And 88% of respondents in the traditional format also expected their courses to be difficult. Details of student course difficulty expectation by learning format are given in Table 7.

Courses that used synchronous format include Systems Analysis and Design, Project Management, and Sociology of Violence. For example, the Systems Analysis and Design course is a required course for all information systems and computer science students, and a prerequisite for all upper division core courses. A term project was used to practice the course content and students had to work in groups to complete the project. As part of the project, students were required to select an organization for their project, identify requirements and develop a proposed informa-

Table 2. Course format

		Frequency	Percent	Valid Percent	Cumulative Percent
Valid	Traditional format	106	64.2	64.2	64.2
	Synchronous format	59	35.8	35.8	100.0
Total		165	100.0	100.0	

Table 3. Courses enrolled

Courses	Frequency	Percent	Cumulative %
Education	2	1.2%	
Electronic Commerce	5	3.0%	
General Psychology	44	26.7%	
History of English Language	7	4.2%	
Information Technology Resources & Policy	16	9.7%	
Life-Span Developmental Psychology	28	17.0%	
Project Management	11	6.7%	
Sociology of Violence	12	7.3%	
Systems Analysis and Design	40	24.2%	
Total	165	100%	

Table 4. Age distribution

		Frequency	Percent	Valid Percent	Cumulative Percent
Valid	Under 19	10	6.1	6.1	6.1
	19 – 23	75	45.5	45.5	51.5
	24 – 29	41	24.8	24.8	76.4
	30 – 35	19	11.5	11.5	87.9
	36 – 40	15	9.1	9.1	97.0
	41 – 45	1	.6	.6	97.6
	46 – 50	2	1.2	1.2	98.8
	Over 50	2	1.2	1.2	100.0
	Total	165	100.0	100.0	

Table 5. Gender distribution

		Frequency	Percent	Valid Percent	Cumulative Percent
Valid	Male	83	50.3	50.6	50.6
	Female	81	49.1	49.4	100.0
	Total	164	99.4	100.0	
Missing	System	1	.6		
Total		165	100.0		

Table 6. Expectation on course difficulty

		Frequency	Percent	Valid Percent	Cumulative Percent
Valid	Very easy	6	3.6	3.7	3.7
	Easy	12	7.3	7.3	11.0
	Somewhat difficult	93	56.4	56.7	67.7
	Difficult	47	28.5	28.7	96.3
	Very difficult	6	3.6	3.7	100.0
	Total	164	99.4	100.0	
Missing	System	1	.6		
Total		165	100.0		

tion system. The modeling language used was Unified Modeling Language (UML). Four major outputs were expected from the term projects: an activity diagram, class diagram, sequence diagram and method specifications. A take-home midterm and final exam were administered for the course. The exams consisted of a case study which required students to create the four major outputs specified above.

The mean rating for overall satisfaction with the learning formats was not significantly different between the groups ($F=0.491$, $p=0.484$); four

Table 7. Expected course difficulty by learning format

Expected Course Difficulty		Learning Format			
		Traditional	Percent	Synchronous	Percent
Valid	Very easy	6	5.7%	1	1.7%
	Easy	7	6.6%	5	8.5%
	Somewhat difficult	56	52.8%	37	62.7%
	Difficult	35	33.0%	12	20.3%
	Very difficult	2	1.9%	4	6.8%
Total		106	100%	59	100%

Table 8. Overall satisfaction with the course delivery format

Course Format	Mean	N	Std. Deviation
Traditional format	4.20	104	.768
Synchronous format	4.29	58	.838
Total	4.23	162	.792

Table 9. Mean differences for overall satisfaction between learning formats (ANOVA)

	Sum of Squares	df	Mean Square	F	Sig.
Between Groups	.310	1	.310	.491	.484
Within Groups	100.777	160	.630		
Total	101.086	161			

Table 10. Overall satisfaction with the course

Course Format	Mean	N	Std. Deviation
Traditional format	4.06	105	.918
Synchronous format	4.21	56	.889
Total	4.11	161	.908

Table 11. Mean differences for overall course satisfaction between learning formats (ANOVA)

	Sum of Squares	df	Mean Square	F	Sig.
Between Groups	.902	1	.902	1.094	.297
Within Groups	131.086	159	.824		
Total	131.988	160			

Table 12. Intention to enroll in the same learning format

Course Format	Mean	N	Std. Deviation
Traditional format	4.23	104	.839
Synchronous format	4.35	57	.813
Total	4.27	161	.829

Table 13. Mean differences for intention to enroll again for learning formats (ANOVA)

	Sum of Squares	df	Mean Square	F	Sig.
Between Groups	.531	1	.531	.772	.381
Within Groups	109.444	159	.688		
Total	109.975	160			

cases excluded. The average rating for traditional format was 4.20 compared to 4.29 for synchronous format; three cases excluded. The details are given in Tables 8 and 9.

Hypothesis 2: Overall Course Satisfaction with Synchronous Format

We compared overall satisfaction between students that used synchronous format with those in asynchronous and traditional formats. The result is shown in Table 10 and 11.

The mean rating for overall course satisfaction was not significantly different between the groups (F=1.094, p=0.297); five cases excluded. The average rating for traditional format was 4.06 compared to 4.21 for synchronous format; four cases excluded.

Hypothesis 3: Expected Future Enrollment

The second hypothesis looked at student intention to enroll in similar formats in future semesters. We asked students if they would enroll in other courses using the current format, i.e. students in the synchronous format to take more courses again using the same format, the same for traditional format. Responses are shown in Tables 12 and 13.

The mean rating for intention to enroll in future courses using the same learning formats was not significantly different between the groups (F=0.772, p=0.381); five cases excluded. The average rating for traditional format was 4.23 compared to 4.35 for synchronous format; four cases excluded.

DISCUSSION

For the purpose of this study students were classified as traditional classroom students or synchronous hybrid eLearning students. The traditional classroom students were those students that attended all classes in a face-to-face format. The synchronous hybrid e-Learning students were those students that attended some of the classes in the synchronous hybrid eLearning format. Fifty-nine respondents (36%) participated in the synchronous hybrid eLearning format; one hundred and six respondents (64%) participated in the traditional classroom format.

The first hypothesis (H1) stated that students will be less satisfied with online VLEs when learning unfamiliar courses. Our results with synchronous VLEs on overall student satisfaction with the learning environment, comparing synchronous and traditional learning formats, found no significant difference between the two formats. As stated earlier prior VLE studies used asynchronous learning format while this study used synchronous format. Our result shows that students learning unfamiliar courses are satisfied with VLEs; the difference is the students in our study used synchronous VLEs.

The second hypothesis (H2) evaluated student overall satisfaction with the course using synchronous and traditional learning formats. Challenges with synchronous learning formats were expected to reduce student overall course satisfaction. Our results showed no significant difference in student overall course satisfaction between synchronous and traditional formats.

The third hypothesis (H3) looked at overall student satisfaction, learning format and course, by evaluating student intention of enrollment for future courses. Our results indicated students that used the synchronous format were as interested in enrolling using the same format for future courses. No significant difference in student intention to enroll in future courses was found between the two formats.

When asked whether they would take another e-Learning class on a 5-point Likert scale, ninety-one percent of the respondents indicated they would by selecting a 4 or 5 on the scale. Eighty-seven percent of the respondents said they did not regret enrolling in this online class, and eighty-three percent said they would recommend this online class format to their friends.

LIMITATIONS AND FUTURE RESEARCH

This course evaluated many courses that students expected to be difficult. While using multiple course is the strength of this study it also raises some limitations. The limitations are that the study did not control for the difference in courses. Future studies may evaluate the difference in the formats using the same courses. The results of this study may also be limited to the specific courses examined in this study and may not be generalizable to other courses, universities, or environments.

Additional research still needs to be undertaken to gain a clearer understanding of synchronous hybrid eLearning versus traditional classroom environments. Research could also be undertaken to compare the levels of satisfaction and self-efficacy of participants in courses in VLEs and traditional environments that are more focused on teamwork as opposed to individual coursework. Additionally, research could also be conducted on this study using larger sample sizes. Finally, researchers are encouraged to conduct studies that compare hybrid synchronous and asynchronous VLEs.

CONCLUSION

VLEs using synchronous hybrid eLearning were examined in this study. Synchronous VLEs provide real time interaction in the classroom. Prior research using asynchronous VLE found differences in how VLEs support unfamiliar and less familiar courses, indicating that students who take unfamiliar courses in VLEs are less satisfied. Many of the difficulties reported by students in asynchronous VLE such as: difficulty managing the high degree of control, overburdened by the shift of responsibility and control, feelings of isolation, experiencing anxiety and difficulty in time management are addressed by synchronous VLEs.

This study provides preliminary evidence to support that VLEs are ready for teaching unfamiliar courses. It is believed that the difference in the results from this study and prior research emanate from the differences between synchronous and asynchronous VLEs. The results of this study show no significant difference between synchronous and traditional student satisfaction when taking unfamiliar course.

REFERENCES

Alavi, M., & Leidner, D. E. (2001). Research commentary: Technology mediated learning - a call for greater depth and breadth of research. *Information Systems Research, 12*(1), 1-10.

Alavi, M., Marakas, G. M., & Yoo, Y. (2002). A comparative study of distributed learning environments on learning outcomes. *Information Systems Research, 13*(4), 404-415.

Brown, B. W., & Liedholm, C. E. (2002). Can web courses replace the classroom in principles of microeconomics? *The American Economic Review, 92*(2), 444-448.

Dagada, R., & Jakovljevic, M. (2004). *Where have all the trainers gone? E-learning strategies and tools in the corporate training environment.* Paper presented at the 2004 Annual Research Conference of the South African Institute of Computer Scientists and Information Technologists on IT Research in Developing Countires, Stellenbosch, Western Capc, South Africa.

Hodges, C. B. (2005). Self-regulation in web-based courses: A review and the need for research. *The Quarterly Review of Distance Education, 6*(4), 375-383.

McCray, G. E. (2000). The hybrid course: Merging on-line instruction and the traditional classroom. *Information Technology and Management, 1,* 307-327.

Newkirk, R.L. (1973). A comparision of learner control and machine control strategies for computer assisted instruction. *Programmed Learning and Educational Technology, 10*(2), 82-91.

Piccoli, G., Ahmad, R., & Ives, B. (2001). Web-based virtual learning environments: A research framework and a preliminary assessment of effectiveness in basic it skills training. *MIS Quarterly, 25*(4), 401-426.

Sauers, D., & Walker, R. C. (2004). A comparison of traditional and technology-assisted instructional methods in the business communication classroom. *Business Communication Quarterly, 67*(4), 430-442.

Seng, L. C., & Al-Hawamdeh, S. (2001). New mode of course delivery for virtual classroom. *Aslib Proceedings, 53*(6), 238-242.

Webb, H. W., Gill, G., & Poe, G. (2005). Teaching with the case method online: Pure versus hybrid approaches. *Decision Sciences Journal of Innovative Education, 3*(2), 223-250.

Wilson, B. G. (1996). *Constructivist learning environments: Case studies in instructional design.* Englewood Cliffs, NJ: Educational Technology Publications.

ENDNOTE

[1] WebCT is a learning management system that supports online learning environments. URL: http://webct.com/

Chapter XX
Understanding the Effectiveness of Collaborative Activity in Online Professional Development with Innovative Educators through Intersubjectivity

Diane Hui
The University of Hong Kong, Hong Kong

Donna L. Russell
University of Missouri-Kansas City, USA

ABSTRACT

Effectiveness of professional development is affected by the quality of social interaction. This study examines how online collaborative dialogues might influence teachers' decisions in their classrooms —sometimes hurting when not appropriated well. This study extends principal sociocultural approaches to cognitive concepts of intersubjectivity and activity through illustrations of empirical data. Part of a larger innovative professional development involving four classroom locations across Missouri, synchronous chatroom dialogues comprising teachers and researchers, and pre- and post-unit interviews underwent qualitative discourse and focused microanalyses. We argue that teachers purposefully used their dynamic intersubjective spaces and strategies in the management of meaning-making negotiations within an online interactive environment. The findings reveal two novel variable forms of intersubjectivity: (a) temporary suspension, and (b) resistance and disagreement. These findings provide useful implications for advanced applications and developments with information communication technology in innovations for enhanced learning and teaching as they relate to the evaluation of teacher effectiveness in implementing collaborative online problem-based activities.

INTRODUCTION, LITERATURE REVIEW, AND SIGNIFICANCE OF THE STUDY

Reform advocates in education have increasing interests and hopes for incorporating information communication technology in reforming both instructional and teacher education. These interests and hopes are important and significant. However, the outcomes of their implementation varied in the field. While some implementing groups sustained their learning, other groups did not. Drawing on the cognitive and sociocultural concepts of intersubjectivity and activity, this study examines the specific ways in which teacher understanding and learning were developed (or not) in online collaborative dialogues and the extent to which these collaborative dialogues might impact on teachers' decision-making when implementing innovative constructivist-based professional development, between four teachers and two researchers involving four classroom locations across Missouri, USA.

Previous studies of professional development from a dialogic perspective implementing similar reforms have proven to benefit innovative teachers. As Zeichner and Liston (1996) wrote, "The challenge and support gained through social interaction is important in helping teachers clarify what they believe and in gaining the courage to pursue their beliefs" (p. 76). To facilitate optimal learning for students with technology, teachers need considerable knowledge, effort, persistence, and self-regulation to devise, implement and assess instructional plans and complex learning environments. In such processes, teachers' collaborative professional development plays a critical role as they construct new understandings, through participation in their "community of practice" (Lave & Wenger, 1991, p. 29). The production of these communities often involved a shared practice that reflected the pursuit of learning through interacting, both with each other, and with the world.

Interest in reforming education through technology has steadily increased in recent years (NCTAF, 2003). Technology has been described as "a fact of American life" (OTA, 1995, p. 2) and the Internet as providing the "fabric of our lives" (Castells, 2001, p. 1). As such, it affects our culture, work, and communication (e.g., Hui, 2003). As the availability of technology in education has become increasingly ubiquitous, research has shown the promising potential of technology in improving student and teacher learning (e.g., Bransford, Brown, & Cocking, 1999). Research has also affirmed the importance of connecting teachers and technology (e.g., Marx, Blumenfeld, Krajcik, & Soloway, 1998). This is crucial for the success of standards-based reform in the American schools (e.g., Brunvand & Fishman, 2005) and has the potential to change the future of education (e.g., Tyack & Cuban, 2004; see also Dede, 1996), given the powerful role of communication technology for mediating teacher education reform. The list of often stated goals includes: (1) sharing information and new pedagogy (e.g., Berge & Collins, 1998), (2) facilitating teacher competencies (e.g., Kabilan, 2005), (3) fostering collaborative professional development (e.g., Bober & Dennen, 2001; Riel & Fulton, 2001; Zhao & Rop, 2001); and (4) building reflective communities (e.g., Berge & Collins, 1998; Borthwick et al., 2004; de Vries, Naidu, Jegede, & Collis, 1995; DiMauro & Jacobs, 1995; Salmon, 2004; Schlager, Fusco, & Schank, 2002). It has proven to be a viable alternative strategy for the development of teachers (e.g., Brunvand, Fishman, & Marx, 2003) and teacher professional development e-communities (e.g., Collison, Elbaum, Haavind, & Tinker, 2000).

Despite its growing prevalence and promise, questions have been raised concerning the pedagogical impact of dialogues as mediated by information communication technology on professional development from conceptual, methodological, and practice perspectives (e.g., Wade & Fauske, 2004; see also Wallace, 2004), for two recurring reasons: (1) the prevalence of shallow online discourse (e.g., Putnam & Borko, 2000; for professional development in general, see Borko, 2004); and (2) teachers' reticence to use networked peer communication (e.g., Zhao & Rop, 2001; see also Zhao & Frank, 2003). Although teachers have reported positive attitudes and values concerning technology in a recent national teacher survey (e.g., NetDay, 2004), they did not consistently use the online resources available for innovative professional development (e.g., Brunvard & Fishman, 2005). As a result of these limitations claims about the contribution of communication technology remain unsubstantiated (e.g., Thompson, Bull, & Bell, 2005), or as described as overblown or "exaggerated" in reality (Selwyn, 2000, p. 750). Calls have also been made for research to provide guidance and support in engaging teachers in productive technology-based discourse (e.g., Gordin, Gomez, Pea, & Fishman, 1996; Putnam & Borko, 2000; Wade & Fauske, 2004).

Extending our previous study (Hui & Russell, 2007), this study touches on these issues by examining the specific ways in which teacher online understanding might occur (or not) through cognitive and sociocultural concepts such as intersubjectivity and activity. These concepts give researchers holistic tools to understand how collaboration can support innovative teachers as they implement new instructional units with the use of technology. More specifically, we use intersubjectivity to understand the nature and the quality of the dialogues of innovative teachers and subsequently the notion of cultural historical activity to draw inferences on how these collaborative dialogues were used by the teachers and their effectiveness at implementing on their new units of study in their classrooms.

Intersubjectivity is the key concept underlying this study. The accomplishment of intersubjectivity is an important step leading to a new solution for solving problems. An example of intersubjectivity can be shown in the agreement between social participants on inquiry- and problem-based activities. In this study, intersubjectivity is defined as the common understanding as shared between social participants concerning their goals, context, action, operation of actions, use of objects, and evaluation of actions and outcomes.

Previous studies of intersubjectivity have often suggested a binary approach to understanding, meaning that the participants either maintained a shared intersubjectivity or understanding, or they did not. In contrast, the current study provides a novel dialogic approach to understanding the variable patterns, or dynamic nature of intersubjectivity, within the zone of proximal development (ZPD) in an online interactive environment as mediated by advanced applications of information communication technology (see further elaboration in the "Theoretical Perspectives" section). The ZPD here refers to the learning difference between guided and independent performance in problem-solving activity (e.g., Vygotsky, 1986).

Activity theory is used in this study to make inferences about the relationship between the dialogic processes of the innovative teachers and their responses to the collaborative processes in their local classrooms. According to Jonassen, Peck, and Wilson (1999), constructivist learning such as problem-solving involves knowledge that is constructed, not transmitted, but embedded in activity, and anchored in the context of the activity. Previous studies have examined reform processes in professional development in educators (e.g., Korthagen, 1993; Shulman, 1986; Schön, 1983). These studies have identified the importance of collaborative professional development for teachers implementing innovative methods. Activity systems are historically conditioned systems of interrelated contacts among individuals and the "proximal culturally organized environments" (Salomon, 1993, p. 8). In this study, activity theory is used to develop inferences on the collaborative dialogues among these innovative teachers and their effectiveness at implementing their new methods in their classrooms.

In this study, we argue for a dynamic nature of understanding that can be appropriated as a mechanism for managing intersubjectivity in a purposeful way within the ZPD, online. Dynamic nature of intersubjectivity does not refer to something that a participant has or does not have, but rather something that varies from one moment to the next depending on the context or activity. Against this background, we address three key research questions:

1. What characteristics or patterns of intersubjectivity occur in the context of online professional development?
2. How might the computer-mediated context enhance or constrain the nature of intersubjectivity?
3. Would, and in what ways might, the dynamic nature of intersubjectivity facilitate or hinder the ultimate effectiveness of online collaborative activity?

The findings of this study reveal two variable forms of intersubjectivity by two participant cases: (1) a temporary suspension, and (2) resistance and disagreement. They show that learning did not occur uniformly among the teachers within an interactive environment as mediated by advanced information communication technology for enhanced learning and teaching. Its effectiveness is dependent largely upon the ways in which collaborative dialogues as constructed in professional development activity were productively appropriated by the teachers and used to solve problems in their classrooms. This process can be tensional at times. These findings do not only increase our awareness of the conditions that might account for the variable patterns of intersubjectivity, but they also make available insightful information into the specific ways in which intersubjectivity may affect the effectiveness of collaborative activity in professional development. These issues provide a formulation for future larger-scale explorations. Moreover, this new knowledge of intersubjectivity through a novel online professional development activity has useful implications for future inquiries of teacher learning (e,g., Cochran-Smith, 2003; Nieto, 2003) by that: cognitive approaches and procedure to online collaborative learning would involve both individual and group or community learning with advanced communication technology; their communicative activities would use both online problem-solving and group collaborative activities; and sociocultural and motivational processes are important in fostering self-directed learning and reflective participation online. All these will inform educational policy-makers, educators and practitioners, and software designers working within teacher development communities, as they design more cost-effective, sustainable models of professional development, using advanced communication technology, therefore having a positive effect on student learning and school outcomes.

THEORETICAL PERSPECTIVES

Conceptually grounded in semiotic and discourse theories, this study's theoretical framework has roots in sociocultural perspectives (e.g., Vygotsky, 1986; Wertsch, 1984, 1985, 1991) and communities of practice (e.g., Lave & Wenger, 1991; Wenger, 1998). Specifically, our argument is grounded in two principal approaches to intersubjectivity: (1) the stable notions (e.g., Bruner, 1996, 1990; Rogoff, 1990; Rommetveit, 1979; Wertsch, 1984, 1985, 1991, to name just a few examples); and (2) the variable patterns or dynamic nature of intersubjectivity (e.g., Mortimer & Wertsch, 2003; see also Hui, 2003; Matusov, 1996; Sawyer, 2003; Topper, 1995) with reference to technology research (e.g., Koschmann, 1996; see also Black, Levin, Mehan, & Quinn 1983; Henri, 1992; Hermann, 1995; Quinn, Mehan, Levin, & Black, 1983; Wilkins, 1991). These approaches will be discussed briefly below.

Conventional Stable Notion of Intersubjectivity

First appropriated from the Scottish philosopher John MacMurray by cognitive developmentalist Colwyn Trevarthen in the 1960s, intersubjectivity has long been recognized as an important concept across social sciences concerning human understanding and transformative learning within the ZPD, both within and between learners (e.g., Rogoff, 1990, 1997; Matusov, 1996), including education (e.g., Bober & Dennen, 2001).

For example, Bruner (1996) defines intersubjectivity as "how people come to know what others have in mind and how they adjust accordingly" (p. 161). Moreover, in adult-child problem-solving situations, Wertsch (1985) suggests that intersubjectivity will be present when "interlocutors share some aspect of their situation definitions" (p. 159). Situation definition is "the way in which a setting or context [objects and events] is represented – that is, defined – by those who are operating in that setting" (Wertsch, 1984, p. 8). A redefinition of the situation is thus an indication of growth. Furthermore, in communicative action, Rommetveit (1979) argues for the "transcendence of the private worlds of the participants" (p. 94) into a temporarily shared social world (see also Tomasello's, 1999, discussion that linguistic symbols as used in communicative events are inherently, socially intersubjective and perspectival). In addition, in mother-infant communication, Rogoff (1990) emphasizes the understanding of a situation shared between people, as providing "a common focus of attention and some shared presuppositions that form the ground for communication" (p. 71). In teacher-learner knowledge construction activity, Bober and Dennen's (2001) notion of intersubjectivity entails a "shared understanding that helps us relate one situation to another" (p. 241).

However, the majority of these intersubjectivity accounts assume that social participants share a stable notion of understanding, reflecting binary (i.e., present or absent) and unidirectional (i.e., a solely forward marching path) characteristics (see also, Davidson, 1992; Göncü, 1993; Forman, 1992). Although a few researchers have considered less stable notions of intersubjectivity, their concerns for intersubjectivity have remained restricted.

The Dynamic Nature of Intersubjectivity

Resembling the dynamic role of intersubjectivity in meaning-making negotiations and extending beyond Ragnar Rommetveit's notion of intersubjectivity, Mortimer and Wertsch (2003) proposed a richer level of intersubjectivity through a concept involving "resist[ance]" (p. 241) in explaining the eighth grade students' reluctance to use a perspective being introduced to them by their science teachers in their group discussion. This notion of resistance was also considered by Matusov (1996) who argued for a participatory notion of intersubjectivity which entailed both agreement and disagreement in the development of sociocultural activity within a play writing class, resembling a "dynamic unity of individual contributions in the joint activity" (p. 33).

Through adult-child activity in a museum, Hui (2003) argued for the dynamic nature of intersubjectivity. In other words, the communicative pattern of the participant was reflected by his/her retreating from and re-approaching towards social interaction as a means for scaffolding an alternative path in regulating and purposefully managing problem-solving behaviors through self-regulative speech. This resonates with the cognitive impact suggested by Azmitia's (2000) "time-outs" (p. 191) for managing cognitive and affective needs within collaborative activity and the strategic use of self-regulative action through discourse in problem-solving activities (e.g., Vygotsky, 1986), but it cannot be easily generalized to interpret collaborative learning.

Drawing from Matusov, Sawyer (2003) explored intersubjectivity in ensemble creativity as "not how performers come to share identical representations, but rather, how a coherent interaction can proceed even when they do not" (p. 10). Sawyer further argued that the "meaning of an individual acts [therefore] emerges from the collective creativity of the group" (p. 11).

Topper (1995) was one of the few scholars to discuss intersubjectivity in an online setting. Drawing predominately from Rogoff (1990) and Matusov (1996), Topper's notion of intersubjectivity appeared to be implicitly dynamic which involved the "transformation of participation through joint activity, which may or may not be symmetrical in terms of the knowledge or understanding of the individuals" (p. 3). However, Topper later associated intersubjectivity explicitly with "shared understanding" (p. 1). Such an association was inconsistent with his earlier description.

Research examining the variable forms of intersubjectivity in collaborative dialogues as mediated by advanced communication technology in innovative professional activities remains uncommon. This study aims to rectify this gap. The analysis here reveals that teachers purposefully used their dynamic intersubjective spaces and strategies in meaning-making negotiations, through exhibiting "peripheral experiences" or participation (Wenger, 1998, p. 117) or "lurking behavior" (e.g., Herrmann, 1995, p. 166). Peripheral participation here refers to the developmental progression of a participant's partial recognition of membership in terms of status, skill, and knowledge within a community of practice. Such a process involves a participant's temporary but strategic suspension from intersubjectivity, as reflected in his/her communicative orientation first moving away from, then back towards the intersubjective space within the problem-solving process. Doing so would thus enhance the ultimate effectiveness of ongoing activity upon re-entering intersubjectivity in social interaction.

Premised upon the presence of intersubjectivity in social interaction, the dynamic nature of intersubjectivity in this study was characterized by the following four features:

1. Variable forms of intersubjectivity through time (i.e., not binary, but may be partial).
2. Participants' temporary suspension from intersubjectivity.
3. Either agreement or disagreement, or both.
4. Purposeful management of intersubjectivity resulting in successful problem-solving as identified using activity theory.

METHODS, RESEARCH DESIGN, CONTEXTS, AND DATA

Methods and Research Design

The online discourse literature continues to call for robust theory and methodology to examine interaction quality (e.g., Cazden, 2001; for online context, see Romiszowski & Mason, 1996; Zhao & Rop, 2001) and to inform best practice (Thompson, 2005) through significant longitudinal studies (e.g., Flynn & Polin, 2003; Rourke, Anderson, Garrison, & Archer, 2001). Research methods, although useful, have remained focused upon structural and discourse content analyses (for an example, see Fahy, Crawford, & Ally, 2001; Henri, 1992) and covered relatively short time frame (for an example, see Gunawardena, Lowe, & Anderson, 1997), thus failed to explain how dialogues may influence decisions. This study aims to fill this gap by making a case for using intersubjectivity for examining online collaborative dialogues and then using the concept of activity to understand how the teachers use the dialogues to make decisions in their classrooms and the outcome or effectiveness of these decisions.

The theoretical grounding for cultural historical activity theory is the sociocultural theory of human learning and development (e.g., Vygotsky, 1986, Bruner, 1990) with an emphasis on understanding the processes of mediated activity (e.g., Wertsch, 1985, 1991). The researchers used activity theory (e.g., Engeström, Miettinen, & Punamaki, 1999) in order to design a systems-based framework for understanding the teachers' conscious goal-directed work activity in context (e.g., Cole & Engeström, 1993; Bereiter, 2002) and to define the interactions and evaluate the consequences of the teachers' work activity. By pairing the use of intersubjectivity with activity theory the researchers are able to draw conclusions concerning the impact of the online dialogues on the classroom decisions and evaluate the effectiveness of their decision-making processes.

Conversational and discourse analyses have been commonly used in analyzing online discourse data (e.g., An & Levin, 2003), for example, through speech acts (e.g., Beals, 1992) and Conversational Analysis theories (e.g., Parrish, 1998), and discourse and ethnographic theories (e.g., Hui, 2006).

This study aims to provide a working analysis of intersubjectivity through the implementation of innovative professional development and activity as mediated by advanced information communication technology. Its analysis integrated both qualitative discourse (e.g., Schegloff, 1999; Gee, 1999) and focused microanalyses (Vygotsky, 1986; Wertsch, 1985, 1991). Such combined methods and analyses provided in-depth information regarding the psychological processes involved in complex collaborative activities. We focused on rich and follow-through qualitative analysis, not content generalizability.

This study was both descriptive and analytical. To examine the development of intersubjectivity between participants, we created a discursive environment through synchronous chatroom within which participants could discuss authentic issues concerned with the design and implementation of an innovative problem-solving unit, based around constructivist approach to professional development and activity. These issues involved the participant's goals, operation of actions, use of tools such as technology, and outcome evaluations, as reflected in the definition of intersubjectivity provided above.

Context and Participants

Part of a three-year collaborative research, this study evaluated the effectiveness of innovative professional development through the examination of online collaborative dialogues and professional learning activities and experiences of two action researchers and four elementary teachers with fourth and fifth graders in four cities throughout Missouri (Linda, Carol, Janet, and Helen, and researchers R1 & R2 – all pseudonyms). The students represented inner city, small city, suburban, and rural populations.

These teachers were voluntary participants in an innovative program in Missouri entitled enhancing Missouri's Instructional Networked Teaching Strategies (eMINTS) in 2001. eMINTS was a program developed by Missouri's Department of Elementary and Secondary Education. It developed technology use through inquiry-based instruction. For readers interested in eMINTS, please refer to the homepage of eMINTS in Missouri at http://missouri.emints.org. These four eMINTS teachers participated in a pilot project at the Missouri Research and Education Network called Pioneers Program that involved using a new online productivity software, Shadow Net Workspace™ in developing constructivist-based problem inquiry. The teachers shared the same three years of technology training, and had similar inquiry experiences and technological access in their classrooms. For instance, they all received the same access to the professional development processes, online and in-person, and materials. However, both Carol and Janet had relatively less experience in collaboration in innovative practice.

In this study, the inquiry on problem-solving activity concerned with the implementation of an innovative unit for elementary students. The

Table 1. Classroom descriptors

Teachers	Grade	Community	Students	Technology Access
Linda	4th	suburban	12 boys and 10 girls, all Caucasian	As a part of their participation in the eMINTS program, each teacher has 12 to 14 Pentium3 LCD computers, a teacher workstation, laptop, a Smartboard and projector, a scanner, a color printer, and a digital camera.
Helen	4th	rural	12 boys and 12 girls, all Caucasian	
Janet	5th	urban	7 boys and 10 girls, all African-American	
Carol	4th	mid-size city	9 boys, 10 girls; 11 Caucasian, 8 African-American	

unit implemented was an authentic design- and problem-based unit entitled "Improving I-70." I-70 is an interstate in Missouri that is in need of repair. Students were to design their solutions in response to the problem. Table 1 shows the classroom settings.

Data

Analyses of this study were based on multiple data sources such as synchronous chatroom discourse among teachers and researchers, pre- and post- unit interviews and researchers' field notes.

The collaborative professional development occurred through seven weekly online chatrooms to help design and develop an online unit. These chats were the only professional development forum scheduled for the teachers who did not meet in person before or during the collaborative processes. The researchers scheduled an initial chat before the unit. Thereafter, with the researchers' facilitation, the teachers took up their next six weekly hour-long chatroom discussion. The teachers identified four common goals for their chats (i.e., scheduling the collaborative unit, scheduling online collaborations among the students, discussing the instructional process, and asking questions and getting feedback from the group) and eight investigative areas (i.e., traffic flow and growth, socio-economic, design/engineering, public affairs, financing, construction, natural, and human environments).

A total of seven chatroom conferences that occurred prior and during the unit implementation were examined. Each chatroom conference was coded for speaking turns and content topics. Of these chatroom discourse, Chat 4, representing an average level of synchronous online conversation, was selected for critical microanalysis as illustrations of the dynamics of intersubjectivity. By Chat 4, teacher participants had already begun the first phase of Unit 1. Informed consent was obtained.

Unit of Analysis

The unit of analysis was the topic. Topic cohesion and progression informed the development of intersubjectivity (or not) among participants through conversational moves, in particular where conflicts, cruxes or turning points occurred. A topic might contain multiple utterances (Bakhtin, 1986) and was thematically classified according to discussion subject. A single speaking turn could conceivably contain multiple topic units. Coding practice was discussed and agreed among researchers.

Using Chat 4 as an illustration, nine main topic areas were discussed throughout the selected transcript. Of these topics, the sixth topic area (i.e., defining expert areas) took up the largest segment. Table 2 shows the overall distribution of topic units and speaking turns produced by each participant.

Table 2. Topic units and speaking turns of each research participant (Chat 4)

Participant	Speaking Turns	Topic Units
R1	29 (30%)	43 (33%)
Carol	20 (21%)	32 (24%)
Linda	24 (25%)	30 (23%)
Janet	13 (14%)	16 (12%)
Helen	10 (10%)	11 (8%)
Total	96 (100%)	132 (100%)

Table 2 illustrates that R1 produced the highest occurrences of both speaking and topic unit turns, Helen the least, and Carol, Linda and Janet scoring in mid range. Such a speaking pattern could be attributable to role specificity (i.e., researcher versus participant roles) by that researchers' interaction aims to facilitate collaborative conversations and orchestrate unit planning activities, whereas teachers' interaction provides information and discusses details in response to the topic at issue.

ANALYSIS AND DISCUSSION

Selected segments of the chatroom transcript will be used to illustrate the dynamics of intersubjectivity (i.e., a temporary suspension phenomenon and different intersubjectivity levels through time). In what follows, original utterances produced by the participants will be presented in italic format or in direct reported speech as bounded by quotation marks within the texts, so as to differentiate them from the authors' analytical narratives.

A Temporary Suspension of Intersubjectivity

The communication pattern, in particular of Helen's, will be presented to illustrate the occurrence of a "temporary suspension" with her retreating from and later re-approaching towards social interaction as contributing to the ultimate effectiveness of the ongoing intersubjective activity upon re-entering intersubjectivity.

Of the eleven topic units Helen produced throughout the chat conference, the majority closely followed from topics that took place in the local discourse context, with an exception of three utterances (i.e., Utterances 83, 118, and 117 in particular) which were then regrouped in two episodes. These episodes demonstrated the temporary suspension from intersubjectivity. Episode 1 will now be discussed in some detail.

The first episode took place in Utterance 83. The group was discussing the importance of defining the expert areas (Topic discussion 6). It was also the longest topic discussion. Helen's last topic turn was shown in Utterance 66. After an extended period of inactivity in the chat conference or "lurking" behavior (entailing a difference of sixteen topic unit turns), Utterance 83 saw Helen express her support for Linda's suggestion for providing additional information to better define the expert areas previously mentioned in Utterances 63, 64, 68 and 75. By such means, she explained, there would be no need to rename the expert groups as initiated by Carol in Utterances 65, 70 and 77. In Utterance 83, Helen wrote,

83	Helen	*I think if we get a good description of the job we won't have to rename them*

Unlike the majority of her topic units, in this episode, Helen summarized topics from a wider discourse area spreading from Utterances 63 through 77. Her peripheral participation was noted in terms of her inactivity or lurking behavior from Utterances 66 through 83 (within Topic discussion 6). Despite this inactivity in terms of writing, her ability to express her position in support for Linda's suggestion to define the expert areas, thus making Carol's suggestion to rename the expert groups unnecessary (Utterance 83), showed that Helen was indeed actively listening or reading during the chatting process. It was thus evident that she was closely following the development of the discussion. Despite the lurking behavior, her acute sense of synthesizing different threads of discourse (i.e., Linda and Carol's different suggestions) and her active participation upon re-entering intersubjectivity contributed to the ultimate effectiveness of the ongoing activity. This lurking behavior and re-participation exemplify the occurrence of a temporary suspension, as reflected by the participant's retreating from and re-approaching towards social interaction.

In summary, Helen's temporary inactivity or peripheral participation was characteristic of her online collaborative behavior. It exemplified the occurrence of a temporary suspension which may facilitate a strategic communicative function.

Different Levels of Intersubjectivity Through Time

Two interesting points will be reported: (1) the divergent aspects of intersubjectivity, featuring resistance and disagreement, and (2) the maintenance of intersubjectivity through group dynamics. Selected discourse of a participant, Carol, will be presented to illustrate these points.

Resistance and Disagreement

Unlike the rest of the participants, throughout the chatroom conference, Carol demonstrated a relatively low level of intersubjectivity compared to that shown by the group's intersubjective pattern. Carol initiated the topic unit to discuss the expert groups in Utterance 1 (Topic discussion 1), but her initiation was quietly ignored. The researcher, R1, shifted the topic unit towards discussion of learning goals (Utterance 3) instead. Moreover, following Linda's expression of concern regarding scheduling online chats with the students (Utterance 18), Carol magnified her own apprehension and wrote (Utterance 19),

19	Carol	*And, with that thought Linda, a point that makes me very nervous is having different groups chatting at a time and I can't see them all at one time. I know I am paranoid, but I like to keep tabs on what is being said.*

When expressing her frustration regarding the expert areas, Carol made notable use of graphic representations such as capital letters and punctuations. For examples:

40	Carol	*BUT I DON'T KNOW WHAT THE AREAS ALL MEAN!!!!!!!!!!!!*
61	Carol	*How would THEY know if it positively or negatively impacts traffic/???*
89	Carol	*I can't agree if I don't understand what they all mean!!!*
91	Carol	*I could swear I was the one saying keep all 8 and let the kids choose 4 and YOU guys said narrow it down NOW and lead them to four!!!! You guys said 4 groups would be more manageable!!!!!!!!!!!!!!!!!*
92	Carol	*Can you tell I'm very frustrated here?????*

Furthermore, Carol would also make use of rhetoric questioning with punctuation (Utterance 25), for example,

25	Carol	*About the 8 experts: I thought we had said last time we were going to "gear" them to a specific 4 that we picked here???*

Carol's frustration level was expressed in an explicitly depersonalized and inflamed manner. On one occasion, Carol left and re-entered the online chat (Utterances 43 to 44, as shown on data log).

In summary, to all intents and purposes, Carol would appear not to be sharing the same intersubjective level with fellow participants. Her level of intersubjectivity seemed to be divergent from that of the rest of the group, consistent in a way more with Mortimer and Wertsch's (2003) notion of resistance and Matusov's (1996) notion of disagreement. However, this proposition might be confounded by the time lag during the electronic transmission process.

Maintaining Intersubjectivity Through Group Dynamics

In addition to the different levels of intersubjectivity within the group, a group dynamic functioned to maintain the group's intersubjectivity level was also noted.

As the chatroom conference developed, all three teacher participants took on the researcher role in responding to Carol's questions and concerns. For example, Carol asked if they could rename the expert groups (Utterance 70), Helen appropriated the researcher's voice in her response to Carol, *"I think if we get a good description of the job we won't have to rename them"* (Utterance 83). Another example took place when Carol asked if they could reduce the number of expert groups. Here, Linda, in her researcher's tone when responding to Carol, *"I thought I understood R1 to say they were all important and we needed to look at all 8 areas ..."* (Utterance 79). On another occasion, Carol asked if the information posted by Helen and Janet was located in the *"before the unit discussion board"* (Utterance 112), Janet responded to her question positively (Utterance 120). It is worthy of note here that due to this collective maintenance of intersubjectivity, Carol finally reconciled and wrote (Utterance 123),

123	Carol	*Thanks. I check it often and have not seen anything new posted. I did not check last night, I'll check tonight.*

In summary, levels of intersubjectivity would seem to change through dialogues over time. Contrary to the cooperative action as demonstrated by the rest of the collaborative group, Carol showed a level of intersubjectivity which was different or even opposed to that of the rest of the group's pattern. However, in the presence of dynamic patterns of intersubjectivity (in particular, disagreement), efforts were noted from the group participants in maintaining the intersubjective pattern, thus securing the ultimate effectiveness of the ongoing activity.

Pre- and Post-Unit Concepts of Innovative Professional Development and Impact of Collaborative Dialogues on the Teachers' Unit Implementation

To understand the extent to which teachers' decisions were affected by their participation in online collaborative dialogues, it would be important to first learn about their expectations of the specific inquiry prior to the unit implementation. Despite their similar experiences using technology, pre-unit interviews revealed that teachers had different expectations for their online collaboration and participation. For instance, one teacher, Carol did not want to work *"in lock-step"* with other teachers and did not want to complete the entire unit. Linda and Helen, who had previously worked collaboratively locally, expressed interests in participating in the online professional development. Janet, who had not participated in collaborative teaching locally, hoped to gain from the online collaboration.

Post-unit interviews were conducted with teachers after unit implementation. While all teachers stated that they would participate in online professional development in the future, these interviews also revealed different pedagogical outcomes and experiences of teachers which reflected the tensions and cruxes underlying the collaborative processes. The two teachers, Janet and Carol, who stated that they had little or no previous collaboration accomplished more in their units as a result of their participation in the online dialogues. Both teachers completed more of the unit with advanced student learning assessed. Janet stated that the online collaboration was *"critical"* to her new understanding of advanced unit design and implementation. Carol did not agree, pre and post, with the advanced problem-based unit design or implementation. She did, however, complete the unit in order to *"stay with the others."*

In this study four innovative teachers collaborated while developing and implementing a problem-based unit. They made decisions in their classrooms that were directly related to their online dialogues and contradictory to their pre-unit statements. Their work activity was analyzed to understand how the collaborative processes impacted their decision-making in their classrooms to overcome contradictions and develop their learning goals for their students. The researchers used cultural historical activity theory (e.g., Engestrom, 1987) to define the activity of the teachers as they appropriated the online collaborative dialogues in the development of their innovative units locally (e.g., Russell, 2005a), while using the notion of intersubjectivity to relate the nature and quality of the teachers' collaborative dialogues to their ability to implement their unit fully. In what follows, we will focus on the specific ways in which teachers responded to the contradiction or divergence throughout the implementation process.

Despite the fact that the innovative teachers had similar technological and inquiry experiences and training, their responses to the professional development were quite varied and had implications in their behaviors in the classroom.

For instance, Helen identified well-formed goals for the implementation of the unit in her classroom. She had experience teaching similar units. She had worked collaboratively with other teachers. She was prepared to use the technology in future units. She seemed well-prepared to implement the unit successfully based on her goals and experience. However, she disintegrated her object and did not implement the entire unit. She shortened her schedule and began the unit after MAP testing in order to stay with the other teachers even though her school was the first to get out for the summer and she was not overly concerned with the testing process. She was also unable to overcome a contradiction in technology support. Her characteristic responses to the professional development seem to be reflected by her orientation of peripheral participation and temporary inactivity through her online collaborative dialogues.

In the end, Helen was forced to end the unit because her school, a rural school which began the school year earlier under Missouri state law, ended two weeks prior to the other schools. Her online participation did not directly add to her unit implementation. In fact, her peripheral participation in the online professional development, specifically changing her schedule to match the other schools, meant that she could not finish the unit as she designed it. She stated, *"They [her students] almost had it. They were learning how to solve problems or learning how to work collaboratively. I don't think we quite made it. With another week, we could have."* She explained that she would not make the decision to shorten her unit because of pressures to work collaboratively online. Her inability to communicate her goals in her online collaboration hindered her unit completion which eventually prevented her students benefiting fully from the innovation.

Another innovative teacher, Carol, implemented the unit successfully. Her students completed all three phases including participating in a complex jigsaw solution group activity. However, her initial goal did not include doing all three phases. She felt that Phase 1 was enough and did not intend to do the unit template as it was structured. She did expand her object but it was entirely through the outside influence of the collaborative process.

She did not have a lot of experience working collaboratively with other teachers and she wanted to develop the unit within the schedule set by the other teachers. She seemed to overcome this contradiction, between her own goals and the unit, through this collaboration. She did not develop new goals about the unit and the learning potential of her students that could result from this type of unit. She will not do similar units in the future. Although she implemented this unit successfully and collaboratively, it is not likely that she has changed in her attitudes toward the learning processes of her students and the type of professional development that could aid her in her classroom.

A third innovative teacher, Linda, had very defined learning goals for her unit as a result of implementing the two innovations. She did not achieve her original goals because of two mediational contradictions, one related to her concepts about the complexity of the unit itself and the other related to the relationship between the technology and her goals. She was able to overcome several other contradictions – working out scheduling issues in her building and technology problems in her classroom – without changing her goals. It appears that her community was favorable to the reform processes and the new tools implemented during this unit. She did not seem to change behaviors and overcome contradictions in response to collaborative professional development.

A final innovative teacher, Linda, ended her online participation with the other teachers during Phase 2 and commented that her students' chatroom participation with the other classes, was *"too chaotic."* Linda stated that she would work collaboratively online but she did not benefit from the collaborative process.

Janet initiated the unit with less well-developed goals than Linda or Helen. However, she changed these goals, expanding her object, during the implementation process. She overcame several contradictions related to her teaching beliefs and the unit through collaborative dialogues with the other participating teachers and the researchers. She adapted her classroom to benefit from interaction with experts using their resources to aid her students as they developed strategies and solutions. She benefited from the professional development processes during the implementation. She, however, was unable to overcome contradictions in her building including a lack of technology support and the continued schedule for departmentalization in her grade.

Janet especially benefited from these dialogues as she worked with the more experienced teachers. In her post-unit interview she stated that she would *"never go back"* to teaching long division for six weeks again. Online collaborative dialogues empowered her to develop an advanced problem-based unit for her urban students (Russell, 2005b).

In summary, using intersubjectivity to understand the nature of the dialogues among the teachers and subsequently using activity theory to make inferences about the impact of collaborative dialogues on the teachers' pedagogical decision-making in their classrooms throughout the development of these innovative units, allows the researchers to understand the types and qualities of collaboration that are supportive of innovative teachers.

CONCLUDING REMARKS

The study has provided a novel exploration of the dynamics of intersubjectivity in online professional development and further clarified the otherwise ambiguous postulation of different levels of growth within the ZPD (Wertsch, 1984). One possible reason for the absence of this type of analysis is that researchers perhaps overestimate the clear knowledge of participants in problem-solving situations, or underestimate the underlying vigorous processes of collaborative learning. The three research questions will be responded in order below.

First, in contrast to previous studies which focused instead upon predominantly stable notions of intersubjectivity, this study reveals new evidence for the management of two variable forms of intersubjectivity: (a) temporary suspension, as reflected by the participant's communicative orientation first moving away from, then re-entering back towards social interaction, and (b) resistance and disagreement. The divergent aspects of two participant cases (i.e., Helen's temporary suspension behavior and Carol's communicative frustrations) illustrated a more dynamic form of intersubjectivity; one that is more partial and managed purposefully at several different levels and facilitated a strategic communicative function, so leading to the ultimate effectiveness of problem-solving activity, supportive of notions closer to, but extending, the sociocultural positions of Mortimer and Wertsch's (2003), Hui (2003), and Matusov's (1996).

Second, the dynamic temporal and spatial parameters of the computer-mediated environment both constrained (e.g., Wallace, 2004) and fostered conditions for creative communicative action (e.g., Herring, 1999). For example, synchronous online discourse seemed to be more suitable for gathering information and promoting socializing dialogues, whereas asynchronous discussion was more useful for task-oriented activities (e.g., Im & Lee, 2003). However, such time lag would also disrupt the flow to sustain interaction (e.g., Brunvand, Fishman, & Marx, 2003).

Finally, resembling a similar discourse suspension phenomenon (Hui, 2003) but in an online environment, for example, Helen's peripheral participation or lurking was construed as providing a strategic communicative function in her online discourse. One might further postulate that this occurred to facilitate her management of ongoing intersubjectivity with chatroom participants, but also to create or transform new intersubjective spaces, thus contributing to the ultimate effectiveness of collaborative learning. Viewed as a strategic act (e.g., Vygotsky, 1986, Wertsch, 1985), this moves us forward beyond Lave and Wenger's (1991) more descriptive "legitimate peripheral participation" and the cognitive impact of Azmitia's (2000) "time-outs" from collaborative activity. In her post-unit interview, Helen identified her responses to the collaboration online as *"letting down"* her students. She reported that she benefited from online dialogues but would not allow decisions made online to negatively impact her classroom integration in the future. By using activity theory to explore the impact of the collaborative discourse on Helen's pedagogical decisions, it has revealed that Helen has experienced difficulty in overcoming her contradictions involving technology support, despite her strategic peripheral participation in maintaining high level of intersubjectivity with other teachers.

In addition, levels of intersubjectivity would appear to change through dialogues over time. Instead of the binary (i.e., share-not shared) understanding assumed by previous studies of intersubjectivity, the findings here illustrate (in particular, Carol's frustration), both support, and extend, Mortimer and Wertsch's (2003) notion of resistance, and Matusov's (1996) notion of disagreement. Moreover, collective efforts leading to Carol's resolution demonstrate the maintenance of an intersubjectivity through role reciprocity (Sawyer, 2003), thus securing the effectiveness of ongoing collaborative activity, a theoretical scenario closer to both Rogoff's (1990, 1997) and Matusov's (1996) conceptions of individual contribution to joint activity. In her post-unit interview, Carol identified the collaboration online as being an important aspect of her unit completion, but did not change her concepts about student learning and technology integration. She did not want to *"let the others down"* by dropping out of the unit. However, she found the problem-based unit a *"waste of time"* and would not implement it again in the future. By using activity theory, it has made explicit how dialogues helped Carol, despite her divergent intersubjectivity because of her own goal to remain collaborative.

In conclusion, collaboration is an important aspect of innovative teachers' professional development. The current analysis and novel identification of variable forms of intersubjectivity has clarified the creative, effective monitoring and management of intersubjectivity as being crucial to success in collaborative activity. Furthermore, such can be seen to occur even despite each participant having putatively distinct representations and communicative goals for him/herself. This form of dialogues through online collaboration can be a beneficial aspect of innovative teachers' professional development. However it can also negatively impact the effectiveness of the teachers' efforts at implementation if the teachers do not appropriate the online dialogues to advance their capabilities and meet their goals. In most cases, online chats allow teachers to acquire information but not to solve problems or initiate new ideas to support the design and implementation of their innovative classrooms (Russell, 2005c). Only one teacher, Janet, used the online professional development as an opportunity to advance her knowledge of this innovative learning environment. To ensure effective integration of innovative technologies into educational settings, online professional development models should provide an interactive platform or medium which encourages multiple forms of dialogues for teachers to proactively communicate their goals and needs and to evaluate and advance their current individual-group efforts at innovation (Russell, 2005b), an important inquiry for future further exploration.

In terms of practice, teachers implementing innovative educational programs do need support but that support should be tailored to their needs by matching the support to their level of expertise and their levels of collaboration. In this study those teachers that were collaborative in their local context did not benefit from the collaboration and actually made decisions that hindered them from meeting their educational goals (Russell, 2005b). Those teachers that did not describe their local context as collaborative did benefit from the online dialogues by developing their innovative units beyond the goals they stated in the pre-unit interviews. Collaborative professional development forums, such as this online forum, should be modified to meet the collaborative needs and experiences of the teachers. If the teachers are highly collaborative in their local context their online dialogues need to be structured and monitored to develop more advanced teaching abilities and concepts. If the teachers are isolated in their local context, the collaboration can be more open and less structured as these teachers will be more likely to benefit as a result of the open dialogues. If possible, online professional development forums should identify the teachers' goals for the dialogues and develop a structure that mentors the teachers (Russell, 2005b).

Using intersubjectivity to examine the nature and the quality of the dialogues of teachers and subsequently using cultural historical activity theory to explore how those dialogues were used by these innovative teachers to develop their units of study and draw inferences on the impact of the collaborative dialogues on teachers' decision-making in their classrooms, gives researchers holistic tools to understand how collaboration can support innovative teachers as they implement new units. New knowledge about how to support innovation in education can lead to more successful change efforts and ultimately may lead to an increase in overall productivity for teachers as they attempt to develop new educational processes for their students.

ACKNOWLEDGMENT

The research reported here was supported by the enhancing Missouri's Instructional Networked Teaching Strategies (eMINTS). We would like to express our thanks to the participating teachers and students contributing to the research process and data corpus; and to Jim Wertsch, Keith

Sawyer, Rebecca Rogers, and Tony Dickinson, for constructive comments on early drafts of this manuscript.

REFERENCES

An, J. H., & Levin, J. A. (2003, April). *Online discourse patterns: Building an instructional framework for designing educational discourses on networks*. Paper presented at the 84th AERA Annual Meeting, Chicago, IL.

Azmitia, M. (2000). Taking time out from collaboration: Opportunities for synthesis and emotion regulation. In R. Joiner, K. Littleton, D. Faulkner and D. Miell (Eds.), *Rethinking collaborative learning* (pp. 179-195). London, UK: Free Association Books.

Bakhtin, M. M. (1986). The problem of speech genres. In M. M. Bakhtin, *Speech genres and other late essays* (V. McGee, Trans. & C. Emerson & M. Holquist, Eds.) (pp. 60-102). Austin, TX: University of Texas Press.

Beals, D. (1992). Computer networks as a new data base. *Journal of Educational Computing Research, 8*(3), 327-345.

Berge, A. L., & Collins, M. (Vol. Eds.). (1998). *Wired together: The online classroom in K-12: Vol. 3. Teacher education and professional development*. Cresskill, NJ: Hampton Press, Inc.

Bereiter, C. (2002). *Education and mind in the knowledge age*. Mahwah, NJ: Lawrence Erlbaum Associates.

Black, S. D., Levin, J. A, Mehan, H., & Quinn, C. N. (1983). Real and non-real time interaction: Unraveling multiple threads of discourse. *Discourse Processes, 6*, 59-75.

Bober, M. J., & Dennen, V. P. (2001). Intersubjectivity: Facilitating knowledge construction in online environments. *Education Media International, 38*(4), 241-250.

Borko, H. (2004). Professional development and teacher learning: Mapping the terrain. *Educational Researcher, 33*(8), 3-15.

Borthwick, A., Pierson, M., Anderson, C., Morris, J., Lathem, S., & Parker, H. (2004). Building learning communities to increase faulty and preservice teacher use of technology. *Journal of Computing in Teacher Education, 21*(1), 23-32.

Bransford, J. D., Brown, A. L., & Cocking, R. R. (1999). *How people learn: Brain, mind, experience, and school* [Electronic version]. Washington, D. C.: National Academy of Sciences.

Bruner, J. (1990). *Acts of meaning.* Cambridge, MA: Harvard University Press.

Bruner, J. (1996). Psychology's next chapter. In J. Bruner, *The culture of education* (pp. 160-85). Cambridge, MA: Harvard University Press.

Brunvand, S., & Fishman, B. (2005). Aligning the design of online professional development with the needs and expectations of teachers. *Proceedings of the Society for Information Technology and Teacher Education (SITE), USA*, 2171-2176.

Brunvand, S., Fishman, B., & Marx, R. (2003). *Moving Professional Development Online: Meeting the Needs and Expectations of All Teachers* [Online]. Paper presented at the Annual Meeting of the National Association of Research on Science Teaching, Philadelphia, PA. Retrieved from http://www-personal.umich.edu/~fishman/papers/papers…

Cochran-Smith, M. (2003). Learning and unlearning: The education of teacher educators. *Teaching and Teacher Education, 19*, 5-28.

Castells, M. (2001). *The Internet galaxy.* Oxford, UK: Oxford University Press.

Cazden, C. (2001). *Classroom discourse: The language of teaching and learning* (2nd ed.). Portsmouth, NH: Heinemann.

Cole, M., & Engeström, Y. (1993). A cultural-historical approach to distributed cognition. In G. Salomon (Ed.), *Distributed cognition: Psychological and educational considerations*. Cambridge, MA: Cambridge University Press.

Collison, G., Elbaum, B., Haavind, S., & Tinker, R. (2000). *Facilitating online learning: Effective strategies for moderators*. Madison, WI: Atwood Publishing.

Davidson, P. M. (1992). The role of social interaction in cognitive development: A propaedeutic. In L. T. Winegar & J. Valsiner (Eds.), *Children's development within social context* (Vol. 1, pp. 19-37). Hillsdale, NJ: Lawrence Erlbaum Associates.

de Vries, L., Naidu, S., Jegede, O., & Collis, B. (1995). Online professional development: An evaluation study. *Distance Education, 16*(1), 157-173.

Dede, C. (1996). Distance learning – Distributed learning: Making the transformation [Electronic version]. *Learning and Leading with Technology, 23*(17), 25-30.

DiMauro, V., & Jacobs, G. (1995). *Filling in the professional gaps: Active teacher participation on a telecommunications network*. Retrieved October 9, 2002, from http://www.terc.edu/papers/labnet/Articles/gaps.html

Engeström, Y. (1987). *Learning by expanding: An activity-theoretical approach to developmental research*. Helsinki, Finland: Orienta-Konsultit.

Engeström, Y., Miettinen, R., & Punamaki, R. (Eds.). (1999). *Perspectives on activity theory*. Cambridge, UK: Cambridge University Press.

Fahy, P. J., Crawford, G., & Ally, M. (2001). Patterns of interaction in a computer conference transcript. *International Review of Research in Open and Distance Learning, 2*(1), 1-18.

Flynn, T., & Polin, L. (2003, April). *Making sense of online learning: Frames, rubrics, tools and coding systems for analyzing asynchronous online discourse*. Paper presented at the 84th AERA Annual Meeting, Chicago, IL.

Forman, E. A. (1992). Discourse, intersubjectivity, and the development of peer collaboration: A Vygotsky approach. In L. T. Winegar & J. Valsiner (Eds.), *Children's development within social context* (Vol. 1, pp. 143-159). Hillsdale, NJ: Lawrence Erlbaum Associates.

Gee, J. (1999). *An introduction to discourse analysis: Theory and method*. London, UK: Routledge.

Göncü, A. (1993). Development of intersubjectivity in social pretend play. *Human Development, 36*, 185-198.

Gordin, D. N., Gomez, L. M., Pea, R. D., & Fishman, B. J. (1996). Using the world wide web to build learning communities in K-12 [Online]. *Journal of Computer Mediated Communication, 2* (3), 1-27. Available: http://www.ascusc.org/jcmc/vol2/issue3/gordin.html

Gunawardena, C. N., Lowe, C. A., & Anderson, T. (1997). Analysis of a global online debate and the development of an interaction analysis model for examining social construction of knowledge in computer conference. *Journal of Educational Computing Research, 17*(4), 397-431.

Henri, F. (1992). Computer conferencing and content analysis. In A. R. Kaye (Ed.), *Collaborative learning through computer conferencing: The Najaden Papers* (pp. 117-136). New York, NY: Springer-Verlag.

Herring, S. (1999). Interactional coherence in CMC [Online]. *Journal of Computer-Mediated Communication, 4*(4), 1-24. Retrieved from http://www.ascusc.org/jcmc/vol4/issue4/herring.html

Herrmann, F. (1995). Listserver communication: The discourse of community-building. *Proceedings of the first international conference on computer support for collaborative learning, October 17-20, 1995, Indiana University, Bloomington, Indiana, USA*, 165-168.

Hui, D. (2003). Managing intersubjectivity in the context of a museum learning environment. *International Journal of Learning, 10*, 1533-1549.

Hui, D. (2006). *Engagement in supporting new teachers: A role for computer-mediated communication in teacher learning within informal professional communities.* Unpublished doctoral dissertation, Washington University in St Louis, Missouri.

Hui, D., & Russell, D. (2007). Understanding innovative professional development for educators through the analysis of intersubjectivity in online collaborative dialogues. *International Journal of Information Communication Technology Education, 3*(3), 25-38.

Im, Y., & Lee, O. (2004). Pedagogical implications of online discussion for pre-service teacher training. *Journal of Research on Technology in Education, 36*(2), 155-170.

Jonassen, D. H., Peck, K. L., & Wilson, B. G. (1999). *Learning with technology: A constructivist perspective.* Upper Saddle River, NJ: Merrill Prentice Hall.

Kabilan, M. (2005). Online professional development: A literature analysis of teacher competency. *Journal of Computing in Teacher Education, 21*(2), 51-57.

Koschmann, T. (Ed.). (1996*). CSCL: Theory and practice of an emerging paradigm*. Mahwah, NJ: Lawrence Erlbaum Associates.

Korthagen, F. (1993). Two modes of reflection. *Teacher and Teacher Education, 9(*3), 317-326.

Lave, J., & Wenger, E. (1991). *Situated learning: Legitimate peripheral participation.* Cambridge, UK: Cambridge University Press.

Marx, R. W., Blumenfeld, P. C., Krajcik, J. S., & Soloway, E. (1998). New technologies for teacher professional development. *Teaching and Teacher Education, 14*(1), 33-52.

Matusov, E. (1996). Intersubjectivity without agreement. *Mind, Culture, and Activity, 3*(1), 25-45.

Mortimer, E. F. & Wertsch, J. V. (2003). The architecture and dynamics of intersubjectivity in science classrooms. *Mind, Culture, and Activity, 10*(3), 230-244.

National Commission on Teaching and America's Future. (2003, January). *No dream denied: A pledge to America' children.* Washington, D.C.: National Commission on Teaching and America's Future.

NetDay. (2004, June). *Insights and ideas of teachers on technology.* Irvine, CA: NetDay Speak Up Day 2004. Retrieved January 30, 2006, from http://www.netday.org/sud4teachers_2004_report.htm

Nieto, S. (2003). *What keeps teachers going?* New York, NY: Teachers College Press.

Office of Technology Assessment, U.S. Congress. (1995, April). *Teachers & technology: Making the connection* (Report No. IR 017 275). Washington, D. C.: Office of Technology Assessment. (ERIC Document Reproduction Service No. ED386154)

Parrish, R. (1998). *Conversation analysis of internet chat rooms* [Online]. Available: http://polisci.wisc.edu/~rdparrish/academic_works.htm

Putnam, R. T., & Borko, H. (2000). What do new views of knowledge and thinking have to say about research on teacher learning? *Educational Researcher,* 29 (1), 4-15.

Quinn, C. N., Mehan, H., Levin, J., & Black, S. (1983). Real education in non-real time: The use of electronic message systems for instruction. *Instructional Science, 11*, 313-327.

Riel, M., & Fulton, K. (2001). The role of technology in supporting learning communities. *Phi Delta Kappan, 82*(7), 518-523.

Rogoff, B. (1990). *Apprenticeship in thinking: Cognitive development in social context.* New York, NY: Oxford University Press.

Rogoff, B. (1997). Evaluating development in the process of participation: Theory, methods, and practice building on each other. In E. Amsel & A. Renninger (Eds.), *Change and development* (pp. 265-285). Hillsdale, NJ: Lawrence Erlbaum Association.

Romiszowski, A. J., & Mason, R. (1996). Computer-mediated communication. In D. H. Jonassen (Ed.), *Handbook of research for educational communications and technology: A project of the association for educational communications and technology* (pp. 438-456). New York, NY: Macmillian Library Reference USA.

Rommetveit, R. (1979). On the architecture of intersubjectivity. In R. Rommetveit & R. M. Baker (Eds.), *Studies of language, thought, and verbal communication* (pp. 93-107). London, UK: Academic Press.

Rourke, L., Anderson, L., Garrison, D. R., & Archer, W. (2001). Methodological issues in the content analysis of computer conference transcripts. *International Journal of Artificial Intelligence in Education, 12*(1), 8-22.

Russell, D. (2005a). Implementing an innovation cluster in educational settings in order develop constructivist-based learning environments. *Educational Technology and Society, 8*(2).

Russell, D. (2005b). Transformation in an urban school: Using systemic analysis to understand an innovative urban teacher's implementation of an online problem-based unit. *International Journal of Instructional Technology & Distance Learning, 2*(2).

Russell, D. (2005c). Understanding innovation in education using activity theory. *Educational Technology in Society, 8*(1), 38-53

Salmon, G. (2004). *E-moderating: The key to teaching and learning online* (2nd ed.). London, UK: Routledge-Falmer.

Salomon, G. (1993). No distribution without individuals' cognition: A dynamic interactional view. In G. Salomon (Ed.), *Distributed cognitions: Psychological and educational considerations.* New York, NY: Cambridge University Press.

Sawyer, R. K. (2003). *Group creativity: Music, theater, collaboration.* Mahwah, NJ: Lawrence Erlbaum Associates.

Schegloff, E. (1999). Talk and social structure. In A. Jaworski, & N. Coupland (Eds.), *The discourse reader* (pp. 107-120). London, UK: Routledge.

Schlager, M. S., Fusco, J., & Schank, P. (2002). Evolution of an on-line education community of practice. In K. A. Renninger & W. Shumar (Eds.), *Building virtual communities: Learning and change in cyberspace* (pp. 129-158). New York, NY: Cambridge University Press.

Schön, D. A. (1983). *The reflective practitioner: How professionals think in action.* New York, NY: Basic Books.

Selwyn, N. (2000). Creating a "connected" community? Teachers' use of an electronic discussion group. *Teachers College Record, 102*(4), 750-778.

Shulman, L. (1986). Those who understand: Knowledge growth in teaching. *Educational Researcher, 15*(2), 4-14.

Thompson, A. (2005). Scientifically based research: Establishing a research agenda for the technology in teacher education community.

Journal of Research on Technology in Education, 37(4), 331-337.

Thompson, A., Bull, G., & Bell, L. (2005). A proactive approach to a research agenda: A call to action (editor's remarks). *Journal of Computing in Teacher Education, 21*(3), 74, 76.

Tomasello, M. (1999). *The cultural origins of human cognition.* Cambridge, MA: Cambridge University Press.

Topper, A. (1995). *Intersubjectivity and educational computer-mediated communication.* Unpublished manuscript.

Tyack, D., & Cuban, L. (2004). Tinkering toward utopia: A century of public school reform. In J. H. Ballantine & J. Z. Spade (Eds.), *Schools and society: A sociological approach to education* (2nd Ed., pp. 459-478). Belmont, CA: Wadsworth/ Thomson Learning.

Vygotsky, L. (1986). *Thought and language* (A. Kozulin, Ed. & Trans.). Cambridge, MA: The MIT Press. (Original work published in 1934).

Wade, S. E., & Fauske, J. R. (2004). Dialogue online: Prospective teachers' discourse strategies in computer-mediated discussions. *Reading Research Quarterly, 39*(2), 134-160.

Wallace, R. M. (2004). A framework for understanding teaching with the internet. *American Educational Research Journal, 41*(2), 447-488.

Wenger, E. (1998). *Communities of practices.* Cambridge, UK: Cambridge University Press.

Wertsch, J. V. (1984). The zone of proximal development: Some conceptual issues. In B. Rogoff & J. Wertsch (Eds.), *Children's learning in the "zone of proximal development"* (pp. 7-18). San Francisco, CA: Jossey-Bass Inc., Publishers.

Wertsch, J. V. (1985). *Vygotsky and the social formation of the mind.* Cambridge, MA: Harvard University Press.

Wertsch, J. V. (1991). *Voices of the mind: A sociocultural approach to mediated action.* Cambridge, MA: Harvard University Press.

Wilkins, H. (1991). Computer talk. *Written Communication, 8*(1), 56-78.

Zeichner, K., & Liston, D. (1996) *Reflective teaching; An introduction.* Mahwah, NJ: Lawrence Erlbaum Association.

Zhao, Y., & Frank, K. A. (2003). Factors affecting technology uses in schools: An ecological perspective. *American Educational Research Journal, 40*(4), 807-840.

Zhao, Y., & Rop, S. (2001). A critical review of the literature on electronic networks as reflective discourse communities for inservice teachers. (ERIC Document Reproduction Service No. ED459483)

Chapter XXI
Effective Questioning to Facilitate Dynamic Online Learning

Silvia Braidic
California University of Pennsylvania, USA

ABSTRACT

Teaching is a complex activity that involves careful preparation, delivery and reflection. As an educator, it is essential to create a sense of community in which students feel significant and are truly engaged as learners. A central focus of the educator is to maximize the capacity of each learner. How does this happen in an online learning environment? This chapter addresses the needs of learners for a learning community that promotes effective discussion; specifically, the practice of questioning that lies at the heart of classroom practice. Just as in a face to face classroom, questioning occurs in a variety of ways for online learners. The chapter shares ideas for effective questioning strategies in an online environment.

INTRODUCTION

In the book, *Technology Literacy Applications in Learning Environments* edited by David Carbonara, Tomei states "Technology plays a significant role in changing the instructional environment by promoting the role of the teacher as a guide in educational discovery, serving as a resource to the student-as-information gatherer." In an online environment, just as in a traditional classroom, you have a spectrum of learners. As an instructor, how do you begin to address the needs of the spectrum of learners in your classroom and create a learning community that promotes effective discussion? Different instructional practices help students learn in meaningful ways. One particular teaching strategy that is utilized in both traditional and online courses is discussion.

Questioning is a significant instructional design element for the promotion of effective discussion (Muilenburg & Berge, 2000). Research on online education consistently finds that high and consistent interaction levels between students and the professor, and high interaction levels between the students themselves, is often seen as a positive variable (Johnson, Aragon, Shaik, & Palmas-Rivas, 2000; Berge & Collins, 1996; Tu, 2000; Muirhead, 2001; Blignaut & Trollip, 2003; Vonderwell, 2003). Akin and Neal (2007) state, "Most online instructors, aware of how important student participation is to online learning, will realize that s/he must produce solid educational discussion questions that also engage. These good questions must also be sound in terms of learning theory, be big enough to engage online classes with possibly 30 or more learners, and long enough to last a module." Questioning provides students with an opportunity to challenge their thinking. As teachers, we are constantly asking questions. Asking questions that require higher level thinking is not an easily acquired skill. Good questioning takes thinking time, planning ahead, and experience. Using effective questioning strategies, teachers restructure their online classroom to engage students in higher level thinking. Questioning can not only help students meet course goals and objectives, but it also engages all students, and improves the quality of teaching and learning at all levels. An excellent first step in differentiating online is to increase the challenge and variety of your class discussions, activities, and assignments through questioning. By paying attention to the kinds of questions you ask, you can stimulate learning with a wide range of learners in your online classroom based on their readiness, interests, and learning style. In order to use the discussion method effectively, it is critical to understand how to design and maintain an online discussion so that all learner needs are met. In order to do so, questioning is an integral focus.

Why Question?

Questions are an important part of communication. It is probably safe to say that questioning is at the heart of classroom practice. Research in classroom behavior indicates that cueing and questioning might account for as much as eighty percent of what occurs in a given classroom on a given day (Marzano, 2001). In Marzano's book, *Classroom Instruction that Works: Research-Based Strategies for Increasing Student Achievement,* he indicated four generalizations as it related to questioning(p.113-114):

1. Cues and questions should focus on what is important as opposed to what is unusual.
2. "Higher level" questions produce deeper learning than "lower level" questions.
3. "Waiting" briefly before accepting responses from students has the effect of increasing the depth of students' answers.
4. Questions are effective learning tools even when asked before a learning experience.

I would propose that these generalizations also hold true for an online learning environment. All students need to be accountable for thinking at higher levels. Some students will be challenged by a more basic question, while others will need more. As in a traditional face to face setting, all students can hear and learn through a wide range of responses and questions; so it is also true in an online setting where students may engage in oral and written responses and discussion.

Who Needs to Ask Questions?

When considering who needs to ask questions in an online classroom, I would include both teachers and students in the **I.Q.** or **I Q**uestion process. Whether in a traditional face to face classroom or in an online environment, teachers must ask questions. Just as important in engaging students to answer questions at various levels of difficulty, students must also engage in opportunities that elicit them to generate and ask questions.

How Do You Question in an Online Environment?

Just as in a face to face classroom, questioning can occur in a variety of ways online to engage all kinds of learners. First, teachers ask questions to promote discussions. Second, teachers pose questions on learning activities and assignments as they work individually or in small or large groups. Finally, teachers find ways to engage students in question-asking.

The remainder of this chapter will share ideas for using the **I.Q. – I Q**uestion strategy in an online environment.

I.Q. – I QUESTION WITH BLOOM'S AT-A-GLANCE

The Taxonomy of Educational Objectives by Dr. Benjamin Bloom brought a structure to higher mental thought processes. It is still utilized today as a way to develop questions that will probe the widest possible range of intellectual activity. In Gridley's book, *Asking Better Classroom Questions*, he states that "by building classroom questions upon Bloom's taxonomy, teachers can systematically cover as wide a range of intellectual processes as they wish, from rote memory through sophisticated and creative manipulations of knowledge." These levels build upon each other as the learner gains knowledge and expertise therefore leading the student to complex understandings and knowledge" (Christopher, Thomas, & Tallent-Runnels, 2004). In the 1990's, Bloom's was revisited and some changes were made. The six categories were changed from noun to verb forms. Bloom's Revised Taxonomy of Educational Objectives (Anderson, et al., 2001) is designed to represent qualitatively different levels of cognition. In the hands of teachers, it can be used to plan for the quality of thinking that an instructor would like to create.

The **I.Q. – I Q**uestion with Bloom's At-A-Glance is a strategy that may be utilized for readings, articles, and other activities. It is important that we model the kinds of questions that stimulate higher level thinking. As teachers, when we design learning environments and experiences, we must be careful to plan for the type of cognitive processing that we hope to foster in an online environment. Incorporating a variety of questions,

Figure 1.

Bloom's	**Bloom's Revised Taxonomy**
Evaluation	Creating
Synthesis	Evaluating
Analysis	Analyzing
Application	Applying
Comprehension	Understanding
Knowledge	Remembering

Knowledge/Remembering – recalling information
Comprehension/Understanding – interpreting; explaining ideas or concepts
Application/Applying – using information in another familiar situation
Analysis/Analyzing – breaking information into parts to explore understanding and relationships
Synthesis/Creating – generating new ideas, products, or ways of viewing things
Evaluation/Evaluating – justifying a decision or course of action

will work to create an environment to address the diverse needs of your learners. The more we can actively engage the learner in the process of thinking and manipulating information, the deeper the processing and the more meaningful the learning (Jonassen, Peck, & Wilson, 2000). One way to accomplish this, is by using a list of prompts, adapted from Gregory and Chapman's book *Differentiated Instructional Strategies: One Size Doesn't Fit All* to springboard discussions in an online class.

Figure 2.

Knowledge
Who, what, when, where, how, ____?
Describe ____________________
Comprehension
Retell _____in your own words
What is the main idea of _____?
Application
How is __an example of ___?
Why is ___significant?
Analysis
What are the parts or features of ___?
Classify ___according to ___?
How does ___compare/contrast with ___?
What evidence can you present for___?
Synthesis
What would you infer from ___?
What ideas can you add to ___?
What might happen if you ___?
What solutions would you suggest for ___?
Evaluation
Do you agree with ___? Why?
What is the most important ___?
Prioritize ___according to ___?
What criteria would you use to assess ___?

As the teacher, it is my role to consider how to most effectively create and ask questions at various levels of complexity so that all students are challenged. **I.Q. – I Q**uestion with Bloom's At-A-Glance serves as a way to help do so. This is most effective when engaging in ongoing threaded discussions. Often, students will post a response, and at times it is appropriate to challenge a students thinking. The **I.Q. – I Q**uestion with Bloom's At-A-Glance may also be shared with students. This may be done to engage students in asking questions at various levels of Bloom's. By modeling the questioning process throughout the course and sharing some guidelines to help students ask questions, will make students more active in the learning process. Now, we not only want students to respond to various levels of questions, but we also want them to ask questions of the instructor and of their classmates. It takes students out of only the response mode and encourages them to set their own agenda for exploration.

I.Q. – I QUESTION BY MOVING UP BLOOM'S

Another approach in utilizing Bloom's Taxonomy to engage students in answering a variety of questions and also asking various questions is the technique **I.Q**. – **I Q**uestion by Moving Up Bloom's Taxonomy. This is an extension of how to use Bloom's prompts for an assignment such as various article readings, cases, or the like. Below is a worksheet titled "Moving Up Bloom's Taxonomy." This may be utilized by instructors with a particular course reading in preparation for a threaded discussion or even a chat session. The purpose is to engage the learners in various levels of questioning.

Figure 3.

Moving Up Bloom's Taxonomy

____

(Title of Article/Case Study/Other)

Knowledge:

What is the definition for ____________________________________?

List words in the article that are new to you.

Recall the main facts of the passage you read. List these.

Make a time line of events.

COMPREHENSION

Explain the main idea of the passage in your own words.

Why was ______________________?

Prepare a flow chart to show...

Write a paragraph explaining...

APPLICATION

What questions would you ask as it relates to the passage?

Demonstrate the way to...

What is another instance of...

How could the information from this article be applied to your situation?

ANALYSIS

If _______________ then _________________?

What is the relationship between __________ and _______?

Compare and contrast ____________________ with __________

SYNTHESIS

What if ____?

How would you deal with _______________________________?

Propose a method to __________________________________.

Based on your reading of the article, create ...

EVALUATION

In your opinion ____?

What solution do you favor and why?

Grade or rank the _____________________________________.

Defend or support the position presented in the article...

(Adapted from Nancy Johnson's Active Questioning, 1995)

I.Q. – I QUESTION WITH DIVERGENT THINKING MODELS AT-A-GLANCE

In addition to Bloom's Taxonomy, divergent questioning models also contribute to creating an effective online learning environment. Divergent questions allow students to explore different avenues and create many different variations and alternative answers or scenarios. These types of questions often require students to analyze, synthesize or evaluate a knowledge base and then project or predict different outcomes. Frequently the intention of these types of questions is to stimulate imaginative and creative thought, or investigate cause and effect relationships, or provoke deeper thought or extensive investigations. And, one needs to be prepared for the fact that there may not be right or definitely correct answers to these questions. Divergent questioning:

- Challenges thinking abilities
- Encourages several answers, or possibilities and stimulates idea creation
- Requires unusual or multiple answers
- Extracts information which is already learned and stimulates productive and divergent thinking
- Includes a large percentage of open questions
- Provides an ideal opportunity to ask thought provoking and probing questions
- Encourages students to ask divergent questions of their peers and of their teacher

The **I.Q. – I Q**uestion with Divergent Thinking Models At-A-Glance below can be used similarly to Bloom's At-A-Glance.

Adapted from the information provided in Dr. Roger Taylor's workshop on *Current, Best Instructional Strategies for Your Gifted and Highly Capable Students* that utilizes the work of Torrance, teachers and students can utilize this strategy in an online classroom.

Figure 4.

Brainstorming Model
(encourages fluent thinking)

List all of the ___

List as many ___ as you can think of

How many ways can you come up with ___

Viewpoint Model
(encourages seeing from a different point of view)

How would this look to a ___

What would a ___mean from the viewpoint of a ___?

How would ___view this?

Involvement Model
(encourages associative thinking)

How would you feel if you were __?

If you were a ___ what would you (see, taste, smell, feel)?

Conscious Self-Deceit Model
(encourages students to imagine that something is true or not true and consider the consequences)

Suppose you could have anything you wanted. What ideas could you produce if this were true?

You have been given the power to___. How will you use it?

Forced Association Model
(encourages making comparisons between different things)

How is a ___like a ___?

Get ideas from __to improve ___.

Reorganization Model
(encourages reorganization of information)

What would happen if ___ were true?

Suppose ___ happened, what would be the consequences?

What would happen if there were no ___?

I.Q. – I QUESTION: AT-A-GLANCE QUESTIONING (BLOOM'S AND DIVERGENT THINKING)

The **I.Q.** – **I Q**uestion At-A-Glance Questioning Sheet is a resource for instructors that was compiled by the author to provide instructors with a tool when creating their online learning activities. This tool provides a simple approach for instructors to help create a balance of questions in order to best address the needs of all kinds of learners in the class.

I.Q. – I QUESTION A PICTURE

In an online environment, you will have students that learn best through auditory, visual or kinesthetic modes at various levels of readiness. Utilizing the I.Q. – I Question a Picture strategy can be a wonderful way to stimulate active questioning by students. Almost more than any other source, a photograph provides an incentive to dig, to burrow, to stretch, to tease out, to investigate and follow up leads (Nuffield Primary History).

Figure 5.

At-A-Glance Questioning - Bloom's Taxonomy	At-A-Glance Questioning - Divergent Thinking
Knowledge – Identification and recall of Information Who, what, when, where, how _____? Describe ________________________	**Brainstorming Model** – encourages fluent thinking List all of the __________. List as many _______ as you can think of. How many ways can you come up with _____.
Comprehension – Organization and selection of facts and ideas Retell ______ in your own words What is the main idea of ________	**Viewpoint Model** – Encourages seeing from a different point of view How would this look to a _____. What would a ______ mean from the viewpoint of a _____? How would _____ view this?
Application – Use of facts, rules, principles How is _____ an example of _____ Why is _____ significant?	**Involvement Model** – Encourages associative thinking How would you feel if you were ________? If you were a ______what would you (see, taste, smell, feel)?
Analysis – Separation of a whole into component parts What are the parts or features of _____ Classify ____ according to __________. Outline/diagram/web _______________. How does _____ compare/contrast with __? What evidence can you present for ___?	**Conscious Self-Deceit Model** – Encourages students to imagine that something is true or not true and consider eth e consequences Suppose you could have anything you wanted. What ideas could you produce if this were true? You can have all of the ____ in the world. How could you use it to ___? You have been given the power to ___. How will you use it?
Synthesis – Combination of ideas to form a new whole What would you predict/infer from _____? What ideas can you add to ____________? How would you create/design a new ____? What might happen if you combined _____ with ___? What solutions would you suggest for ___?	**Forced Association Model** – Encourages making comparisons between different things How is a ____like a _____? Get ideas from ___to improve ___.
Evaluation – Development of opinions, judgments, or decisions Do you agree _____________________? What is the most important _________? Prioritize _____ according to ________? How would you decide about ________? What criteria would you use to assess ___?	**Reorganization Model** – Encourages reorganization of information What would happen if ___ were true? Suppose ___ happened, what would be the consequences? What would happen if there were no ___?

Consider presenting students with a picture. Ask students to think about what they "see" and what they "don't see." By utilizing a picture, students are thinking differently. Students might consider such questions as a starting point for discussion. As they engage in asking and answering questions, a new question may be formed.

- Where did this take place?
- What happened before the picture?
- What happened after?
- What might have caused this situation?
- Who is involved?
- When did this take place?
- Why is this happening?

I.Q. – I Question a Picture provides a questioning technique that will address those students who learn best visually. A similar technique may be utilized for audio where students are presented with similar prompts that are addressed to a musical piece or audio clip of a speech.

Taking Time to Reflect on Questioning

Not only is it important to engage learners with answering and asking questions, but it is also important to take time to reflect upon questioning. Consider these questions for reflection as it relates to your online course and your utilization of questions.

Setting the Stage for Learning

- What questions did you use to focus or introduce the students thinking before the lesson was taught?

Questions

- When reflecting upon your online course, how many questions did you post for a particular learning activity? For a unit?
- What question(s) did I ask that did not elicit student responses at the level I seeked? Why?
- What kinds of questions did I ask most? Least?
- What question(s) do I need to think more about before responding?
- What question(s) should I have asked?

Goal Setting

- My goal is to ask more of the following kinds of questions:

CONCLUSION

Learning to ask good questions is a valuable skill to acquire, and our students will become good at questioning if we build in opportunities for them to ask their own questions. In my experience, when students are properly encouraged, they will ask a wide range of sensible questions, better than those we might have asked them. When we invite students to not only answer questions but also to ask questions about a topic, we must treat their questions with respect: by recording them, encouraging investigation of the topic, and at the end, reviewing them to check whether we have been able to address them and if not, why. Questions allow us to make sense of the world. Used purposefully, questioning can help instructors achieve well defined goals. Meulinburg and Berge (2000) state, "In a constructivist learning environment, the instructor always needs to keep in mind that when facilitating online discussion, asking the right questions is almost always more important than giving the right answers." Good questioning generates good discussion. It is a skill. Questioning is at the heart of learning. Discussions will be more effective when questions are well planned and aligned to the purposes of the class. To learn effectively, students need to learn actively, and one way to encourage this is through questions. Whether in a traditional classroom or in an online learning environment, instructors must develop a place where students feel comfortable with questions. As Albert Einstein said, "The important thing is not to stop questioning."

REFERENCES

Akin, L., & Neal, D. (2007). CREST + Model: Writing Effective Online Discussion Questions. Journal of Online Learning and Teaching, Vol. 3. No. 2, June 2007. Retrieved September 26, 2007 from http://jolt.merlot.org/vol3no2/akin.htm

Anderson, L.W., & Krathwohl (Eds.). (2001). *A Taxonomy for Learning, Teaching, and Assessing: A Revision of Bloom's Taxonomy of Educational Objectives*. New York: Longman.

Benjamin, A. (2005). Differentiated instruction using technology. Larchmont, NY: Eye on Education, Inc.

Berge, Z., & Collins, M. (1996). Where interaction intersects time. *MC Journal: The Journal of Academic Media Librarianship,* 4(1). Retrieved September 26, 2007, from http://wings.buffalo.edu/puublications/mcjrnl/v4n1/berge.html

Blignaut, S., & Trollip, S. (2003). Developing a taxonomy of faculty participation in asynchronous learning environments. An exploratory investigation. *Computers and Education,* 41(2), 149-172.

Bloom, B. (1956). Taxonomy of educational objectives. New York, NY: David McKay Company, Inc.

Carbonara, D.(editor) 2005. Technology literacy applications in learning environments. Hershey, PA: Idea Group, Inc.

Cruz, E. (2003). Bloom's revised taxonomy. In B. Hoffman (Ed.), *Encyclopedia of Educational Technology*. Retrieved September 25, 2007, from http://coe.sdsu.edu/eet/articles/bloomrev/start.htm

Gregory, G. and Chapman, C. (2002). Differentiated instructional strategies. Thousand Oaks, Ca: Corwin Press, Inc.

Gridley, R. Asking better classroom questions: A teachers' mini-workbook.

Davidson, K. and Decker, T.(2006). Bloom's and Beyond. Pieces of Learning Publishing.

Johnson, N. (1995). Active questioning. Pieces of Learning Publishing.

Johnson, S., Aragon, S., Najmuddin, S., & Palma-Rivas, N. (2000). Comparative analysis of learner satisfaction and learning outcomes in online and face-to-face learning environments. *Journal of Interactive Learning Research,* 11(1), 29-48.

Jonassen, D., Peck, K., and Wilson, B. (2000). Learning With Technology: A Constructivist Perspective. Merrill: Upper Saddle, NJ.

Marzano, R., Pickering, D., and Pollock, J. (2001). Classroom instruction that works: research-based strategies for increasing student achievement. Alexandria, VA: Association for Supervision and Curriculum Development.

Muelinburg, L. and Berge, Z. (2000). The moderators' homepage: A framework for designing questions for online learning (website). Retrieved from http://www.emoderators.com/moderators/muilenburg.html

Muirhead, B. (2001). Enhancing social interaction in computer-mediated distance education. *Ed at a Distance,* 15(40). Retrieved September 26, 2007, from http://www.usdla.org/html/journal/APR01_Issue/articel02.html

Nuffield Primary History (website). Retrieved from http://www.primaryhistory.org/aboutus

Taylor, R. (2001). Current, best instructional strategies for your gifted and highly capable students (workshop). Presented by Bureau of Education and Research.

Tu, C. (2000). Strategies to increase interaction in online social learning environments. Paper presented at the Society for Information Technology and Teacher Education International Conference, San Diego, CA.

Vonderwell, S. (2003). An examination of asynchronous communication experiences and perspectives of students in an online course: A case study. *The Internet and Higher Education, 6*, 77-90.

Chapter XXIX
Transitioning from Face-to-Face to Online Instruction:
How to Increase Presence and Cognitive/Social Interaction in an Online Information Security Risk Assessment Class

Cindy S. York
Purdue University, USA

Dazhi Yang
Purdue University, USA

Melissa Dark
Purdue University, USA

ABSTRACT

This article briefly reviews two important goals in online education: interaction and presence. These are important goals in online education because they are linked to learning and motivation to learn. The article provides guidelines and an extended example of how to design an online course in information security in a manner that will enhance interaction and presence. This article's contribution is to provide guidelines with a corresponding extended and concrete example for those who are tasked with designing and delivering online courses. Although the guidelines and example were targeted to the field of information security, they can be readily adopted by other disciplines.

INTRODUCTION

Although online education can offer convenience and flexibility for learners, it is not without challenges. Frequently, online education is no more than instructor notes and lecture materials posted on a Web site, perhaps with some required discussion. Much online instruction is designed, developed, and delivered without careful consideration of foundational instructional design principles. Research has shown that online courses that lack substantive and meaningful interaction, coupled with a sense of presence (feeling as though belonging in a virtual environment), contribute to a sense of isolation, unsatisfying learning experiences, and high dropout rates (Aragon, 2003; Bennett, Priest, & Macpherson, 1999; Glickman, 2003; Moore & Kearsley, 1996). The goal of this article is to provide a set of online course design guidelines based on research findings and best practices to enhance interaction and sense of presence, which are two critical factors that impact learning and motivation to learn in online courses (Moore, 1992; 1993; Muirhead, 1999; Richardson & Swan, 2003). Finally, an example is provided for applying the guidelines to transition a face-to-face class to an online class, using an information security risk assessment class. In order for these guidelines to make sense, we start with a brief discussion of interaction and presence.

Interaction

Moore (1989) identified three major types of interaction: a) learner-content, b) learner-instructor, and c) learner-learner. Learner-content interaction refers to the amount of substantive interaction occurring between the learner(s) and the content. Content could be in the form of text, radio, television, and/or audiotape. Participant interaction (learner-learner and learner-instructor) refers to the engagement of the learners and instructor in the learning and teaching process. It also refers to dialogue between and/or among different participants in online learning environments. Thus, interaction is more than a communication exchange; interaction occurs when objects, actions, and events mutually influence one another (Wagner, 1994). Instructional interaction is meaningful communication that challenges learners' thinking, shapes the acquisition of knowledge in meaningful ways, and changes learners, moving them toward achieving their goals. Effective interaction is not necessarily more interaction, rather it is interaction resulting in learners thinking in new and more profound ways. While the literature and research confirmed the importance of interaction in the learning process (Muirhead, 2001), online learners frequently do not interact at sufficient levels and/or in substantive ways with the instructor or other learners in online courses. The lack of appropriate and deep interactions is a common inadequacy of current online courses (Bennett et al., 1999).

Presence

Closely related to interaction is the concept of presence. From the learner's perspective, presence is the "sense of being in and belonging in a course and the ability to interact with other students and an instructor although physical contact is not available" (Shin, 2002, p. 22). Presence also refers to the "involvement, warmth, and immediacy" (Danchak, Walther, & Swan, 2001, p. 1) learners experience during communication and interaction with others in the online learning environment. According to Picard (1997), an online course that conveys affective or emotional information to learners will lead to a higher sense of social presence and interaction. Leh (2001) found lack of interaction, originally due to lack of physical and face-to-face contact, in online learning environments leads to a sense of isolation (or lack of social presence). On the other side, an appropriate level of interaction promotes a better sense of social presence (Rovai, 2001). Research also has shown social presence is positively related to learner satisfaction, perceived learning (Richardson & Swan, 2003), and learning success (Rifkind, 1992; Tu, 2000). In other words, a good sense of social presence influences interaction and interaction influences students' sense of social presence. Together, appropriate interaction and presence lead to increased cognitive activity and also cognitive activity at higher levels, resulting in more meaningful learning in online learning environments. While these relationships are known, many designers, developers, and instructors of online courses do not consciously implement instructional methods and techniques that will effectively increase interaction and social presence. We asked ourselves why. The answer, we believe, is that they have not had access to pedagogically content-based guidelines grounded in research. With this information, we turn to the guidelines.

GUIDELINES FOR PROMOTING INTERACTION AND PRESENCE IN AN ONLINE COURSE

There are four main components to consider when transitioning a traditional face-to-face course to an online version: (a) introductions, (b) organization, (c) instruction, and (d) feedback. There are techniques to use for all four of these components that will allow students to be more socially and cognitively interactive and present in an online course.

Introductions

Much of the current literature on online courses emphasizes the value of creating a learning community among the online participants. According to Hanna, Glowacki-Dudka, and Conceiçào-Runlee (2000), "a learning community is a group of people who have come together to form a culture of learning in which everyone is involved in a collective effort of understanding" (p. 14). This sounds great, but as an instructor you are probably asking what techniques can be used to accomplish this task. "You need to build a climate that will foster professional learning or collaboration by crafting communications that support a sense of safety in the discussion areas" (Collison, Elbaum, Haavind, & Tinker, 2000, p. 30). You want students to share their experiences with each other, but this is difficult unless they feel comfortable with each other. There are a number of strategies that can be used to foster this feeling of community.

In the course content discussion area, start with a social icebreaker for students to introduce themselves. This should be a non threatening type of interaction that "breaks the ice of using technology to communicate," (Conrad & Donaldson, 2004, p. 47) is participant focused versus academic content focused, and requires reading and responding to other postings (Conrad & Donaldson, 2004). Conrad and Donaldson (2004) list and describe a number of different types of ice breakers. For example, BINGO requires everyone to post a short biography on the discussion board. The instructor then e-mails everyone a bingo card with something from everyone's posting in a box. Students must then determine which box belongs to which student and fill in the correct name. Another possible icebreaker is TWO TRUTHS AND A LIE. Students post two truths about themselves and one lie. Other students then try to determine which is the lie. This is most fun when the truths are so outrageous it is hard to distinguish them from a lie.

Another method to help foster the feeling of community is to have a page dedicated to the biographical sketches and pictures of the students. This could also be a social space with a title, such as lounge, hallway, or water cooler, where students can discuss any off-content topics. Students need a space provided just for "*social dialogue* or simple chitchat" (Collison et al., 2000, p. 20). This helps prevent clutter in the content-discussion area and encourages students to contact each other via e-mail, instant messaging, or chat.

As the facilitator of this community, you will want to send an opening message to each student or post one on the content-discussion board. It should be a warm, welcoming message, perhaps with a friendly photo. The opening message should include a question requiring a response from students. This first message will set the tone for the class; it also can serve as a model for online discussions.

Organization

There are a number of organizational strategies to use to help increase interaction and presence in an online class. As the instructor, you will want to hold "online" office hours. This can either be a specified time when you will be answering e-mail or using instant messenger to "chat" with students either synchronously and privately. Another strategy is to provide job aids on how to use the technology employed in the course, which allows the technology to become invisible as students become more familiar with using it.

The course syllabus should consist of more than taking the face-to-face paper copy and making it digital. In an online course, the syllabus needs to include things such as guidelines for discussions, definition of roles, and so forth, and to function as a contract between instructor and students. In addition to content traditionally included in a syllabus, you should include contact information for student technical difficulties. Instead of listing"participation" or "online attendance" and the point value, define what participation entails. For example, we suggest the participation grade be based on the quality of the postings and not just the quantity. Participation could include posting on an asynchronous discussion board, showing up for a synchronous chat, working on a team project, and so forth.

One strategy to help foster quality asynchronous postings is to ask the students to send private e-mails for single comments, such as "I liked your last posting" or "I agree." According to Moore and Kearsley (1996), this helps prevent cluttering the discussion board; they encourage only postings that will contribute to the "community's pool of knowledge (p. 151)." Do not discourage positive comments like this, because they contribute to the social presence of the community.

Another strategy is to group students into teams of three or four and have them write up what they believe are good ground rules for discussions and participation. For example, what to do if someone reads all the postings, but does not post any. These could be posted in a forum that explains discussion board procedures and guidelines. Guidelines that include a posting with "emoticons" for students unfamiliar with how to express text-based emotions are helpful (e.g., ☺ means smiling or happy). In addition, some students might use abbreviations that are now common in Internet-based chat, such as LOL or "laughing out loud." These small additions can add personality to the text-based "voices" of the participants. In a face-to-face classroom, physical presence is displayed through "voice, body language, intonation, expressions, [and] gestures" and helps communication (Ko & Rossen, 2004, p. 12). In the online environment, participants rely solely on text-based communication and need to avoid words that could be misinterpreted, such as sarcasm, inappropriate jokes, and so forth. Thus everyone in the community must demonstrate a culture of respect, so participants "feel what they say matters and is valued by the other members of the community" (Collison et al., 2000, p. 30).

An additional aspect of the syllabus that needs to be addressed is the schedule. The schedule is the lifeline of the online course. Students will look to this to effectively manage time. Therefore, it is critical to present course content in a consistent manner, either all up front or on a regular schedule. This will reduce confusion and promote consistent checking of the site by students. Keep in mind not everyone is logged on when an assignment is posted. Give approximately a week for assignments, to allow for those who log in later in the week. This is one of the advantages to online learning: the ability to log on anytime of the day or night. You could require everyone to log in every two to three days to ensure they have the most up-to-date information. Supplying information, such as due dates, in more than one location on the course Web site is also a good idea as some students might look in different locations for information.

Instruction

There are a number of different instructional techniques to use when attempting to increase presence and interaction in an online class. Collaboration can be fostered in small or large groups of students. If you choose to have large class projects, there are grouping considerations. Before grouping the students into teams they will work with for the large class project, consider pairing them up for a smaller assignment, such as an article critique or peer review. This helps foster feelings of comfort when learning how to work with others over distance. When assembling teams for a large class project, groups of four are typically the optimum number. Encourage collaboration to prevent the group from splitting up the work, then putting it together to turn in; you want them to "construct their learning together" (Palloff & Pratt, 2005, p. 39). Also, explain to the students why it is important that they work collaboratively and that it is a requirement.

Palloff and Pratt (2005) discuss the importance of collaboration, saying it promotes critical thinking skills and helps to foster the feeling of community. There are a number of ways to do this online. Students first can do the work individually and bring it to the group for critique and to certify the correctness of the papers. Then the instructor can pick one student randomly to answer the questions studied by the group or choose one paper from the group to grade with everyone in the group receiving that grade. A second technique is to provide different team members with the charge of finding different information. This is called information interdependence, or the jigsaw strategy, where students have the different pieces needed to complete the puzzle. In order for all the team members to do well on the assignment, they need to rely on the information the rest of the team members have learned. Hence, the students are held accountable for teaching the material to their team members. Another technique is to have the team devise a "charter" or team agreement delineating the different roles individuals will play, how they will interact, and different project deadlines. It is helpful to provide a sample charter, so students know what is expected of them. Some possible roles are secretary, liaison to the instructor, organizer, discussion board poster, and so forth. These roles might change during the project's phases. Have the team create a team name; this helps with team identification on the discussion board and also with a sense of community. To encourage team buy-in, give the team choices in determining project topic. Monitor the team's progress and intervene if there are participation problems. Johnson, Johnson, and Holubec (1991) state that there are three reasons an instructor should intervene:

1. To correct misunderstandings or misconceptions about task instructions and the academic assignments they are completing.
2. To correct the absence, incorrect use, or inappropriate use of collaborative skills.
3. To reinforce the appropriate or competent use of collaborative skills. (pp. 6:29)

In addition, tips for online conflict resolution could be included as a job aid (Palloff & Pratt, 2005). When assessing the collaborative assignment, perhaps include peer evaluations in the grading criteria. This can help prevent noncontributing team members. In addition, ask each team member to write a reflection on what they did to contribute to the project, and how they tackled their role and task throughout the process.

As many online courses are taught using mainly asynchronous discussion boards, there are discussion strategies and activities that encourage interaction and a sense of presence online. A main goal is to ensure there is a high level of interaction and dialogue. This can be facilitated by using different types of questions, activities, and presentations.

Questions

When posting discussion questions, the instructor does not always need to be the initiator. After the instructor has modeled question facilitation, allow students, or pairs of students, to take turns facilitating different discussion topics. This allows students to see that each participant in the community is as valuable as the instructor because every participant shares personal experiences to help the community learn. It also allows participants to see multiple perspectives. During online discussion, it is important to provide the discussions with a distinct beginning and end to prevent information overload and frustration among students (Conrad & Donaldson, 2004). Different types of questions can help encourage critical thinking, such as questions asking for more evidence, questions asking for clarification, open-ended questions, hypothetical questions, cause and effect questions, and summary and synthesis questions (Palloff & Pratt, 2005). In addition, you or the discussion topic facilitator should write a wrap-up paragraph summarizing the main points of the discussion, including students' names and the different points they made, which contributes to the feeling of presence.

Activities

Different types of activities can take place on a discussion board. The typical threaded discussion can get boring to students who read numerous posts. Some activities to increase interaction are the following: a) role playing, b) debates, c) simulations, d) case studies, e) outside experts, f) sharing related personal/professional experiences, and g) electronic virtual field trips. In addition to these asynchronous activities, consider having a few required synchronous discussions. It should be noted, however, synchronous discussions tend to be more social; therefore, they are usually more effective at fostering social interaction than cognitive interaction. Guest lecturers via audio or video conferencing and synchronous large group sessions, where the instructor uses a whiteboard to demonstrate a problem, also can be used. If the instructor must present some sort of lecture to provide information to the class, include meaningful interactive links, discussion threads, and other activities to make the lecture interactive.

Presentations

Online course technologies often allow for different types of presentations. Individuals can post papers, PowerPoint presentations, and other documents in discussion threads. However, how do you have a group do a presentation to the entire class? If the students have access to software, such as Breeze, Camtasia, or Articulate, they can create multimedia presentations the class can watch. If this type of software is not available, students can prepare a discussion thread led by the team to present their project. Teams also can create simple Web sites to showcase their projects.

Feedback

The final component we are going to discuss is the use of feedback, which is essential to fostering interactivity. Online feedback should consist of both instructor-to-student and student-to-student (or peer) feedback. Responding to individual e-mails asking the same questions can get redundant; therefore, encourage students to post questions on the discussion board, so everyone will benefit from the response. In addition, provide a discussion forum that allows students to provide feedback about the course; perhaps these are recommendations for improvement or lessons learned that can be shared with future classes.

No matter the activity students are involved in, provide opportunities for individual as well as group practice and feedback; this may be the first online class they have taken. In addition, the instructor should respond to all student queries. Make sure responses are prompt if it is a technical question. If there is a delay in responding, explain the reason. Instructor feedback should offer detailed analysis of student work and suggestions for improvement, along with additional hints and information to supplement learning. These can be private, via e-mail, or public to a team via the discussion board. If a student is not accessing the site enough, the instructor can send informal e-mails to see if the student is having problems in terms of the technology. The instructor should send encouraging supportive e-mails to individuals on an ongoing basis. Include questions that require the student to respond, thus drawing them into active participation.

Students should complete peer reviews for student-to-student feedback. This provides the reviewer the opportunity to focus on others' interpretations and the original writer to receive multiple perspectives. Provide guidelines and the rubric to be used for grading. Both the instructor and the students should use "track changes" in Microsoft Word documents to provide feedback, so everyone can see changes made, comments, or notes that include questions. Also try to get feedback from participants about their progress. This can be done through direct questions, assignments, quizzes, polls, and questionnaires.

THE ONLINE INFORMATION SECURITY CLASS EXAMPLE

This section of the article begins with a brief overview of how to introduce and organize the online information security course. Next, there is an in-depth focus on three weeks of instruction, explaining how the course was transitioned from face-to-face instruction to an online format. While this specific example focuses on an information security course, the purpose of the example is simply to enact the guidelines. The guidelines can be generalized to other topics and fields in technology education.

Introductions for Building a Learning Community

When building a learning community in a face-to-face security assurance class, the instructor tends to have class introductions and perhaps an ice breaker activity. For an *online* security assurance class, the instructor needs to facilitate a learning community in a similar, yet different way. The instructor needs to provide a Web-based orientation to both the online environment and the course materials. An opening message should be sent by instructors, including a question requiring a response from students. In order to allow students to get to know each other early in the course, online ice breaker activities that are via discussion boards as well as having students work in virtual teams to produce a visual presentation about the team are useful. In addition, a space is provided on the class Web site for students to post their digitized images. The instructor also should encourage students to contact each other via e-mail and chat.

Course Organization

When organizing a face-to-face security assurance class, the instructor has a syllabus that includes a schedule, required textbook, and office hours. In the online version of the security assurance class, the instructor needs to go further. In addition to the online syllabus, the instructor needs to post a schedule with hyperlinks to that day's information as well as discussion and participation guidelines and requirements. Links to online readings, in addition to the listed textbook, are included. Furthermore, the instructor needs to hold online office hours when the students can be sure to reach him or her immediately.

Instruction—Week One

Perform Asset Identification and Classification

In the first week of the face-to-face security assurance class, the instructor provides the students with readings on the risk assessment process and various models. She also presents a lecture to provide them with additional information. A discussion ensues about asset identification. They look at the different authors and different information provided in order to compare and contrast what each author said. They also discuss the purpose in the risk assessment process. In addition, the class brainstorms assets in the k-12 setting. The instructor assembles small groups and has the students apply asset identification to the k-12 setting. The groups then compare their new list to the other groups' lists. As a class, they then group information assets (types of data, part of classification). FIPS 199 is discussed as a classification scheme for sensitivity of assets. For homework, students are asked to apply FIPS 199 to their list and write a critique of the usability of FIPS 199. They can revisit the first readings to discuss their classifications.

In the online security assurance class, the instructor provides the students with links to readings and Web sites about the risk assessment process and various models. She also posts a short lecture (approximately 10 minutes) with a PowerPoint presentation via Breeze to provide them with additional information. A discussion forum is started in which the instructor poses an initial question about asset identification. The students have two to three days to respond. The instructor assigns different students to read different authors to gain different information about risk assessment models. Concurrently, the instructor creates a new discussion forum for students to a) post a summary of their article, b) then compare their article to other postings, and c) discuss the purpose of models in the risk assessment process. The students again have two to three days to respond. The instructor creates a new discussion The instructor creates a new discussion thread to brainstorm information assets in the k-12 setting and posts an initial questions. She has students individually apply asset identification to the k-12 setting then post their responses. She organizes students into groups of three or four and provides them group discussion forums. She has each group create one new list and has groups compare/contrast lists with each other. She also has groups apply a classification scheme to their list as well as write a group critique of FIPS 199. Groups also discuss their classification, according to the first readings they did. The students have four to five days to respond.

Instruction—Week Two

Perform Threat Identification

In the second week of the face-to-face security assurance class, the instructor provides the students with readings on information security threat analysis and classifications of threat types. As a class, they discuss how different threats might correlate to different assets. This is done first in small groups, then together in one large group discussion. The instructor also presents them with information on methods for identifying types of threats. She provides them with existing reports (FBI CIC Survey to Crime Data) and tells them where they can get this type of information for typical threats in other organizations. She asks the students if they can try to generalize to their organization, and how they would monitor their own employees/network/system. She poses the question, "How are you going to get clients to think about modeling their threats before we go out to the client? Where we do actual threat identification?"

In the online security assurance class, the instructor provides the students with links to readings on information security threat analysis as well as Web sites about classifications of threat types. The students have two to three days to read this information. The instructor creates a new discussion forum about how different threats might correlate to different assets. Students are first assigned to small group discussion areas to answer a posted question. Then students discuss their findings in a large group discussion area. Students have two to three days to respond. Another discussion thread is created about methods for identifying types of threats. The instructor posts open-ended questions about the following: existing reports (FBI CIC Survey to Crime Data); where to get information; typical threats in other organizations; how the students could generalize to their organization; and how to monitor their own employees/network/system. In addition the instructor posts a fourth question: "How are yhou going to get clients to think about modeling their threats before we go out the the client, where we do actual threat identification?" Students have the same two to three days to respond.

Instruction—Week Three

Perform Vulnerability Identification

In the third week of the face-to-face security assurance class, the instructor provides a lecture and PowerPoint presentation on the three types of vulnerabilities— people, policy, and technology—and about establishing criteria for assessing vulnerability. She asks the students to individually develop an evaluation checklist (for policy) to take into a company. The instructor presents information about technical vulnerability. For example, she discusses the reporting tools companies and schools are likely to have as well as passive scanning tools. The class goes to a computer lab as a group and experiments with a variety of these tools. The students are provided criteria to evaluate different types of scanning tools: purpose, when to use, cost, and advantages/disadvantages; this is done in small groups. For homework, students are put into small groups and asked to select a tool. They are then provided with a flawed system with known vulnerabilities to run their tool against. They must then take, analyze, and report their findings.

In the online security assurance class, the instructor provides the students with links to readings and Web sites on the three types of vulnerabilities—people, policy, and technology—and about establishing criteria for assessing vulnerability. She also posts a short lecture (approximately 10 minutes) with PowerPoint presentation via Breeze to provide them with additional information. She asks the students to individually develop an evaluation checklist (for policy) to take into a company and submit this to the online assignment drop box. Students have two to three days for this. Students are then put into small groups, where each student presents his/her checklist to the other group members. The small group is then tasked with coming up with a "Best of Breed" checklist, using their individual checklists. The instructor presents information about technical vulnerability via an audio presentation. For example, she discusses the reporting tools and passive scanning tools that companies and schools are likely to have. The instructor provides links to demonstration software for students to experiment with different types of these tools. She posts a list of criteria along with an example for students to evaluate different types of scanning tools: purpose, when to use, cost, and advantages/disadvantages; she assigns this to be done in small groups and posted within three days. The students also are requested to select one tool per small group. They are then provided a flawed system with known vulnerabilities to run their tool against. This system is accessed via a virtual private network (VPN). The groups of students must then take, analyze, and report their findings on the discussion board within three days.

SUMMARY

The goal of this article was to provide guidance to faculty who are tasked with transitioning face-to-face instruction into distance learning. More specifically, these guidelines for an online course and the example of one are meant to provide readers with action steps that can be taken to improve the level and nature of interaction as well as students' sense of presence. The ultimate goal, of course, is to produce equally, if not more effective, results from online learning. Our hope is that faculty who attempt to use these guidelines will see increased learning and motivation to learn among their distance learning students.

REFERENCES

Aragon, S. R. (2003). Creating social presence in online environments. In S. R. Aragon (Ed.), *Facilitating learning in online environments* (pp. 57-68). San Francisco: Jossey-Bass.

Bennett, S., Priest, A., & Macpherson, C. (1999). Learning about online learning: An approach to staff development for university teachers. *Australian Journal of Educational Technology, 15*(3), 207-221.

Collison, G., Elbaum, B., Haavind, S., & Tinker, R. (2000). *Facilitating online learning: Effective strategies for moderators.* Madison, WI: Atwood.

Conrad, R-M., & Donaldson, J. A. (2004). *Engaging the online learner: Activities and resources for creative instruction.* San Francisco: Jossey-Bass.

Danchak, M. M., Walther, J. B., & Swan, K. P. (November 2001). *Presence in mediated instruction: Bandwidth, behavior, and expectancy violations.* Presented at the annual meeting on Asynchronous Learning Networks Conference, Orlando, FL.

Glickman, N. W. (2003). *The veterinary distance learning program at Purdue University: Factors associated with persistence and dropout. Digital Dissertations, 64(11), 4018B.* (UMI no. 3113802): Purdue University, West Lafayette, IN.

Hanna, D. E., Glowacki-Dudka, M., & Conceiçào-Runlee, S. (2000). *147 practical tips for teaching online groups: Essentials of web-based education.* Madison, WI: Atwood.

Johnson, D. W., Johnson, R. T., & Holubec, E. J. (1991). *Cooperation in the classroom.* Edna, MN: Interaction Book Co.

Ko, S., & Rossen, S. (2004). *Teaching online: A practical guide* (2nd ed.). New York: Houghton Mifflin.

Leh, A. S. (2001). Computer-mediated communication and social presence in a distance learning environment. *International Journal of Educational Telecommunications, 7*(2), 109-128.

Moore, M. G. (1989). Editorial: Three types of interaction. *The American Journal of Distance Education, 3*(2). Retrieved January 24, 2005, from http://www.ajde.com/Contents/vol3_2.htm#editorial

Moore, M. G. (1992). Distance education theory. *The American Journal of Distance Education, 5*(3), 1-6.

Moore, M. G. (1993). Theory of transactional distance. In D. Keegan (Ed.), *Theoretical principles of distance education* (pp. 22-38). New York: Routledge.

Moore, M., & Kearsley, G. (1996). *Distance education: A systems view.* Belmont, CA: Wadsworth Publishing Company.

Muirhead, B. (1999). *Attitudes toward interactivity in a graduate distance education program: A qualitative analysis.* Parkland, FL: Dissertation.com.

Muirhead, B. (2001). Practical strategies for teaching computer-mediated classes. *Educational Technology & Society, 4*(2). Retrieved September 12, 2005, from http://ifets.ieee.org/periodical/vol_2_2001/discuss_summary_jan2001.html

Palloff, R. M., & Pratt, K. (2005). *Collaborating online: Learning together in community.* San Francisco: Jossey-Bass.

Picard, R. W. (1997). *Affective computing.* Cambridge, MA: MIT Press.

Richardson, J. C., & Swan, K. (2003). Examining social presence in online courses in relation to students' perceived learning and satisfaction [Online]. *Journal of Asynchronous Learning Networks, 7*(1), 68-87.

Rifkind, L. J. (1992). Immediacy as a predictor of teacher effectiveness in the instructional television classroom. *Journal of Interactive Television, 1*(1), 31-38.

Rovai, A. P. (2001). Building and sustaining community in asynchronous learning network. *Internet and Higher Education, 3*(2000), 285-297.

Shin, N. (2002). Beyond interaction: The relational construct of 'Transactional Presence.' *Open Learning, 17*, 121-137.

Tu, C. H. (2000). Strategies to increase interaction in online social learning environments. In *SITE 2000.* Proceedings from the Society for Information Technology and Teacher Education International conference. Norfolk, Va.: AACE, 2000. (ED 444 550).

Wagner, E. D. (1994). In support of a functional definition of interaction. *The American Journal of Distance Education, 8*(2), 6-29.

This work was previously published in International Journal of Information and Communication Technology Education, Vol. 3, Issue 2, edited by L. A. Tomei, pp. 41-50, copyright 2007 by IGI Publishing, formerly known as Idea Group Publishing (an imprint of IGI Global).

Compilation of References

"Technology and Motivation: Can Computers Motivate Students to Read?" Business Wire. April 13, 2007. FindArticles.com. 25 Nov. 2007. http://findarticles.com/p/articles/mi_m0EIN/is_2007_April_13/ai_n19001667

AAUW. (2000). *Tech-savvy: Educating girls in the new computer age*. Washington, DC: AAUW Educational Foundation.

Abrams, D., & Hogg, M. A. (1999). *Social identity and social cognition*. Malden, MA: Blackwell.

Agarwal, R., & Atish, S. P. (1996). Cognitive fit in requirements modeling: a study of object and process methodologies. *Journal of Management Information Systems, 13*(2), 137-162.

Ahn, J., Han, K. , and Han, B. (2005). Web-based education: characteristics, problems, and some solutions. *International Journal of Innovation and Learning 2005, 2*(3), 274 - 282.

Ahuja, M. K., & Thatcher, J. B. (2005). Moving Beyond Intentions and Toward the Theory of Trying: Effects of Work Environment and Gender on Postadoption Information Technology Use. *MIS Quarterly, 29*(3), 427–459.

Akin, L., & Neal, D. (2007). CREST + Model: Writing Effective Online Discussion Questions. Journal of Online Learning and Teaching, Vol. 3. No. 2, June 2007. Retrieved September 26, 2007 from http://jolt.merlot.org/vol3no2/akin.htm

Alavi, M., & Leidner, D. E. (2001). Research commentary: Technology mediated learning - a call for greater depth and breadth of research. *Information Systems Research, 12*(1), 1-10.

Alavi, M., Marakas, G. M., & Yoo, Y. (2002). A comparative study of distributed learning environments on learning outcomes. *Information Systems Research, 13*(4), 404-415.

Aldrich, D.,Bell, B. and Batzel, T. (2006). Automated podcasting solution expands the boundaries of the classroom, Proceedings of SIGUCCS'06, Nov 5-8, 2006, Edmonton.

Alfaro,R.(1999).The Technology-Reading Connection. *Educational Leadership, March*. 48-50.

Allan, B. (2007). Time to learn?: E-learners' experiences of time in virtual learning communities. *Management Learning, 38*(5), 557-572.

Alty, J. (2002). Dual coding theory and computer education: Some media experiments to examine the effects of different media on learning. In P. Kommers & G. Richards (Eds.), *Proceedings of World Conference on Educational Multimedia, Hypermedia, and Telecommunications 2002* (pp. 42-47). Chesapeake, VA: AACE.

An, J. H., & Levin, J. A. (2003, April). *Online discourse patterns: Building an instructional framework for designing educational discourses on networks*. Paper presented at the 84th AERA Annual Meeting, Chicago, IL.

Anandarajan, M., Igbaria, M. and Anakwe, U. (2002). IT Acceptance in a Less-developed Country: A Motivational Factor Perspective, *International Journal of Information Management, 22*, 47-65.

Anderson, J.R., Corbett, A.T., Koedinger, K.R., & Pelletier, R. (1995). Cognitive tutors: Lessons learned. *Journal of the Learning Sciences, 4*(2), 167-207.

Anderson, K. J., & Leaper, C. (1998). Meta-analyses of Gender Effects on Conversational Interruption: Who, What, When, Where, and How. *Sex Roles, 39*(3/4), 225-252.

Anderson, L.W., & Krathwohl (Eds.). (2001). *A Taxonomy for Learning, Teaching, and Assessing: A Revision of Bloom's Taxonomy of Educational Objectives*. New York: Longman.

Andrews, G. (2006). Laptops + Challenging Curriculum = Student Success. *Techniques, May.* 42-44.

Apple (2006). Today's Top Podcasts—Education Genre, www.apple.com, Sep 28, 2006.

Arbaugh, J.B. & Hiltz, S. (2005). Improving quantitative research on aln effectiveness. In S. Hiltz & R. Goldman (Eds.), *Learning together online: Research on asynchronous learning networks* (pp 81-102). London: Lawrence Erlbaum Associates.

Armitage, S., Brown, T. and Jenkins, M. (2001). Management and Implementation of Virtual Learning Environments: a UCISA funded Survey, *UCISA*

Armstrong, L. Berry, M. and Lamshed, R. (2004). Blogs as electronic learning journals. *e-Journal of Instructional Science and Technology,* 7(1). Retrieved from http://www.usq.edu.au/electpub/e-jist/docs/Vol7_No1/CurrentPractice/Blogs.htm

Azmitia, M. (2000). Taking time out from collaboration: Opportunities for synthesis and emotion regulation. In R. Joiner, K. Littleton, D. Faulkner and D. Miell (Eds.), *Rethinking collaborative learning* (pp. 179-195). London, UK: Free Association Books.

Bakhtin, M. M. (1986). The problem of speech genres. In M. M. Bakhtin, *Speech genres and other late essays* (V. McGee, Trans. & C. Emerson & M. Holquist, Eds.) (pp. 60-102). Austin, TX: University of Texas Press.

Banâtre, J.-P., & Le Metayer, D. (1990). The Gamma model and its discipline of programming. *Science of Computer Programming, 15*, 55-77.

Banâtre, J.-P., & Le Metayer, D. (1993). Programming by multiset transformation. *CACM, 36*(1), 98-111.

Bandura, A. (1997). Self-efficacy: The exercise of control. New York: W.H. Freeman.

Barger, J. (1997). Weblog resources FAQ, Robot Wisdom Werblog. Retrieved from http://www.robotwisdom.com/weblogs

Barnett, R. (2004). Learning for an unknown future. *Higher Education Research and Development, 23*(3), 247-260.

Barnett, S.M., & Ceci, S.J. (2002). When and where do we apply what we learn? A taxonomy for far transfer. *Psychological Bulletin, 128*, 612-637.

BBC News (2005). Wikipedia survives research test. BBC News, Dec 15, 2005.

Beacham, N. A., Elliot, A. C., Alty, J. L., & Al-Sharrah, A. (2002). Media combinations andlearning styles: A dual coding approach. In P. Kommers & G. Richards (Eds.), *Proceedings of World Conference on Educational Multimedia, Hypermedia, and Telecommunications 2002* (pp. 111-116). Chesapeake, VA: AACE.

Beals, D. (1992). Computer networks as a new data base. *Journal of Educational Computing Research, 8*(3), 327-345.

Beck, P., Kung, M., Park, Y., and Yang, S. (2004). 'E-learning architecture: challenges and mapping of individuals in an internet-based pedagogical interface,' *International Journal of Innovation and Learning 2004, 1*(3) 279-292.

Belbin, R. (1981). *Management teams: Why they succeed or fail*. Oxford: Butterworth-Heinemann.

Bell, P., Hoadley, C.M., & Linn, M.C. (2004). Design-based research in education. In M.C. Linn, E.A. Davis, & P. Bell (Eds.). *Internet environments for science education* (pp. 73-88). Laurence Erlbaum Associates.

Bellefeuille, G. L. (2006). Rethinking reflective practice education in social work education: A blended constructivist and objectivist instructional design strategy for a web-based child welfare practice course. *Journal of Social Work Education, 42*(1), 85-103.

Benbunan-Fich, R., Hiltz, S., and Harasim, L. (2005). The online interaction learning model: An integrated theoretical framework for learning networks. In S.Hiltz & R.Goldman (Eds.), *Learning together online: Research on asynchronous learning networks* (pp. 19-37). London: Lawrence Erlbaum Associates.

Benjamin, A. (2005). Differentiated instruction using technology. Larchmont, NY: Eye on Education, Inc.

Bereiter, C. (2002). *Education and mind in the knowledge age*. Mahwah, NJ: Lawrence Erlbaum Associates.

Berge, A. L., & Collins, M. (Vol. Eds.). (1998). *Wired together: The online classroom in K-12: Vol. 3. Teacher education and professional development*. Cresskill, NJ: Hampton Press, Inc.

Berge, O., & Fjuk, A. (2006). Understanding the roles of online meetings in a net-based course. *Journal of Computer Assisted Learning, 22*(1), 13-23.

Berge, Z. (2001). *Sustaining distance training: Integrating learning technologies into the fabric of the enterprise*. San Francisco: Jossey-Bass.

Berge, Z. L., & Clark, T. (2005). *Virtual schools: Planning for success*. New York: Teachers College Press.

Berge, Z., & Collins, M. (1996). Where interaction intersects time. *MC Journal: The Journal of Academic Media Librarianship,* 4(1). Retrieved September 26, 2007, from http://wings.buffalo.edu/puublications/mcjrnl/v4n1/berge.html

Berge, Z.L. & Clark, T. (2005). *Virtual schools: Planning for success*. Teachers College Press.

Berger, P. L., & Luckmann, T. (1966). The social construction of reality: A treatise in the sociology of knowledge. Garden City, NY: Anchor Books.

Bergin, J. (2002) Teaching on the Wiki Web. *ACM SIGCSE Bulletin*, 1.

Bernard, R.M., Abrami, Lou, Borokhovski, Wade, Wozney, Wallet, Fiset, & Huang (2004). How does distance education compare to classroom instruction? A meta-analysis of the empirical literature. *Review of Educational Research, 74*(3), 379-439.

Betts, J. D. and Glogoff, S. (2004). Instruction models for using weblogs in e-learning. Proceedings of Syllabus 2004, July, San Francisco.

Bhagyavati, Kurkovsky, S., & Whitehead, C. C. (2005). Using asynchronous discussions to enhance student participation in CS courses. In *Proceedings of the 36th SIGCSE Technical Symposium on Computer Science Education* (pp. 111-115). Retrieved December 22, 2005, from the ACM database.

Bhattacherjee, A. (2001) Understanding Information Systems Continuance: An Expectation-Confirmation Model, *MIS Quarterly, 25*(3), 351-370.

Biggs, J. (1999). Teaching for Quality Learning at University, *SRHE*, Open University.

Biggs, J. (2003). *Teaching for quality learning at university* (2nd edition). Buckingham: Society for Research into Higher Education and Open University Press.

Bird, L. (2007). The 3 'c' design model for networked collaborative e-learning: A tool for novice designers. *Innovations in Education and Teaching International, 44*(2), 153-167.

Bjerlöv, M. (1999). *Learning in work based discourse*. Doctoral dissertation. Arbete och Hälsa 1999:1, Solna: National Institute for Working Life. (In Swedish, English summary)

Black, S. D., Levin, J. A, Mehan, H., & Quinn, C. N. (1983). Real and non-real time interaction: Unraveling multiple threads of discourse. *Discourse Processes, 6*, 59-75.

Blignaut, S., & Trollip, S. (2003). Developing a taxonomy of faculty participation in asynchronous learning environments: An exploratory investigation. *Computers and Education,* 41(2), 149-172.

Blood, R. (2004). How Blogging Software Reshapes the Online Community. *Communications of the ACM, 47*(12), 53-33.

Bloom, B. (1956). Taxonomy of educational objectives. New York, NY: David McKay Company, Inc.

Blumer, H. 1933. *Movies and conduct.* New York: MacMillan.

Bober, M. J., & Dennen, V. P. (2001). Intersubjectivity: Facilitating knowledge construction in online environments. *Education Media International, 38*(4), 241-250.

Boeree, C. G. (2007) Donald Snygg and Arthur Combs: Personality theory. Retrieved January, 14, 2007 from http://www.social-psychology.de/cc/click.php?id=40

Boneva, B., Kraut, R., & Frohlich, D. (2001). Using e-Mail for Personal Relationships: The Difference Gender Makes. *American Behavioral Scientist, 45*(3), 530-549.

Bongey, S. Cizadlo, G. and Kalnbach, L. (2006). Explorations in course-casting. *Campus-Wide Information Systems, 23*(5), 350-367.

Bonk, C. and Zhang, K. (2006). Intrpducing the R2D2 Model: Online learning for the diverse learners of this world. *Distance Education, 27*(2), 249-264.

Borja, R. (2006). Researchers Weigh Benefits of One Computer Per Lap. *Education Week, 25(36).* 10.

Borko, H. (2004). Professional development and teacher learning: Mapping the terrain. *Educational Researcher, 33*(8), 3-15.

Borthwick, A., Pierson, M., Anderson, C., Morris, J., Lathem, S., & Parker, H. (2004). Building learning communities to increase faulty and preservice teacher use of technology. *Journal of Computing in Teacher Education, 21*(1), 23-32

Boyce, T.E., & Hineline, P.N. (2002). Interteaching: A strategy for enhancing the user-friendliness of behavioral arrangements in the college classroom. *The Behavior Analyst, 25*, 215-226.

Boyle, T., Bradley, C., Chalk, P., Jones, R., & Pickard, P. (2003). Using blended learning to improve student success rates in learning to program. *Journal of Educational Media, 28*(2-3), 165-178.

Bransford, J. D., Brown, A. L., & Cocking, R. R. (1999). *How people learn: Brain, mind, experience, and school* [Electronic version]. Washington, D. C.: National Academy of Sciences.

Brislin, R. (1986). "The Wording and Translation of Research Instruments." In W.J. Lonner and J.W. Berry (eds.), *Fields Methods in Cross-Cultural Research.* Newbury Park, CA: Sage.

Brown, B. W., & Liedholm, C. E. (2002). Can web courses replace the classroom in principles of microeconomics? *The American Economic Review, 92*(2), 444-448.

Brown, I. (2002). Individual and Technological Factors Affecting Perceived Ease of Use of Web-based Learning Technologies in a Developing Country, *Electronic Journal of Information Systems in Developing Countries, 9*(5), 1-15.

Brown, J.S., Collins, A. & Duguid, P. (1989). Situated cognition and the culture of learning. *Educational Researcher, 18*(1), 32-42.

Bruner, J. (1990). *Acts of meaning.* Cambridge, MA: Harvard University Press.

Bruner, J. (1996). Psychology's next chapter. In J. Bruner, *The culture of education* (pp. 160-85). Cambridge, MA: Harvard University Press.

Bruns, A. and Humphreys, S. (2005). Wikis in Teaching and Assessment: The M/Cyclopedia Project. Proceedings WikiSim '05, Oct. 16-18, San Diego CA.

Brunsdon, C., D'Acci, J., & Spigel, L. (Eds.). (1997). *Feminist television criticism.* Oxford: Oxford University Press.

Brunvand, S., & Fishman, B. (2005). Aligning the design of online professional development with the needs and expectations of teachers. *Proceedings of the Society for Information Technology and Teacher Education (SITE), USA*, 2171-2176.

Brunvand, S., Fishman, B., & Marx, R. (2003). *Moving Professional Development Online: Meeting the Needs and Expectations of All Teachers* [Online]. Paper presented at the Annual Meeting of the National

Association of Research on Science Teaching, Philadelphia, PA. Retrieved from http://www-personal.umich.edu/~fishman/papers/papers...

Bryant, S. M. and Hunton, J. E. (2000). The Use of Technology in the Delivery of Instruction: Implications for Accounting Educators and Education Researchers. *Issues in Accounting Education, 15*(1), 129-163.

Buchanan, E. A., Myers, S. E., & Hardin, S. L. (2005). Holding your hand from a distance: Online mentoring and the graduate library and information science student. *The Journal of Educators Online, 2*(2), 1-18.

Buendia, F., Diaz, P., & Benlloch, J. V. (2002). A framework for the instructional design of multi-structured educational applications. In P. Kommers & G. Richards (Eds.), *Proceedings of World Conference on Educational Multimedia, Hypermedia & Telecommunications 2002* (pp. 210-215). Chesapeake, VA: AACE.

Burke, M. (2004). Instant Messaging and Course Diaries. *CCSC Northeastern Conference proceedings, May 2004, 19*(5), 207-274.

Buschmann, M. (1994). Jungen und Koedukation. Zur Polarisierung der Geschlechterrollen. *Die deutsche Schule*, 86. Jg. 11.2, 192-213.

Bush, G. (2007). Learning about learning: from theories to trends. *Teacher Librarian, 34*(2), 14-18.

Business Roundtable. (2005). *Tapping America's potential: The education for innovation initiative*. Washington, DC: The Business Roundtable.

Butz, C.J., Hua, S., & Mcguire, R.B. (2006). A web-based bayesian intelligent tutoring system for computer programming. *Web Intelligence & Agent Systems, 4*(1), 61-67.

Callison, D. (2007). Portfolio Revisted with Digital Considerations. *School Library media Activities Monthly. Vol.* 23 (6): 43-47.

Camp, T. (1997). The incredible shrinking pipeline. *Communications of the Association of Computing Machinery, 40 (10),* 103-110.

Camp, T. (2002). Forward to women and computing, *SIGSE Bulletin, 34* (2). 1-6.

Cantril, H., & Allport, G. (1935). *The psychology of radio.* New York: Harper & Bros.

Carbonara, D.(editor) 2005. Technology literacy applications in learning environments. Hershey, PA: Idea Group, Inc.

Caspi, A., & Gorsky, P. (2006). Online Deception: Prevalence, Motivation and Emotion. *Cyberpsychology and Behavior, 9*(1), 54-59.

Cassel, E. (January 2003). Teaching and learning law online. *Modern Practice: FindLaw's Practice and Technology Magazine.* Retrieved October 27, 2004 from http://practice.findlaw.com/archives/teaching_0103.html.

Castells, M. (2001). *The Internet galaxy.* Oxford, UK: Oxford University Press.

Cazden, C. (2001). *Classroom discourse: The language of teaching and learning* (2nd ed.). Portsmouth, NH: Heinemann.

Chandra, S. (2007). Lecture Video Capture for the Masses.

Charles, M., & Bradley, K. (2005). *Women and information technology: Research on the reasons for under representation.* Paper presented at the 100th Annual Meeting of the American Sociological Association, Philadelphia, PA, August 13.

Chawner, B. and Gorman, GE (2002).Wikis are wicked. *New Library World, 103*(1182/1183), 483.

Chawner, B. and Lewis, P. (2006). Wiki wiki webs: new ways to communicate in a web environment. *Information Technology and Libraries, March, 25*(1), 33-43.

Chenoweth, K. (2001). Keeping Score. *School Library Journal, 47(9).* 48-51.

Cherry, N. (1998). *Action Research: A Pathway to Action, Knowledge, and Learning.* RMIT Publishing.

Cheung, K., Lee, F., Ip, R. and Wagner, C. (2005). The development of successful online communities. *International Journal of the Computer, the Internet, and Management, 13*, 1.

Chin, W. (1998). Issues and opinion on Structural Equation Modeling, *MIS Quarterly, 22*(1), vii-xvi.

Cho, S.K. & Berge, Z.L. (2002). Overcoming barriers to distance training and education. *Education at a Distance, 16*(1). Retrieved November 18, 2007 from http://www.usdla.org/html/journal/JAN02_Issue/article01.html

Chou, S.W. and Liu C.H. (2005). Learning Effectiveness in a Web-based Virtual Learning Environment: A Learner Control Perspective. *Journal of Computer Assisted Learning, 21.* 65-76.

Chu, H.C., & Hinton, B. E. (2001). *Factors affecting student completion in a distance learning mediated HRD baccalaureate program.* Paper presented at the Academy of Human Resource Development 2001 Conference, Fayetteville, AR. (AHRD Reference No. 021)

Chuang, S., & Tsai, C. (2004). Preferences toward the constructivist internet-based learning environments among high school students in taiwan. *Computers in Human Behavior, 21*(2), 255-272.

Cicciarelli, M. (2006). Telementoring and computer mediated discussions: A description of online instructors' support. *Dissertation Abstracts International,* (UMI No. 3238537)

Cicciarelli, M. (2007). Behavioral theories that guide online course design. *Encyclopedia of Information Technology Curriculum Integration,* Pennsylvania: Idea Group Publishing.

Cicciarelli, M. (2007). Cognitive theories that guide online course design. *Encyclopedia of Information Technology Curriculum Integration,* Pennsylvania: Idea Group Publishing.

Cicciarelli, M. (2007). Humanistic theories that guide online course design. *Encyclopedia of Information Technology Curriculum Integration,* Pennsylvania: Idea Group Publishing.

Cicciarelli, M. (2008). A description of online instructors use of design theory. *International Journal of Information & Communication Technology Education, 4*(1), 25-32.

Clark, J.M., & Paivio, A. (1991). Dual coding theory and education. *Educational Psychology Review, 3,* 149-210.

Clark, T., & Else, D. (2003). Distance education, electronic networking, and school policy. In D. R. Walling & J. F. Jennings (Eds.), *Virtual schooling* (pp. 31-45). Bloomington, IN: Phi Delta Kappa Educational Foundation.

Clay, M. (1999). Development of training and support programs for distance education instructors. *The Online Journal of Distance Learning Administration* (Fall). Retrieved on September 15, 2006, from http://www.westga.edu/~distance/clay23.html

CNN (2006). Apple surpasses beer on college campuses. Retrieved June 7, 2006, from www.cnn.com

Coates, J. (1986). *Women, Men and Languages: Studies in Language and Linguistics.* London, UK: Longman.

Cochran-Smith, M. (2003). Learning and unlearning: The education of teacher educators. *Teaching and Teacher Education, 19*, 5-28.

Cold, S. (2005). Using RSS to Enhance Student Research. *ACM SIGITE Newsletter, 3,* 1.

Cole, M., & Engeström, Y. (1993). A cultural-historical approach to distributed cognition. In G. Salomon (Ed.), *Distributed cognition: Psychological and educational considerations.* Cambridge, MA: Cambridge University Press.

Collins, A., Joseph, D., & Bielaczyc, K. (2004). Design research: Theoretical and methodological issues. *Journal of the Learning Sciences, 13*(1), 15-42.

Collins, M., & Berge, Z. (1996). *Facilitating interaction in computer mediated online courses.* Paper presented at the FSU/AECT Distance Education Conference. Retrieved June 29, 2005, from http://www.emoderators.com/moderators/fleehtml.

Collison, G., Elbaum, B., Haavind, S., & Tinker, R. (2000). *Facilitating online learning: Effective strategies for moderators*. Madison, WI: Atwood Publishing.

Committee on Maximizing the Potential of Women in Academic Science and Engineering. (2007). *Beyond bias and barriers: Fulfilling the potential of women in academic science and engineering*. Washington, DC: The National Academies Press.

Committee on Science, U. S. House of Representatives. (2002). *A review of the Morella Commission report on recommendations to attract more women and minorities in science and engineering.* Serial No. 106-82. Education 6 (10): 1-2. Washington, DC: U. S. Government Printing Office.

Compeau, D. and Higgins, C. A. (1995). Computer Self-efficacy: Development of a Measure and Initial Test, *MIS Quarterly, 19*(2), 189-212.

Conover, W.J. (1971). *Practical nonparametric statistics.* New York, NY: John Wiley & Sons, Inc.

Cook, K. C. (2007). Immersion in a digital pool: Training prospective online instructors in online environments. *Technical Communication Quarterly, 16*(1), 55-83.

Copley, J. (2007). Audio and video podcasts of lectures for campus-based students. *Innovations in Education and Teaching International, 44*(4), 387-399.

Costa, P. T. J., Terracciano, A., & McCrae, R. R. (2001). Gender Differences in Personality Traits across Cultures: Robust and Surprising Findings. *Journal of Personality and Social Psychology, 81*(2), 322-331.

Course-Management Systems. (2005). *Library technology reports, 41*(3), 7-11.

Cradler, J., McNabb, M., Freeman, M., & Burchett, R. (2002). How does technology influence student learning? *Learning and Leading with Technology, 29, 46-52.*

Cramer, S. and Smith, A. (2002). Technology's Impact on Student Writing at the Middle School Level. *Journal of Instructional Psychology, 29 (1).* 3-14.

Crawford, M. (1995). *Talking Difference. On Gender and Language.* London, UK.: Sage.

Cruz, E. (2003). Bloom's revised taxonomy. In B. Hoffman (Ed.), *Encyclopedia of Educational Technology.* Retrieved September 25, 2007, from http://coe.sdsu.edu/eet/articles/bloomrev/start.htm

D'Ambrosio, J. (2003) *E-Teaching: Creating Web Sites and Student Web Portfolios Using Microsoft PowerPoint.* Columbus: Linworth.

Dagada, R., & Jakovljevic, M. (2004). *Where have all the trainers gone? E-learning strategies and tools in the corporate training environment.* Paper presented at the 2004 Annual Research Conference of the South African Institute of Computer Scientists and Information Technologists on IT Research in Developing Countires, Stellenbosch, Western Cape, South Africa.

Davidson, K. and Decker, T.(2006). Bloom's and Beyond. Pieces of Learning Publishing.

Davidson, P. M. (1992). The role of social interaction in cognitive development: A propaedeutic. In L. T. Winegar & J. Valsiner (Eds.), *Children's development within social context* (Vol. 1, pp. 19-37). Hillsdale, NJ: Lawrence Erlbaum Associates.

Davis, F. D. (1989). Perceived Usefulness, Perceived Ease of Use, and User Acceptance of Information T Technology, *MIS Quarterly, 13*(3), 319-339.

de Vries, L., Naidu, S., Jegede, O., & Collis, B. (1995). Online professional development: An evaluation study. *Distance Education, 16*(1), 157-173.

Dearstyne, B. (2005). Blogs: The New Information Revolution. *Information Management Journal, Sep/Oct*, 38-44.

Dearstyne, B. (2007). Blogs, Mashups, & Wikis, Oh My! *Information Management Journal, 41*(4), 25-33.

Dede, C. (1996). Distance learning–Distributed learning: Making the transformation [Electronic version]. *Learning and Leading with Technology, 23*(17), 25-30.

DeHaan, R.L. (2005). The impending revolution in undergraduate science education. *Journal of Science Education and Technology, 14*(2), 253-269.

DeMarco, T. (1978). *Structured Analysis and System Specification*. Englewood Cliffs, NJ: Prentice-Hall.

Dennen, V. P. (2005). From message posting to learning dialogues: Factors affecting learner participation in asynchronous discussion. *Distance Education, 26*(1). Retrieved December 23, 2005, from the ProQuest database.

Dennen, V. P. (2007). Presence and positioning as components of online instructor persona. *Journal of Research on Technology in Education, 40*(1), 95-108.

Descy, D. (2005). Podcasting: Online media Delivery... with a Twist. *TechTrends, Sep/Oct, 49*(5), 4-6.

Design-Based Research Collective (2003). *Educational Researcher, 32*(1), 5-8.

Desilets, A., Paquet, S. and Vinson, N. (2005). Are wikis usable? *Proceedings WikiSim'05, Oct.* 16-18, San Diego.

Dewey, J. (1938). *Experience and education.* New York: Macmillan.

Dewiyanti, S., Brand-Gruwel, S., Jochems, W., & Broers, N. J. (2004). Students' experiences with collaborative learning in asynchronous computer-supported collaborative learning environments. *Computers in Human Behavior, 23*(1), 496-514.

Dickey, M. (2004). The impact of web-logs on student perceptions of isolation and alienation in a web-based distance learning environment. *Open Learning, 19*(3), 279-291.

Diener, K. (2005). Principal, Waukesha *IQ Academies*. Interview conducted via the phone on April 14, 2005.

Dillon, A., and Morris, M. G. (1996). User Acceptance of Information Technology: Theories and Models, *Annual Review of Information Science and Technology, 31,* 3-32.

DiMarco, J. (2006). *Web Portfolio Design and Applications.* Hershey: Idea Group.

DiMarco, J. (2007). Web Portfolio Design for Teachers and Professors. *Proceedings of the 2007 Information resources management Association International Conference.* IGI Global: New York.

DiMauro, V., & Jacobs, G. (1995). *Filling in the professional gaps: Active teacher participation on a telecommunications network.* Retrieved October 9, 2002, from http://www.terc.edu/papers/labnet/Articles/gaps.html

Dixon, N. (1994). *The organizational learning cycle. How we can learn collectively.* London: McGraw-Hill.

Doering, Beach, and O'Brien (2007). Infusing Multimodal Tools and Digital Literacies into an English Education Program, *English Education, 40,* 1.

Donohue, C. (2002). It's a small world: Taking your first steps into online teaching and learning, *Childcare Information Exchange, 9,* 20-25.

Döös, M. (1997). *The qualifying experience.* Doctoral dissertation. Arbete och Hälsa 1997:10, Solna: National Institute for Working Life. (In Swedish, English summary)

Döös, M. (2007). Organizational learning: Competence-bearing relations and breakdowns of workplace relatonics. In L. Farell & T. Fenwick (Eds.), *World Year Book of Education 2007. Educating the global workforce* (pp. 141-153). London: Routledge.

Döös, M., & Ohlsson, J. (1999). Relating theory construction to practice development – Some contextual didactic reflections. In J. Ohlsson & M. Döös (Eds.), *Pedagogic interventions as conditions for learning* (pp. 5-13). Stockholm: Department of Education, Stockholm University.

Döös, M., & Wilhelmson, L. (2005). Collective learning - on the importance of interaction in action and a common action arena. *Journal of Swedish Educational Research, 10*(3/4), 209-226. (In Swedish; English summary)

Dringus, L., & Ellis, T. (2005). Using data mining as a strategy for assessing asynchronous discussion forums.

Computers & Education, 45(1), 141-160. Retrieved December 8, 2005, from the Science Direct database.

Dron, J (2005) Control Termites and E-Learning, *IADIS International Conference Web Based Communities*, 23-25.

Drucker, P.F. (1964). The big power of little ideas. *Harvard Business Review, 42*(3), 6-8.

Drucker, P.F. (1993). *Post-capitalist society*. New York: Harper Collins.

Dryburgh, H. (2000). Underrepresentation of girls and women in computer science: Classification of 1990s research. *Journal of Educational Computing Research, 23*, 181-202.

Dubofsky, M. (1975). *Industrialism and the American worker*. New York: Crowell.

Ducate, L. and Lomicka , L. (2005). Exploring the Blogosphere: Use of Web Logs in the Foreign Language Classroom. *Foreign Language Annals, Fall, 38*(3), 410-421.

Dunlap, J. C. (2005). Workload reduction in online course: Getting some shuteye. *Performance Improvement, 44*(5). Retrieved December 8, 2005, from the ProQuest database.

Eastmond, D. V. (1995). *Alone but Together: Adult Distance Study through Computer Conferencing*. Cresskill, NJ: Hampton Press.

ECOT. (2006). High school. Retrieved November 18, 2007 from http://www.ecotohio.org/highschool.html

Edelsky, C. (1993). Who's Got the Floor? In D. Tannen (Ed.), *Gender and Conversational Interaction* (pp. 189-227). New York, NY: Oxford University Press.

Edelson, D., Pea, R., & Gomez, L. (1996). Constructivism in the collaboratory. In B. G. Wilson (Ed.), *Constructivist learning environments: Case studies in instructional design* (pp. 151-164). Englewood Cliffs, NJ: Educational Technology Publications.

Edelstein, S., & Edwards, J. (2002). If you build it they will come: Building learning communities through threaded discussions. *ELearn, 2002*(4). Retrieved December 12, 2005, from the ACM database.

Efimova, L. and Fiedler, S. (2004). Learning Webs: Learning in Weblog Networks. In Kommers, Isaias and Nunes (eds.) Proceedings of the IADIS Intl. Conference on Web Based Communities 2004. Lisbon: IADIS Press.

Electronic Classroom of Tomorrow (ECOT). (2004). *High school*. Retrieved September 15, 2006, from http://www.ecotohio.org/highschool.html

Ellis, Kristine. (December 2000). A model class. *Training*, 37(12).

Emurian, H.H. (2001). The consequences of e-learning (Editorial), *Information Resources Management Journal*, April-June, 3-5.

Emurian, H.H. (2004). A programmed instruction tutoring system for Java: Consideration of learning performance and software self-efficacy. *Computers in Human Behavior, 20*(3), 423-459.

Emurian, H.H. (2005). Web-based programmed instruction: Evidence of rule-governed learning. *Computers in Human Behavior, 21*(6), 893-915.

Emurian, H.H. (2006). A Web-based tutor for Java□: Evidence of meaningful learning. *Journal of Distance Education Technologies, 4*(2), 10-30.

Emurian, H.H. (2006). Assessing the effectiveness of programmed instruction and collaborative peer tutoring in teaching Java . *International Journal of Information and Communication Technology Education, 2*(2), 1-16.

Emurian, H.H. (in press). Applications of behavior analysis to ICT education: Teaching java with programmed instruction and interteaching. *Encyclopedia of Information Technology Curriculum Integration*. Hershey, PA: IRM Press.

Emurian, H.H. (in press). Managing programmed instruction and collaborative peer tutoring in the classroom: Applications in teaching Java™. *Computers in Human Behavior*.

Emurian, H.H., & Durham, A.G. (2003). Computer-based tutoring systems: A behavioral approach. In J.A. Jacko

& A. Sears (Eds.), *Handbook of human-computer interaction* (pp. 677-697). Mahwah, NJ: Lawrence Erlbaum & Associates.

Emurian, H.H., Hu, X., Wang, J., & Durham, A.G. (2000). Learning Java: A programmed instruction approach using applets. *Computers in Human Behavior, 16*, 395-422.

Emurian, H.H., Wang, J., & Durham, A.G. (2003). Analysis of learner performance on a tutoring system for Java. In T. McGill (Ed.), *Current Issues in IT Education* (pp. 46-76). Hershey, PA: IRM Press.

Engeström, Y. (1987). *Learning by expanding: An activity-theoretical approach to developmental research.* Helsinki, Finland: Orienta-Konsultit.

Engestrom, Y. (2002). Cultural Historical Activity Theory. Retrieved from http://www.helsinki.fi/~jengestr/activity/6a.htm

Engeström, Y., Miettinen, R., & Punamaki, R. (Eds.). (1999). *Perspectives on activity theory.* Cambridge, UK: Cambridge University Press.

Engstrom, M. E. and Jewitt, D. (2005). Collaborative Learning the Wiki Way. *TechTrends, Nov/Dec 2005, 49*(6), 12-15.

Estonian eUniversity (2004). The UNIVe Project. Retrieved September 10, 2004, from http://www.e-uni.ee/Minerva/

Estonian eUniversity (2004). Estonian e-University NEEDS Analysis. Retrieved May 5, 2004, from http://www.e-uni.ee/doc/uuring/E-ylikool_vajaduste_analyys_mp2.pdf

Fåhræus, E. R. (2000). *Growing knowledge - How to support collaborative learning e-discussions in forum systems.* No. 00-005. Dept. of Computer and Systems Sciences, Stockholm University/KTH. Retrieved from http://www.dsv.su.se/~evafaahr/lic/index.html

Fåhræus, E. R. (2003). *A triple helix of learning processes - How to cultivate learning, communication and collaboration among distance-education students.* Doctoral dissertation. Dept. of Computer and Systems Sciences, Stockholm University/KTH, Report No. 03-015. Retrieved from ftp://ftp.dsv.su.se/users/jpalme/eva-f-doktorsavhandling.pdf

Fåhræus, E. R., & Döös, M. (2007). Competent web dialogue: Thoughts linked in digital conversations. *The International Journal of Information and Communications Technology Education, 3*(3), 14-24.

Fahy, P. J., Crawford, G., & Ally, M. (2001). Patterns of interaction in a computer conference transcript. *International Review of Research in Open and Distance Learning, 2*(1), 1-18.

Fallows, D. (2005, 12/28). How Women and Men Use the Internet. Retrieved December 28, 2005, from http://www.pewinternet.org/PPF/r/171/report_display.asp

Felder, R. M., Felder, G. N., Mauney, M., Hamrin, C. E. J., & Dietz, E. J. (1995). A Longitudinal Study of Engineering Student Performance and Retention. III. Gender Differences in Student Performance and Attitudes. *Journal of Engineering Education, 84*(2), 151-174.

Femtec. (2002). *Introduction to Femtec: University-based career center for women Berlin, Inc.* Berlin, Germany: Femtec.

Ferrán-Urdaneta, C. (1999). *Teams or communities? Organizational structures for knowledge management.* Proceedings of SIGCPR 99, New Orleans, LA.

Ferris, S.P. and Wilder, H. (2006). Uses and Potentials of Wikis in the Classroom. *Innovate, 1*(5).

Figueroa, S., & Huie, C. (2001). *The use of blackboard in computer information systems courses.* New York: City University of New York Hostos Community College. (ERIC Document Reproduction Service No. ED469883)

Finkelstein, J. E. (2006). *Learning in real time: Synchronous teaching and learning online.* San Fransisco: Jossey-Bass.

Finkelstein, J. E. (2007). *Back to school: Virtual eye contact.* October 7th, 2007. Retrieved from http://www.learninginrealtime.com/minute/

Fishbein, M. and Ajzen, I. (1975). Belief, Attitude, Intention, and Behavior: An Introduction to Theory and Research. Reading, Mass: Addison-Wesley.

Fisher, M. and Baird, D. (2005). Online learning design that fosters support, self-regulation, and retention. *Campus-Wide Information Systems, 22*(2), 88-107.

Fjermestad, Hiltz, S. & Zhang, Y. (2005). Effectiveness for students: Comparisons of "in-seat" and aln courses. In S.Hiltz & R.Goldman (Eds.), *Learning together online: Research on asynchronous learning networks* (pp. 39-79). London: Lawrence Erlbaum Associates.

Flatley, M. (2005). Blogging for Enhanced Teaching and Learning. *Business Communication Quarterly, 68*, 2005.

Fleming, N. and Mills, C. (1992). Not another inventory, mrater a catalyst for reflection. *To Improve the Academy, 11*, 137-149.

Fleming, N. and Mills, C. (1992). VARK a guide to learning styles. Retrieved from http://www.vark-learn.com/English/index.asp

Flores, R.A., Kremer, R.C., & Norrie, D.H. (2001). An architecture for modeling internet-based collaborative agent systems. in Wagner, T., & Rana, O.F. (Eds.), Infrastructure for Agents, Multi-Agent Systems, and Scalable Multi-Agent Systems (pp. 56-63), LNCS1887, Springer-Verlag, London, UK.

Flynn, T., & Polin, L. (2003, April). *Making sense of online learning: Frames, rubrics, tools and coding systems for analyzing asynchronous online discourse*. Paper presented at the 84th AERA Annual Meeting, Chicago, IL.

Forman, E. A. (1992). Discourse, intersubjectivity, and the development of peer collaboration: A Vygotsky approach. In L. T. Winegar & J. Valsiner (Eds.), *Children's development within social context* (Vol. 1, pp. 143-159). Hillsdale, NJ: Lawrence Erlbaum Associates.

Fornell, C. and Larcker, D. F. (1981). Evaluating Structural Equations Models with Unobservable Variables and Measurement Error, *Journal of Marketing Research, 18*, 39-50.

Fox, M.A., & Hackerman, N. (2003). *Evaluating and improving undergraduate teaching in science, technology, engineering, and mathematics.* Washington, DC: The National Academies of Science Press.

Frame, J.D. (1999). *Project management competence.* US: Jossey-Bass.

Francescato, D., Mebane, M., Porcelli, R., Attanasio, C., & Pulin, M. (2007). Developing professional skills and social capital through computer supported collaborative learning in university contexts. *Int. J. Human-Computer Studies* (65), 140-152.

Frantz, P.L. & Wilson, A.H. (2004). Student performance in the legal environment course: determinants and comparisons. *The Journal of Legal Studies Education, 21*(2) 225.

Freeman, C. E. (2004). *Trends in educational equity of girls & women: 2004* (No. NCES 2005-016). Washington, DC: U.S. Government Printing Office: U.S. Department of Education, National Center for Education Statistics.

Freeman, L. A. (2003). A refresher in data flow diagramming: an effective aid for analysts. *Communications of the ACM, 46*(9), 147-151.

Friedman, T. (2005). *The World is Flat.* New York.

Fuchs-Kittowski, F. and Kohler, A. (2002). Knowledge Creating Communities in the Context of Work Processes. *SIGGROUP Bulletin, 23*, 3.

Fuchs-Kittowski, F. and Kohler, A. (2005). Wiki Communities in the Context of Work Processes. Proceedings WikiSim'05, Oct. 16-18, San Diego CA.

Fulwiler, T. (1987). *The Journal Book*. Portsmouth, NH: Boynton/Cook.

Furhammer, I., & Isaksson, F. (1971). *Politics and film.* New York: Praeger..

Gabbert, P. (2000). Systems analysis: the challenge of integrating two competing technologies. *Proceedings of the fourteenth annual consortium on Small Colleges Southeastern conference*, 193-200.

Gaddis, P.O. (1959). The project manager. *Harvard Business Review, 32*(May-June), 89-97.

Gagne, R. M. (1985). *The conditions of learning and theory of instruction* (4th ed.). New York: Holt, Rinehart, and Winston.

Gagne, R. M., Wager, W. W., Golas, K. C., & Keller, J. M. (2005). *Principles of instructional design* (5th ed.). Belmont, CA: Wadsworth/Thomson Learning.

Gane, C., & Sarson, T. (1979). *Structured Systems Analysis: Tools and Techniques*. Englewood Cliffs, NJ: Prentice-Hall.

Gee, J. (1999). *An introduction to discourse analysis: Theory and method.* London, UK: Routledge.

Gefen, D. (2000). Gender Differences in the Perception and Adoption of E-mail and Computer-Mediated Communication Media: A Sociolinguistics Approach. In A. Kent (Ed.), *The Encyclopedia Of Library And Information Science (Volume 69)* (pp. 139-153). New York: Marcel Dekker, Inc.

Gefen, D. and Straub, D. (2000). The Relative Importance of Perceived Ease of Use in the IS Adoption: A Study of E-Commerce Adoption, *Journal of Association for Information Systems, 1*(8), 1-28.

Gefen, D., & Ridings, C. (2005). If You Spoke as She Does, Sir, Instead of the Way You Do: A Sociolinguistics Perspective of Gender Differences in Virtual Communities. *The DATA BASE for Advances in Information Systems, 36*(2), 78-92.

Gefen, D., & Straub, D. W. (1997). Gender Differences in Perception and Adoption of E-mail: An Extension to the Technology Acceptance Model. *MIS Quarterly, 21*(4), 389-400.

Genuter, B. (1995). *Television and gender representation.* London: John Libbey.

George, S. E. (2001). Learning and the reflective journal in computer science. *Australian Computer Science Communications, 24*(1), 77-86.

Gerbner, G. (1976). *Television and its viewers: What social science sees.* Santa Monica, CA: Rand Corporation.

Gerbner, G., Gross, L., Morgan, M., & Signorielli, N. (1994). Growing up with television: The cultivation perspective. In Bryant, J., & D. Zillman (Eds.). *The cultivation perspective* (pp. 17-41). Hillsdale, N. J.: Erlbaum.

Geri, N., Mendelson, O., & Gefen, D. (2007, August 9-12). *On the Centrality of Social Norms in Determining MBA Program Success.* Paper presented at the Americas conference on Information Systems, Keystone, Colorado, USA.

Gfeller, M. K., Niess, M. L., & Lederman, N. G. (1999). Preservice teachers' use of multiple representations in solving arithmetic mean problems. *School Science and Mathematics, 99*, 250-257.

Giere, R. (2002). Distributed Cognition in Epistemic Cultures. *Philosophy of Science, 69*(4), 637-644.

Gieryn, T.F. (1995). Boundaries of Science, Jasanoff et. al. (Ed.), *Handbook of Science and Technology Studies*, 393-444.

Gilbert, A. (2005). *Teachers leave grading up to the computer.* CNET News.com. Retrieved December 30, 2005, from http://news.com.com/Teachers+leave+grading+up+to+the+computer/2100-1032_3-5659366.html?part=rss&tag=5659366&subj=news

Gilson, C., Pratt, M., Roberts, K. & Weymes, E. (2000). *Peak performance: Business lessons from the world's top sports organizations.* Netley SA: Harper Collins Business.

Gladwell, M. (2000). The tipping point: How little things can make a big difference. London: Little Brown.

Gläser, J. and Laudel, G. (2004). Experteninterviews und qualitative Inhaltsanalyse als Instrumente rekonstruierender Untersuchungen, *Verlag für Sozialwissenschaften.*

Glogoff, S. (2005). Instructional Blogging: Promoting Interactivity, student-centered learning, and peer input. *Innovate, 1,* (5).

Goffe, W.and Sosin, K. (2005). Teaching with Technology: May you Live in Interesting Times. *Journal of Economic Education, 36,* (3).

Goldsby, D., Fazal, M. (2001). Now that your students have created web-based digital portfolios, How do you evaluate them? *Journal of Technology and Teacher Education,* Vol. 9, (4): 607-616. http://proquest.umi.com

Göncü, A. (1993). Development of intersubjectivity in social pretend play. *Human Development, 36,* 185-198.

Gordin, D. N., Gomez, L. M., Pea, R. D., & Fishman, B. J. (1996). Using the world wide web to build learning communities in K-12 [Online]. *Journal of Computer Mediated Communication, 2* (3), 1-27. Available: http://www.ascusc.org/jcmc/vol2/issue3/gordin.html

Grabinger, R. S. (1996). Rich environments for active learning. In D.H. Jonassen (ed.) *Handbook of Research for Educational Communications and Technology.* New York: Macmillan, 665-692.

Grabinger, S. (2004). Design lessons for social education. In Duffy, T. M., & Kirkley, J. R., *Learner-Centered Theory and Practice in Distance Education: Cases from Higher Education 2004* (pp. 49-60).

Grant, R.M. (1997). The knowledge-based view of the firm: Implications for management practice. *Long Range Planning, 30*(3), 450-454.

Gray, J. (1992). *Men are From Mars, Women Are From Venus.* New York, NY: HarperCollins.

Greenberg, G. (2004). The digital convergence: extending the portfolio mode [electronic version]. *Educase 39*(4): 28-37.

Greer, R.D. (2002). *Designing teaching strategies: An applied behavior analysis systems approach.* New York: Academic Press.

Greer, R.D., & McDonough, S.H. (1999). Is the learn unit a fundamental measure of pedagogy? *The Behavior Analyst, 22,* 5-16.

Gregory, G. and Chapman, C. (2002). Differentiated instructional strategies. Thousand Oaks, Ca: Corwin Press, Inc.

Gridley, R. Asking better classroom questions: A teachers' mini-workbook.

Griss, M., & Pour, G. (2001). Accelerating development with agent components. *Computer, IEEE, 34*(5), 37-43.

Gronbeck, B. Farrell, T., and Soukup, P. (1991). *Media Consciousness and Culture,* Newbury Mark CA: Sage.

Grossman, M., Aronson, J. E., & McCarthy, R. V. (2005). Does UML make the grade? Insights from the software development community. *Information and Software Technology, 47,* 383-397.

Guernsey, L. (2003). In the Lecture Hall, a Geek Chorus. *The New York Times,* July 24.

Guillier, J., & Drndell, A. (2006). I Totally Agree with You: Gender Interactions in Educational Online Discussion Groups. *Journal of Computer Assisted Learning, 22*(5), 368-381.

Gunawardena, C. N., & Duphorne, P. L. (2000). Predictors of learner satisfaction in an academic computer conference. *Distance Education, 21*(1), 101-117.

Gunawardena, C. N., Lowe, C. A., & Anderson, T. (1997). Analysis of a global online debate and the development of an interaction analysis model for examining social construction of knowledge in computer conference. *Journal of Educational Computing Research, 17*(4), 397-431.

Gunter, B. (1995). *Television and gender representation.* London: John Libbey

Guri-Rosenblit, S. (2005). Eight Paradoxes in the Implementation Process of E-learning in Higher Education. *Higher Education Policy, 18*(1), 5-29.

Haake, A., Lukosch, S. and Schummer, T. (2005). Wiki Templates. Proceedings WikiSym'05, Oct 16-18, San Diego, 41-51.

Halpern, D.F., & Hakel, M.F. (2003). Applying the science of learning to the university and beyond: Teaching for long-term retention and transfer. *Change*, *35*(4), 37-41.

Hannah, A., & Murachver, T. (1999). Gender and Conversational Style as Predictors of Conversational Behavior. *Journal of Language and Social Psychology, 18*(2), 153-174.

Hara, N. and Kling, R. (1999). Students' Frustrations with a Web-Based Distance Education, *First Monday, 4*(12). Retrieved March 3, 2004, from http://www.firstmonday.org/issues/issue4_12/hara/index.html

Hargis, J. and Wilson, B. (2005). Fishing for Learning with a Podcast Net. Instruction and Research Technology, Univ. of North Florida.

Harrington, D. (1999). Teaching statistics: A comparison of traditional classroom and programmed instruction/distance learning approaches. *Journal of Social Work Education*, *35*(3), 343-352.

Hartge, B. (2005). Personal correspondence.

Hartman, S. and Schecker, H. (2005). Mädchen im Umgang mit Informatik, Technik und Naturwissenschaften—Externe Evaluationsergebnisse zu dem Projekt Roberta, *Zeitschrift für Didaktik der Naturwissenschaften*, ZfDN

Hartmann, S., Schecker, and H., Rethfeld, J. (2005). Mädchen und Roboter—Ein Weg zur Physik? Anja Pitton, (Ed.) *Tagungsband der Jahrestagung in Heidelberg 2004 der Gesellschaft für Didaktik der Chemie und Physik—Relevanz fachdidaktischer Forschungsergebnisse für die Lehrerbildung*. Münster: Lit Verlag.

Heckel, R. W. (1996). Engineering freshman enrollments: Critical and non-critical factors. *Journal of Engineering Education, 85* (1), 15-21.

Heggestad, E.D., & Kanfer, R. (2005). The predictive validity of self-efficacy in training performance: Little more than past performance. *Journal of Experimental Psychology: Applied, 11*(2), 84-97.

Hembrooke, H. and Gay, G. (2003). The laptop and the lecture. *Journal of Computing in Higher Education, 15*(1).

Hemphill, L. and Yew, J. (2007). Twiki and WetPaint. *Proceedings of GROUP'07*, Nov 4-7, Sanibel Island FL, 273-276.

Henri, F. (1992). Computer conferencing and content analysis. In A. R. Kaye (Ed.), *Collaborative learning through computer conferencing: The Najaden Papers* (pp. 117-136). New York, NY: Springer-Verlag.

Herring, S. (1999). Interactional coherence in CMC [Online]. *Journal of Computer-Mediated Communication, 4*(4), 1-24. Retrieved from http://www.ascusc.org/jcmc/vol4/issue4/herring.html

Herring, S. C. (1993, January 20). Gender and Democracy in Computer Mediated Communication. *Electronic Journal of Communication 3(2)* Retrieved July 29, 2006, from http://www.cios.org/www/ejc/v3n293.htm

Herring, S. C. (1996). Posting in a Different Voice: Gender and Ethics in Computer-Mediated Communication. In C. Ess (Ed.), *Philosophical Perspectives on Computer-Mediated Communication* (pp. 115 145). Albany: State University of New York Press.

Herring, S. C. (1996). Two Variants of an Electronic Message Schema. In S. C. Herring (Ed.), *Computer-Mediated Communication Linguistic, Social and Cross-cultural Perspectives* (pp. 81-106). Philadelphia, PA: John Benjamins Publishing Company.

Herring, S., Scheidt, L., Bonus, S. and Wright, E (2004). Bridging the Gap: A Genre Analysis of Weblogs. Proceedings of HICSS 2004, Big Island, Hawaii.

Herrmann, F. (1995). Listserver communication: The discourse of community-building. *Proceedings of the first international conference on computer support for collaborative learning, October 17-20, 1995, Indiana University, Bloomington, Indiana, USA*, 165-168.

Higdon, J. (2006). Pedagogies of Wikis. EDUCAUSE NLII Meeting, San Diego, Educause.

Higgins, C. Reeves, L. and Byrd, E. (2004). Interactive Online Journaling. Proceedings SIGUCCS'04, Oct 10-13, 2004, Baltimore MD.

Hiltz, S. R., & Turoff, M. (2005). Education Goes Digital: The Evolution of Online Learning and the Revolution in Higher Education. *Communications of the ACM, 48*(10), 59-64.

Hirschheim, R. (2005). The Internet-Based Education Bandwagon: Look before You Leap. *Communications of the ACM, 48*(7), 97-101.

Hodges, C. B. (2005). Self-regulation in web-based courses: A review and the need for research. *The Quarterly Review of Distance Education, 6*(4), 375-383.

Hofstede, G. (1980). *Culture's Consequences: International Differences in Work Related Values*. London, UK: Sage.

Holmes, J. (1992). Women's Talk in Public Contexts. *Discourse and Society, 3*(2), 131-150.

Honegger, B. (2005). Wikis—a rapidly growing phenomenon in the German-Speaking School Community. Proceedings WikiSim'05, Oct. 16-18, San Diego CA.

Hong, S. J., & Tam, K. Y. (2006). Understanding the Adoption of Multipurpose Information Appliances: The Case of Mobile Data Services. *Information Systems Research, 17*(2), 162-179.

Hrastinski, S. (2006). Introducing an informal synchronous medium in a distance learning course: How is participation affected? *The Internet and Higher Education* (9), 117-131.

Hrastinski, S. (2006). The relationship between adopting a synchronous medium and participation in online group work: An explorative study. *Interactive Learning Environments 14*(2), 137-152.

Hsia, J.I., Simpson, E., Smith, D., & Cartwright, R. (2005, February 23-27). Taming Java for the classroom. *SIGCSE'05,* St. Louis, MI, 327-331.

Hsu, K. and Tanner, J. (2006). An assessment of classroom technology usage in business. *International Journal of Innovation and Learning 3*(5), 488-498.

Hsu, M-H, Yen, C-H, Chiu, C-M and Chang, C-M (2006). A Longitudinal Investigation of Continued Online Shopping Behavior: An Extension of the Theory of Planned Behavior, *International Journal of Human-Computer Studies, 64*(9), 889-904.

Hu, C. (2006). It's mathematical after all: The nature of learning computer programming. *Education and Information Technologies, 11*(1), 83-92.

Hu, P. J., Chau, P. Y. K., Sheng, O. R. L. and Tam, K. Y. (1999). Examining the Technology Acceptance Model Using Physician Acceptance of Telemedicine Technology, *Journal of Management Information Systems, 16*(2), 91-112.

Huang, C. D., and Behara, R.S. (2007). Outcome Drivem Experiential Learning with Web 2.0. *Journal of Information Systems Education, 18,* (3), 329-335.

Huang, G., Taddese, N., & Walter, E. (2000). *Entry and persistence of women and minorities in college science and engineering.* (No. NCES 2001-601). U.S. Department of Education National Center for Education Statistics. Washington, DC: U. S. Government Printing Office.

Huang, H. M. (2002). Student perceptions in an online mediated environment. *International Journal of Instructional Media 29,* 405-422.

Huang, H. M., & Liaw, S. S. (2004). Guiding distance educators in building web-based instructions. *International Journal of Instructional Media 31*(2), 125-137.

Hui, D. (2003). Managing intersubjectivity in the context of a museum learning environment. *International Journal of Learning, 10,* 1533-1549.

Hui, D. (2006). *Engagement in supporting new teachers: A role for computer-mediated communication in teacher learning within informal professional communities.* Unpublished doctoral dissertation, Washington University in St Louis, Missouri.

Hutchins, E. (1995). *Cognition in the Wild.* Cambridge MA: MIT Press.

IEP. (2007). L.D in depth. *LD Online*. Retrieved November 18, 2007 from http://www.ldonline.org/ld_indepth/iep/iep.html

Ifinedo, P. (2005, September 12-15). E-learning Technology Adoption Factors in an Eastern European Country: An Exploratory Study, *Proceedings of the 9th East-European Conference on Advances in Databases and Information Systems*. Tallinn, Estonia.

Ifinedo, P. (2005). Uncertainties and Risks in the Implementation of e-learning Information Systems Project in a Higher Learning Environment: Viewpoints from Estonia, *Journal of Information and Knowledge Management, 4*(1), 37-46.

Ifinedo, P. (2006). Acceptance and Continuance Intention of Web-based Learning Technologies (WLT) Use Among University Students in a Baltic Country, *Electronic Journal of Information Systems in Developing Countries, 23*(6), 1-20.

Ifinedo, P. (2007). Investigating the Antecedents of Continuance Intention of Course Management Systems Use among Estonian Undergraduates, *International Journal of Information and Communication Technology Education, 3*(4), 76-92.

Ifinedo, P. (2007). E-learning in the Nigerian Higher Education Sector: Opportunities and Challenges, *The African Symposium: An Online Journal of African Educational Research Network, 7*(2), 48-54.

Igbaria, M. (1990). End-User Computing Effectiveness: A Structural Equation Model, *Omega, 18*(6), 637-652.

Igbaria, M. and Parasuraman, S. (1989). A Path Analytic Study of Individual Characteristics, Computer Anxiety, and Attitudes toward Microcomputers, *Journal of Management, 15*(3), 373-388.

Im, Y., & Lee, O. (2004). Pedagogical implications of online discussion for pre-service teacher training. *Journal of Research on Technology in Education, 36*(2), 155-170.

Individualized Education Program (IEP). (1999). L.D in depth. *LD Online*. Retrieved September 15, 2006, from http://www.ldonline.org/ld_indepth/iep/iep.html

Ingvarsson, E.T., & Hanley, G.P. (2006). An evaluation of computer-based programmed instruction for promoting teachers' greeting of parents by name. *Journal of Applied Behavior Analysis, 39*(2), 203-214.

International Society for Technology Education (ISTE). (n.d.). *Educational technology standards and performance indicators for all teachers*. Retrieved September 15, 2006, from http://cnets.iste.org/teachers/t_stands.html

iQ Academies. (2006). *iQ Academies* homepage. Retrieved November 18, 2007 from http://www.gotoiq.com/index.asp

Isaacs, W. (1999). *Dialogue: The art of thinking together*. New York: Doubleday.

ISTE. (2000). Educational technology standards and performance indicators for all teachers. Retrieved November 18, 2007 from http://cnets.iste.org/teachers/pdf/page09.pdf

Ivankova, N. V., & Stick, S. L. (2005). Collegiality and community: Building as a means for sustaining persistence in the computer-mediated asynchronous learning environment. *Online Journal of Distance Learning Administration, 8*(3). Retrieved December 15, 2005, from http://www.westga.edu/%7Edistance/ojdla/fall83/ivankova83.htm

Jacobs, J. A. (1999). The sex segregation of occupations." In Powell, G. N. (Ed.). *Handbook of gender and work*, (pp. 125-139). Thousand Oaks, CA: Sage Publications.

Jay, R. (1995). *Build a great team*. London: Pitman Publishing.

Jehng, J-C. J. (1997). The psycho-social processes and cognitive effects of peer-based collaborative interactions with computers. *Journal of Educational Computing Research, 17*(1), 19-46.

Jewels, T. & Bruce, C. (2003, June 24-27). *Using a case method approach in an IT project management curriculum: A long look over the shoulder of a practitioner at work*. Proceedings of the Informing Science + IT Education Conference, Pori, Finland.

Jewels, T. & Ford, M. (2004). A single case study approach to teaching: Effects on learning & understanding. *Issues in Informing Science and Information Technology, 1*, 359-372.

Jewels, T. & Underwood, A. (2004). The impact of informal networks on knowledge management strategy. In B. Montano (Ed.), *Innovations of knowledge management* (Chapter 1). Hershey, PA: IRM Press.

Joerges, B. (1996). Technik Körper der Gesellschaft. Arbeiten zur Techniksoziologie, Suhrkamp, Frankfurt am Main.

Johnson, D.W. and Johnson, R. T. (1998). *Cooperation and Competition*. Edina MN:Interaction.

Johnson, D.W., Johnson, R.T. and Holubec, E. (1991). *Cooperation in the Classroom* (3/e) Edina MN: Interaction.

Johnson, N. (1995). Active questioning. Pieces of Learning Publishing.

Johnson, R. A., Hardgrave, B. C., & Doke, E. R. (1999). An industry analysis of developer beliefs about object-oriented systems development. *ACM SIGMIS Database, 30*(1), 47-64.

Johnson, R.D. (2005). An empirical investigation of sources of application-specific computer-self-efficacy and mediators of the efficacy-performance relationship. *International Journal of Human-Computer Studies, 62*(6), 737-758.

Johnson, S., Aragon, S., Najmuddin, S., & Palma-Rivas, N. (2000). Comparative analysis of learner satisfaction and learning outcomes in online and face-to-face learning environments. *Journal of Interactive Learning Research,* 11(1), 29-48.

Jonassen, D. (2003). Using cognitive tools to represent problems. *Journal of Research on Technology in Education, 35*, 362-381.

Jonassen, D. H. (1994). Thinking technology: Toward a constructivist design model. *Educational Technology, 34*(3), 34-37.

Jonassen, D. H., Peck, K. L., & Wilson, B. G. (1999). *Learning with technology: A constructivist perspective.* Upper Saddle River, NJ: Merrill-Prentice Hall.

Jonassen, D., Peck, K., and Wilson, B. (2000). Learning With Technology: A Constructivist Perspective. Merrill: Upper Saddle, NJ.

Jones, M. G., Howe, A., & Rua, M. J. (2000). Gender differences in students' experiences, interests and attitudes toward science and scientists. *Science Education, 84*(2), 180-192.

Joyce, Weil, & Calhoun. (2000). *Models of Teaching (6th Ed.).* Needham Heights, MA: Allyn & Bacon.

Kabilan, M. (2005). Online professional development: A literature analysis of teacher competency. *Journal of Computing in Teacher Education, 21*(2), 51-57.

Kanuka, H., Collett, D., & Caswell, C. (2002). University instructor perceptions of the use of asynchronous text-based discussion in distance courses. *The American Journal of Distance Education, 16*(3), 151-167.

Kashihara A., Kinshuk, Oppermann R., Rashev R., Simm H., & Toyoda J. (1998). Towards sharable intelligent assistance for learning-by-exploration. SIG Technical Notes in Japanese Society for Artificial Intelligence, SIG-IES9703, JSAI, (pp. 1-6)Shizuoka, Japan,.

Katila, R. (2002). New product search overtime: Past ideas in their prime? *Academy of Management Journal, 45*(5), 995-1010.

Katzenbach, J. & Smith, D. (1993). *The wisdom of teams.* Boston: Harvard Business School Press.

Keegan, D. (1996). *Foundations of distance education.* (3rd ed). London: Routledge.

Keller, B, Stuve, M (2006). Teacher as Brand: Pursuing Professional Identities in a Digital Domain. In: Sharon Y. Tettegah and Richard C. Hunter, Editor(s), Advances in Educational AdministrationJAI. *Technology and Education: Issues in Administration, Policy, and Applications in K12 Schools.* Vol.8: 57-70.

Keller, C. and Cernerud, L. (2002). Students' Perceptions of E-learning in University Education, *Journal of Educational Media, 27*(1/2), 55-67.

Kilbourne, W., & Weeks, S. (1997). A Socio-economic Perspective on Gender Bias in Technology. *Journal of Socio-Economics, 26*(1), 243-260.

King, K. P. (2001). Playing out the realities of web-based bulletin boards: Enhancing face to face learning. *New Horizons in Adult Education, 15*(1).

Kinzie, M., Whitaker,S. and Hofer, M. (2005). Instructional Uses of Instant Messaging During Classroom Lectures. *Educational Technology and Society, 8*(2), 150-160.

Kinzie, S. (2005). Colleges Land Lines nearing an end. The Washington Post, Feb 12.

Kirkup, G and von Prummen, C. (1990). Support and Connectiveness: The needs of Women Distance Education Particpants, *Journal of distance Education*, Retrieved from http://cade.athabascau.ca.vol5.2/7_kirkup_and_von_prummer.html

Kirschner, P.A., Sweller, J., & Clark, R.E. (2006). Why minimal guidance during instruction does not work: An analysis of the failure of constructivist, discovery, problem-based, experiential, and inquiry-based teaching. *Educational Psychologist, 41*(2), 75-86.

Kish, M.H.Z. (2006). Overview of using vignettes to develop higher order thinking and academic achievement in adult learners in an online learning environment. *International Journal of Information and Communication Technology Education, 2*(3), 60-74.

Kleiner, A. & Farris, E. (2002). *Internet access in U.S. public schools and classrooms: 1994–2001* (NCES 2002-018). Washington, DC: U.S. Department of Education, National Center for Education Statistics.

Koen, B.V. (2005). Creating a sense of "presence" in a web-based PSI course: The search for Mark Hopkins' log in a digital log. *IEEE Transactions on Education, 48*(4), 599-604.

Kolb, D. (1984). *Experiential learning. Experience as the source of learning and development*. Englewood Cliffs, NJ: Prentice-Hall.

Kolling, M., Quig, B., Patterson, A., & Rosenberg, J. (2003). The BlueJ system and its pedagogy. *Journal of Computer Science Education, 14*(Dec), 1-12.

Koper, R. (2006). Current research in learning design. *Journal of Educational Technology & Society, International Forum of Educational Technology & Society, 9*(1), 13-22.

Korthagen, F. (1993). Two modes of reflection. *Teacher and Teacher Education, 9(*3), 317-326.

Koschmann, T. (Ed.). (1996*). CSCL: Theory and practice of an emerging paradigm*. Mahwah, NJ: Lawrence Erlbaum Associates.

Kozlowski, S. (2000). Better Learning with Technology? *The School Administrator, April.*

KPMG (2003). Insights from KPMG's European Knowledge Management Survey 2002/2003. KPMG Knowledge Advisory Servicees, Amsterdam.

Kramarski, B. and Feldman, Y. (2000). Internet in the Classroom: Effects on Reading Comprehension, Motivation and Metacognitive Awareness. *Educational Media International, 37(3).* 149-55.

Krisnamurthy, S. (2002). The Multidimensionality of Blog Conversations. Internet Research 3.0, Maastricht, Netherlands.

Kurbanoglu, N.I., Taskesenligil, Y., & Sozbilir, M. (2006). Programmed instruction revisited: A study on teaching stereochemistry. *Chemistry Education Research and Practice*, 7(1), 13-21.

Lacey, K. (1996). *Feminine frequencies: Gender, German radio, and the public sphere: 1923-1945.* Ann Arbor: University of Michigan Press.

Lamb, B. (2004). Wide Open Spaces Wikis ready or Not. *Educause, Sep/Oct*, 36-48.

Latour, Bruno (1998). Über technische Vermittlung. Philosophie, Soziologie, Genealogie. Rammert, Werner (Ed.)

Technik und Sozialtheorie. Campus-Verlag, 29-83.

Lave, J. and Wenger, E. (1991). *Situated Learning, Legitimate Peripheral Participation*. New York: Cambridge University Press.

Lave, J., & Wenger, E. (1991). *Situated learning: Legitimate peripheral participation*. Cambridge, UK: Cambridge University Press.

Lawlor, C. (2006). Gendered Interactions in Computer-Mediated Computer Conferencing. *The Journal of Distance Education / Revue de l'Éducation à Distance, 21*(2), 26-43.

Lebaron, J., & Miller, D. (2005). The potential of jigsaw role playing to promote the social construction of knowledge in an online graduate education course. *Teachers College Record, 107*(8), 1652-1674.

Lederer, A., Maupin, D., Sena, M. and Zhuang, Y. (2000). The Technology Acceptance Model and the World Wide Web, *Decision Support Systems, 29*(3), 269-282.

Lee, C., Witta, E. L. (2001). Online Students' Perceived self-efficacy: Does It Change? *Annual meeting of the Association for Educational Communications and Technology (AECT)*, Atlanta, GA.

Lee, E.-J. (2003). Effects of "Gender" of the Computer on Informational Social Influence: The Moderating Role of Task Type. *International Journal of Human-Computer Studies, 58*, 347–362.

Lee, J, Cho, H., Gay, G., Davidson, B. and Ingraffea, A. (2003). Technology Acceptance and Social Networking in Distance Learning, *Educational Technology & Society, 6*(2), 50-61.

Lee, J., & Lee, Y. (2006). Personality types and learners' interaction in web-based threaded discussion. *Quarterly Review of Distance Education, 7*(1), 83-94.

Legris, P., Ingham, J. and Collerette, P. (2003). Why Do People Use Information Technology?: A Critical Review of the Technology Acceptance Model, *Information and Management, 40*(3), 191-204.

Leidner, D. E. and Jarvenpaa, S. L. (1993). The Information Age Confronts Education: Case Studies on Electronic Classrooms, *Information Systems Research, 4*(1), 24-54.

Leonard-Barton, D. (1995). *Wellsprings of knowledge: Building and sustaining the sources of innovation*. Boston: Harvard Business School Press.

Leontiev, A. (1978). Activity, Consciousness, and Personality. Englewood Cliffs, NJ: Prentice Hall.

Leuf , B. and Cunningham, W. (2001). *The Wiki Way*. Reading, MA: AW Professional.

Leuf, B. and Cunningham, W. (1999). The WIKI Way: Quick Collaboration on the Web. Reading, MA: Addison Wesley.

Levin, D. and Arafeh, S. (2002). The Digital Disconnect. Pew Charitable Trust, 2004.

Levin, D. & Arafeh, S. (2002). *The digital disconnect: The widening gap between Internet-savvy students and their schools*. Washington DC: Pew Internet & American Life Project.

Levy, Y. (2006). The Top 10 Most Valuable Online Learning Activities for Graduate MIS Students. *International Journal of Information and Communication Technology Education, 2*(3), 27-44.

Levy, Y. (2007). Comparing Dropouts and Persistence in e-learning Courses. *Computers & Education, 45*(2), 185-204.

Liaw, S.-S. (2005). Developing a web assisted knowledge construction system based on the approach of constructivist knowledge analysis of tasks. *Computers in Human Behavior 21*(1), 29-44.

Lim, K. (2005). Now Hear This—Exploring Podcasting as a Tool in Geography Education. Nanyang Technological University.

Limayem, M., Cheung, C. M. K. and Chan, G. W. W. (2003). Explaining Information Systems Adoption and Post-Adoption: Towards An Integrative Model. In *Proceedings of 24th International Conference on Information Systems.*

Lin, H. (2004). A language for specifying agent systems in E-Learning environments. in: Lin, F.O. (ed.), Designing Distributed Learning Environments with Intelligent Software Agents (pp. 242-272). IGI Global, Hershey, PA, USA.

Lin, H., & Nguyen, K. (2005). Classroom simulation of massive parallel computing. Proc. The 16th IASTED International Conference on Modeling and Simulation (MS 2005), Cancun, Mexico, May 18-20 (pp. 45-50). ACTA Press, Calgary, AB, Canada.

Lin, H., & Yang, C. (2006). Specifying Distributed Multi-Agent Systems in Chemical Reaction Metaphor. *The International Journal of Artificial Intelligence, Neural Networks, and Complex Problem Solving Technologies, Springer-Verlag, 24*(2), pp.155-168.

Lindkvist, L. (2005). Knowledge Communities and Knowledge Collectivities. *Journal of Management Studies, 42*(6), 1189-1210.

Lipman, Matthew (1991). *Thinking in Education*. New York: Cambridge University Press.

Lister, R. (2004). Teaching java first: Experiments with a pigs-early pedagogy. *Proceedings of the 6th Conference on Australian Computing Education* Vol. 30 (pp. 177-183), Dunedin: Australian Computer Society, Inc.

Listing of colleges and universities managed by NYS Dept. of Higher Education. http://www.nymentor.edu [retrieved college websites 2/1/2006 through 4/2/2006]

Liu, C. and Chen, S. (2005). Determinants of knowledge sharing of e-learners. *International Journal of Innovation and Learning 2005, 2*(4), 434-445.

Liu, C., Lu, J. and Yu, C. (2003). Learning Style, Learning Patterns, and Learning Performance in a WebCT-based MIS course, *Information and Management, 40*(6), 497-507.

Locke, C. et al. (2000). The Cluetrain Mainfesto: The End of Business as Usual, Cambridge, MA: Perseus.

Louridas, P. (2006). Using wikis in software development. IEEE Software, Mar/Apr, 88-91.

Lu, H-P, Liu, S-H. and Liao, H-L. (2005). Factors Influencing the Adoption of E-learning Websites: An Empirical Study, *Issues in Information, 6*(1), 190-196.

Ludwig, B. (2000, August). *Web-based instruction: Theoretical differences in treatment of subject matter.* Paper presented at the Annual Meeting of the Psychological Association, Washington, DC. (ERIC Document Reproduction Service No. ED453708)

Lum, L. (2006). The Power of Podcasting. *Diverse issues in Higher education, 23*(2), 32.

Lund, A. and Smordal, O. (2006). Is There Space for a Teacher in a Wiki? Proceedings of WikiSym'06, August 21-23, Odense Denmark.

Majeski, R., & Stover, M. (2007). Theoretically based pedagogical strategies leading to deep learning in asynchronous online gerontology courses. *Educational Gerontology, 33*(3), 171-185.

Major, B., & O'Brien, L. T. (2005). The social psychology of stigma. *Annual Review of Psychology, 56*, 393-421.

Makrakis, V. (1998). *Guidelines for the design and development of computer mediated collaborative open distance learning courseware.* Paper presented at the 1998 World Conference on Educational Multimedia and Hypermedia & World Conference on Educational Telecommunications. Frieburg, Germany. (ERIC Document Reproduction Service No. ED428694)

Marcel, K. (2002). Can law be taught effectively online? *JURIST*, December, 2002. Retrieved May 5, 2005 from http://jurist.law.pitt.edu/lessons/lesdeco2.php.

March, J. G. (1991). Exploration and exploitation in organizational learning. *Organization Science, 2*(1), 71-87.

Margolis, J. & Fisher, A. (2002). *Unlocking the clubhouse: Women in computing.* Cambridge, MA: MIT Press, pp. 3-5).

Markel, S. L. (2001). Technology and education online discussion forums: It's in the response. *Online Journal of Distance Learning Administration, 4*(2). Retrieved February 23, 2006, from http://www.westga.edu/~distance/ojdla/summer2001/markel42.pdf

Marshall, A. (1920). *Principles of economics*. Retrieved August 2, 2002, from www.econlib.org/library/Marshall/marP0.html

Martindale, T. and Wiley, D. (2005). Using Weblogs in Scholarship and Teaching. *TechTrends, 49*, 2.

Martyn, M. A. (2004). *The effect of online threaded discussion on student perceptions and learning outcomes in both face-to-face and online courses*. Doctoral dissertation, University of Akron, 2004. (UMI No. 3123389)

Marx, R. W., Blumenfeld, P. C., Krajcik, J. S., & Soloway, E. (1998). New technologies for teacher professional development. *Teaching and Teacher Education, 14*(1), 33-52.

Marzano, R., Pickering, D., and Pollock, J. (2001). Classroom instruction that works: research-based strategies for increasing student achievement. Alexandria, VA: Association for Supervision and Curriculum Development.

Matusov, E. (1996). Intersubjectivity without agreement. *Mind, Culture, and Activity, 3*(1), 25-45.

Maxwell, S.E., & Delaney, H.D. (2004). *Designing experiments and analyzing data: A model comparison perspective (2nd Ed)*. Mahwah, NJ: Lawrence Erlbaum Associates.

Mayer, R.E. (2002). *The promise of educational psychology. Volume II. Teaching for meaningful learning*. Upper Saddle River, NJ: Pearson Education, Inc.

Mayer, R.E. (2004). Should there be a three-strikes rule against pure discovery learning? *American Psychologist, 59*(1), 14-19.

Mayring, P. (2000). Qualitative Inhaltsanalyse. Grundlagen und Techniken, *Deutscher Studien Verlag*.

McAlpine, I., & Ashcroft, B. (2002). Turning points: Learning from online discussions in an off-campus course. In P. Kommers and G. Richards (Eds.), *Proceedings of World Conference on Educational Multimedia, Hypermedia& Telecommunications 2002* (pp. 1251-1257), Chesapeake, VA: AACE.

McCann, D. (2005). *Team learning*. Retrieved March 7, 2005, from www.tms.com.au/tms12-2c.html

McCray, G. E. (2000). The hybrid course: Merging on-line instruction and the traditional classroom. *Information Technology and Management, 1*, 307-327.

McDougall, A., & Boyle, M. (2004). Student strategies for learning computer programming: Implications for pedagogy in informatics. *Education and Information Technologies, 9*(2), 109-116.

McLeod, R. J. (1996). Comparing undergraduate courses in systems analysis and design. *Communications of the ACM, 39*(5), 113-121.

McQueen, D. (1998). *Television: A media student's guide*. London: Arnold Press.

Meisalo, V., Lavonen, J. and Juuti, K. (2005). A Case Study on a group of unqualified Primary Teachers taking a science education course in a web based environment. *IADIS International Conference Web Based Communities*.

Mervis, J. (2000). Diversity: Easier said than done. *Science, 289* (5478), 378-379.

Mezirow, J. (2000). Learning to think like an adult. Core concepts of Transformation Theory. In Mezirow J (ed.) *Learning as transformation. Critical perspectives on a theory in progress* (pp. 3-33). San Fransisco: Jossey-Bass.

Millard, D. and Ross, M. (2006). Web 2.0: Hypertext by any other name? Proceedings of HT'06, August 22-25, Odense Denmark, 27-30.

Miller, K.H.(2004).The law catches up with distance learning. *THE Journal, 31*(7), 31-34.

Millet, I. (1999). A Proposal to Simplify data Flow Diagrams. *IBM Systems Journal, 38*(1), 118-121

Millet, I., & Nelson, R. (2007). Data Flow Diagrams versus Use Cases - Student Perceptions. International Journal of Information & Communication Technology Education, 3(1), 70-78.

Mimirinis, M. & Bhattacharya, M. (2007). Design of virtual learning environments for deep learning. *Journal of Interactive Learning Research, 18*(1), 55-64.

Mindel, J. and Verma, S. (2006). Wikis for teaching and learning. *Communications of the AIS, 18,* 1.

Mock, K. (2002). The Use of Internet Tools to Supplement Communication in the Classroom. *CCSC Northeastern Conference proceedings, 17*(2), 14-21.

Molenda, M. (2002). *A new framework for teaching in the cognitive domain.* Syracuse, NY: ERIC Clearinghouse on Information and Technology (ERIC Document Reproduction Service No. ED470983)

Moonen, J. Tulner, H. (2003). *E-learning and electronic portfolio: Some New Insights.* Universiteit van Twente, The Netherlands. Retrieved Sept 5, 2006, from http://www.connict.nl/pdf/moonentulner-portfolio.pdf.

Moore, M.G. (1980). Independent study, Redefining the Discipline of Adult Education, *Redefining the Discipline of Adult Education,* Boyd and Apps (eds), San Francisco, 16-27.

Moreno, L., Gonzalez, C., Castilla, I., Gonzalez, E., & Sigut, J. (2006). Applying a constructivist and collaborative methodological approach in engineering education, computers & Education, in press.

Moreno, R. (2002). Who learns best with multiple representations? Cognitive theory implications for individual differences in multimedia learning. In P. Kommers and G. Richards (Eds). *Proceedings of World Conference on Educational Multimedia, Hypermedia & Telecommunications 2002* (pp. 1380-1385). Chesapeake, VA: AACE.

Morris, M.G. and Turner, J. M. (2001). Assessing Users' Subjective Quality of Experience with the World Wide Web: An Exploratory Examination of Temporal Changes in Technology Acceptance, *International Journal of Human-Computer Studies, 54*(6), 877-901.

Morss, D. A. (1999). A Study of Student Perspectives on Web-based Learning: WebCT in the Classroom, *Internet Research: Electronic Network Applications and Policy, 9*(5), 393-408.

Mortimer, E. F. & Wertsch, J. V. (2003). The architecture and dynamics of intersubjectivity in science classrooms. *Mind, Culture, and Activity, 10*(3), 230-244.

Muelinburg, L. and Berge, Z. (2000). The moderators' homepage: A framework for designing questions for online learning (website). Retrieved from http://www.emoderators.com/moderators/muilenburg.html

Muilenburg, L.Y. & Berge, Z.L. (2001). Barriers to distance education: A factor-analytic study. *The American Journal of Distance Education.* 15(2), 7-24.

Muirhead, B. (2001). Enhancing social interaction in computer-mediated distance education. *Ed at a Distance,* 15(40). Retrieved September 26, 2007, from http://www.usdla.org/html/journal/APR01_Issue/articel02.html

Mulac, A., Erlandson, K. T., Farrar, W. J., & Hallett, J. S. (1998). Uh-huh. What's That all about? Differing Interpretations of Conversational Backchannels and Questions as Source of Miscommunication Across Gender Boundaries. *Communication Research, 25*(6), 641-668.

Müllerburg, M., Petersen, U., and Theidig, G. (2004). Mit Robotern spielend lernen. VDI (Ed.) *ROBOTIK 2004.* VDI Berichte Nr. 1841, 393-400.

Murphy, E. (1997). *Constructivist checklist.* retrieved August 18, 2005, from http://www.stemnet.nf.ca/~elmurphy/emurphy/cle4.html

Murphy, E. (2004). Recognizing and promoting collaboration in an online asynchronous discussion. *British Journal of Educational Technology, 35,* 421-431.

Nardi, B. and Bradner, E. (2000). Interaction and Outeraction. Proceedings of CSCW'00, Dec. 2-6, 2000, Philadelphia PA.

Nardi, B., Schiabo, D.J. , Gumbrecht, M. and Swartz, L (2004). Why We Blog. *Communications of the ACM, 47*(12), 41-46.

Nardi, B., Schiano, D.J. and Sumbrecht, M. (2004). Blogging as Social activity. Proceedings of CSCW 2004, Chicago, Nov. 6-10, 222-231.

Nass, C., Moon, Y., & Green, N. (1997). Are Computers Gender-neutral? Gender Stereotypic Responses to

Computers. *Journal ofApplied Social Psychology, 27,* 864–876.

National Center for Research on Women. (2001). *Balancing the equation: Where are the women and girls in science, engineering, and technology?* New York: National Council for Research on Women.

National Commission on Teaching and America's Future. (2003, January). *No dream denied: A pledge to America' children.* Washington, D.C.: National Commission on Teaching and America's Future.

National Education Teacher Standards (NETS, 2000). International Society for Technology in Education.

National Science Board. (2000). *Science and engineering indicators, 2000.* Arlington, VA: National Science Foundation.

Navarro, P & Shoemaker, J. (2000). Policy issues in the teaching of economics in cyberspace: research design, course design, and research results. *Contemporary Economic Policy, 18*(3), 359-366.

Nelson, M., Bhagyavati, Miles, G., Settle, A., Shaffer, D., Watts, J., et al. (2005). Online teaching practices (both best and worst). Panel discussion. *Journal of Computing Sciences in Colleges, 21*(2). Retrieved December 14, 2005, from the ACM database.

NetDay. (2004, June). *Insights and ideas of teachers on technology.* Irvine, CA: NetDay Speak Up Day 2004. Retrieved January 30, 2006, from http://www.netday.org/sud4teachers_2004_report.htm

Neuman, W. L. (1997). *Social Research Method.* London: Allyn and Bacon.

Newby, M., & Nguyen, T. (2007). Using the Same Problem with Different Techniques in Programming Assignments. Journal of Information Systems Education, 18(3), 279-282.

Newkirk, R.L. (1973). A comparision of learner control and machine control strategies for computer assisted instruction. *Programmed Learning and Educational Technology, 10*(2), 82-91.

Ng, K. C. (2007). Replacing face-to-face tutorials by synchronous online technologies: Challenges and pedagogical implications. *The International Review of Research in Open and Distance Learning, 8*(1).

Ngai, E. W. T., Poon, J. K. L. and Chan, Y. H. C. (2006). Empirical Examination of the Adoption of WebCT using TAM, *Computer and Education, 48*(2), 250-267.

Nichols, R. and Anderer (2007). Using Video Podcasts Outside the Classroom in *Proceedings of SIGUCCS'07,* Oct 7-10, 2007, Orlando FL, 277-279.

Nieto, S. (2003). *What keeps teachers going?* New York, NY: Teachers College Press.

Nonaka, I. (1991). The knowledge-creating company. *Harvard Business Review, 6*(8), 96-104.

Nortar, C. E., Wilson, J. D., & Montgomery, M. K. (2005). A distance learning model for teaching higher order thinking. *College Student Journal, 31*(1), 17-26.

Northrup, P., Lee, R., & Burgess, V. (2002). Learner perception of online interaction. In P. Kommers & G. Richards (Eds.), *Proceedings of World Conference on Educational Multimedia, Hypermedia, and Telecommunications 2002* (pp. 1462-1467). Chesapeake, VA: AACE.

Nuffield Primary History (website). Retrieved from http://www.primaryhistory.org/aboutus

Nunnally, J. C. (1978). *Psychometric Theory.* New York, NY: McGraw-Hill.

Nussbaum, E. (2004). My so-called blog. *The New York Times,* Jan 11.

O'Neill, M. (2005). Automated Use of a Wiki for Collaborative Lecture Notes. Proceedings SIGCSE'05, Feb 23-27, 2005, 267-271.

O'Reilly, T. (2005). What is Web 2.0. Retrieved September, 2005, from http://www.oreilly.com

Office of Technology Assessment, U.S. Congress. (1995, April). *Teachers & technology: Making the connection* (Report No. IR 017 275). Washington, D. C.: Office of Technology Assessment. (ERIC Document Reproduction Service No. ED386154)

Ohlsson, J. Granberg, O., & Stedt, L. (2004). Collective learning - the bridge between the individual and the learning organization? In *Workplace learning - From the learner's perspective*. LearningLab Denmark. (Conference paper).

Oliverm K. Leveraging Web 2.0 in the redesign of a Graduate-Level Technology Integration Course. *TechTrends, 51*(5), 55-61.

Ong, C.-S., Lai, J.-Y., & Wang, Y.-S. (2004). Factors affecting engineers' acceptance of synchronous e-learning systems in high-tech companies. *Information & Management, 1*(6), 795-804.

Ong, W. (1982). *Orality and Literacy*. New York: Routledge.

Palloff, R.M. & Pratt, K. (1999). *Building learning communities in cyberspace: Effective strategies for the online classroom*. San Francisco: Jossey-Bass.

Pan, C., Siva, S. and Brophy, J. (2003). Students' Attitude in a Web-enhanced Hybrid Course: A Structural Equation Modeling Inquiry, *Journal of Educational Media and Library Sciences, 41*(2), 181-194.

Pan, W., & Hawryszkiewycz, I. (2004, December 5-8). A method of defining learning processes. In R. tkinson, C. McBeath, D. Jonas-Dwyer, & R. Phillips (Eds.), *Beyond the comfort zone: Proceedings of the 21st ASCILITE Conference* (pp. 734-742). Perth,. http://www.ascilite.org.au/conferences/perth04/procs/pan.html

Paquette, G. (2001). Designing Virtual Learning Centers. In H. Adelsberger, B. Collis, J. Pawlowski (Eds). Handbook on Information Technologies for Education & Training (pp. 249-272), Springer-Verlag, Berlin et al.

Parker, J.R., & Becker, K. (2003). Measuring effectiveness of constructivist and behaviourist assigments. In CS102, Proceedings ITiCSE'03 ACM, (pp. 40-44).

Parks, M. R., & Floyd, K. (1995, January 10, 2003). Making Friends in Cyberspace. *Journal of Computer Mediated Communication 1(4)* Retrieved October 3, 2007, from http://www.ascusc.org/jcmc/vol1/issue4/parks.html

Parrish, R. (1998). *Conversation analysis of internet chat rooms* [Online]. Available: http://polisci.wisc.edu/~rdparrish/academic_works.htm

Pecheone, R.L., Pigg, M.J., Chung, R.R., & Souviney, R.J. (2005). Performance assessment and electronic portfolios:Their effect on teacher learning and education. *The Clearing House, 78*, 164–176.

Perone, M. (1999). Statistical inference in behavior analysis: Experimental control is better. *The Behavior Analyst, 22*(2), 109-116.

Phipps, R. & Merisotis, J. (1999). *What's the difference? A review of contemporary research on the effectiveness of distance learning in higher education.* Washington, D.C.: The Institute for Higher Education Policy. *18*(3), 359-366.

Piaget, J. (1960). *The child's conception of the world.* Totowa, NJ: Littlefield, Adams.

Piaget, J. (1975). *The construction of reality in the child.* Ballantine Books.

Piccoli, G., Ahmad, R., & Ives, B. (2001). Web-based virtual learning environments: A research framework and a preliminary assessment of effectiveness in basic it skills training. *MIS Quarterly, 25*(4), 401-426.

Pinar, W. F., Reynolds, W. M., Slattery, P., & Taubman, P. m. (1996). Understanding curriculum. New York, NY: Peter Lang Publishing, Inc.

Pirolli, P., & Recker, M. (1994). Learning strategies and transfer in the domain of programming. *Cognition and Instruction, 12*(3), 235-275.

Potts, R., & Martinez, I. (1994). Television viewing and children's beliefs about scientists. *Journal of Applied Development Psychology, 15*, 287-300.

Prensky, M (2001). Digital Natives, Digital Immigrants. *On the Horizon NCB University Press, 9*(5).

Prestera, G. E., & Moller, L. A. (2001, April). *Facilitating asynchronous distance learning: Exploiting opportunities for knowledge building in asynchronous distance learning environments.* Paper presented at the Annual

Mid-South Instructional Technology Conference, Murfreesboro, TN. (ERIC Document Reproduction Service No. ED463723)

Price, L. (2006). Gender Differences and Similarities in Online Courses: Challenging Stereotypical Views of Women. *Journal of Computer Assisted Learning, 22*(5), 349-359.

Prinsen, F. R., Volman, M. L. L., & Terwel, J. (2007). Gender-Related Differences in Computer-Mediated Communication and Computer-supported Collaborative Learning. *Journal of Computer Assisted Learning, 23*(5), 393-409.

Puntambekar, S. (2006). Analyzing collaborative interactions: Divergence, shared understanding and construction of knowledge. *Computers & Education, 47*(3) 332-351.

Putnam, R. T., & Borko, H. (2000). What do new views of knowledge and thinking have to say about research on teacher learning? *Educational Researcher,* 29 (1), 4-15.

Quible. (2005). Blogs: A Natural in Business Communication Courses.

Quinn, C. N., Mehan, H., Levin, J., & Black, S. (1983). Real education in non-real time: The use of electronic message systems for instruction. *Instructional Science, 11*, 313-327.

Ractham, P. and Zhang, X. (2006). Podcasting in Academia. Proceedings SIGMIS-CPR'06, April 13-15.

Raman, M, Ryan, T. and Olfman, L. (2005). Designing Knowledge Management Systems for Teaching and Learning with Wiki Technology. *Journal of Information Systems Education, 16*(3), 311-320.

Rammert, W. (1999). Weder festes Faktum noch kontingentes Konstrukt: Natur als Produkt experimenteller Interaktivität. *Soziale Welt* 50 (3), 281-296.

Rammert, W. (2002). Technik als verteilte Aktion. Wie technisches Wirken als Agentur in hybriden Aktionszusammenhängen gedeutet werden kann. *Technical University Technology Studies Working Papers*, TUTS-WP-3.

Rammert, W. and Schulz-Schaeffer, I. (2002). Technik und Handeln. Wenn soziales Handeln sich auf menschliches Verhalten und technische Abläufe verteilt. Rammert et.al. (Ed.) *Können Maschinen handeln? Soziologische Beiträge zum Verhältnis von Mensch und Technik.* Campus Verlag, 11-65.

Read, B . (2005). Lectures on the Go. *The Chronicle of Higher Education, 52*(10), A39.

Reiser, R. (2001). A history of instructional design and technology. Part 2: A history of instructional design. *Educational Technology, Research and Development, 49*(2), 57-67.

Rethfeld, J. and Schecker, H. (2005). Roberta—Mädchen erobern Roboter. Ergebnisse der quantitativen Begleitforschung. *3. Qualitätsmanagement-Workshop*, 20-21 Juni, Fraunhofer-Institut AIS, Birlinghoven.

Rieber, L. P., Tzeng, S. C., Tribble, K., & Chu, G. (1996). *Feedback and elaboration within a computer-based simulation: A dual-coding perspective.* Paper presented at the 1996 National Convention of the Association for Educational Communications and Technology, Indianapolis, IN. (ERIC Document Reproduction Service No. ED397829)

Riel, M., & Fulton, K. (2001). The role of technology in supporting learning communities. *Phi Delta Kappan, 82*(7), 518-523.

Rieman, J. (1996). A field study of exploratory learning strategies. *ACM Transactions on computer-Human Interaction, 3*(3), 189-218.

Riley, W., & Anderson, P. (2006). Randomized study on the impact of cooperative learning. *Quarterly Review of Distance Education, 7*(2), 129-144.

Rivera, J.C. & Rice, M.L. (2002). A comparison of student outcomes and satisfaction between traditional & web based course offerings. *Online Journal of Distance Learning Administration, V*(III).

Robins, A., Rountree, J., Rountree, N. (2003). Learning and teaching programming: A review and discussion. *Computer Science Education, 13*(2), 137-172.

Roca, J. C., Chiu, C-M. and Martínez, F. J. (2006). Understanding e-learning Continuance Intention: An extension of the Technology Acceptance Model, *International Journal of Human-Computer Studies, 64*(8), 683-696.

Rogoff, B. (1990). *Apprenticeship in thinking: Cognitive development in social context.* New York, NY: Oxford University Press.

Rogoff, B. (1997). Evaluating development in the process of participation: Theory, methods, and practice building on each other. In E. Amsel & A. Renninger (Eds.), *Change and development* (pp. 265-285). Hillsdale, NJ: Lawrence Erlbaum Association.

Romiszowski, A. J., & Mason, R. (1996). Computer-mediated communication. In D. H. Jonassen (Ed.), *Handbook of research for educational communications and technology: A project of the association for educational communications and technology* (pp. 438-456). New York, NY: Macmillian Library Reference USA.

Rommetveit, R. (1979). On the architecture of intersubjectivity. In R. Rommetveit & R. M. Baker (Eds.), *Studies of language, thought, and verbal communication* (pp. 93-107). London, UK: Academic Press.

Rosenberg, M. (2001). *E-learning: Strategies for delivering knowledge in the digital age*. New York: McGraw-Hill.

Rosser, S. V. (1998). Applying feminist theories to women in science programs. *Signs, 24,* 174-200.

Rourke, L., Anderson, L., Garrison, D. R., & Archer, W. (2001). Methodological issues in the content analysis of computer conference transcripts. *International Journal of Artificial Intelligence in Education, 12*(1), 8-22.

Rourke, L., Anderson, T., Garrison, D. R., & Archer, W. (2001). Assessing social presentation asynchronous text-based computer conferencing. *Journal of Distance Education, 14*(2), 50-71.

Rovai, A. P. (2002). Building Sense of Community at a Distance. *International Review of Research in Open and Distance Learning, 3*(1), www.irrodl.org/index.php/irrodl/article/viewPDFInterstitial/79/153.

Rovai, A. P. (2003). In Search of Higher Persistence Rates in Distance Education Online Programs. *Internet and Higher Education, 6*(1), 1-16.

Rudestam, K.E. & Schoenholtz-Read, J. (2002). The coming of age of adult online education. In K.E. Rudestam & J. Schoenholtz-Read (Eds.), *Handbook of online learning: Innovations in higher education and corporate training* (pp. 3-28). Thousand Oaks, California: Sage Publications.

Russell, D. (2005). Implementing an innovation cluster in educational settings in order develop constructivist-based learning environments. *Educational Technology and Society, 8*(2).

Russell, D. (2005). Transformation in an urban school: Using systemic analysis to understand an innovative urban teacher's implementation of an online problem-based unit. *International Journal of Instructional Technology & Distance Learning, 2***(2).**

Russell, D. (2005). Understanding innovation in education using activity theory. *Educational Technology in Society, 8*(1), 38-53

Russell, T. (1999). *The no significant difference phenomenon.* Office of Instructional Telecommunications, North Carolina State University Chapel Hill, N.C.

Ryan, R.C. (2000). Student assessment comparison of lecture and online construction equipment and methods classes. *THE Journal, 6*(27).

Sackett, R.R., & Mullen, E.J. (1993). Beyond formal experimental design: Towards an expanded view of the training evaluation process. *Personnel Psychology, 46,* 613-627.

Salas, E., & Cannon-Bowers, J.A. (2001). The science of training: A decade of progress. *Annual Review of Psychology, 52*, 471-499.

Sällström, A. (2005). *Best practice net-based meetings - a pilot study.* Report 11:2005. Härnösand: Nätuniversitetet. (In Swedish)

Salmon, G (2000). Computer Mediated Conferencing for Management Learning at the Open University, *Management Learning*, 31 (4), 491-502.

Salmon, G. (2004). *E-moderating: The Key to Teaching and Learning Online* (2nd ed.). London, UK: Routledge-Falmer, Taylor and Francis Group.

Salomon, G. (1993). No distribution without individuals' cognition: A dynamic interactional view. In G. Salomon (Ed.), *Distributed cognitions: Psychological and educational considerations*. New York, NY: Cambridge University Press.

Sanders, M. (2000). Web-based portfolios for technology education: a personal case study. *The Journal of Technology Studies, 11*. Retrieved Sept 5, 2006, from http://scholar.lib.vt.edu/ejournals/JOTS/Winter-Spring-2000/pdf/sanders.pdf.

Sandholtz, J., Ringstaff, C., & Dwyer, D. (1994). Student engagement revisited: Views from Technology Rich Classrooms. Retrieved October 4, 2007, from http://www.apple.com/education/k12/leadership/acot/pdf/rpt21.pdf.

Sauers, D., & Walker, R. C. (2004). A comparison of traditional and technology-assisted instructional methods in the business communication classroom. *Business Communication Quarterly, 67*(4), 430-442.

Saville, B.K., Zinn, T.E., Neef, N.A., Norman, R.V., & Ferreri, S.J. (2006). A comparison of interteaching and lecture in the college classroom. *Journal of Applied Behavior Analysis*, 39, 49-61.

Sawyer, R. K. (2003). *Group creativity: Music, theater, collaboration*. Mahwah, NJ: Lawrence Erlbaum Associates.

Scardamalia, M., & Bereiter, C. (1994). Computer support for knowledge building communities. *Journal of the Learning Sciences, 3*(3), 265-283.

Schaino, D., Nardi, B. Gumbrecht, M. and Swartz, L. (2004). Blogging by the Rest of Us. Proceedings CHI 2004, 1143-1146.

Schegloff, E. (1999). Talk and social structure. In A. Jaworski, & N. Coupland (Eds.), *The discourse reader* (pp. 107-120). London, UK: Routledge.

Schellens, T., & Valcke, M. (2004). Fostering knowledge construction in university students through asynchronous discussion groups. *Computers & Education, 46*(4). Retrieved June 20, 2006, from the ScienceDirect database.

Schlager, M. S., Fusco, J., & Schank, P. (2002). Evolution of an on-line education community of practice. In K. A. Renninger & W. Shumar (Eds.), *Building virtual communities: Learning and change in cyberspace* (pp. 129-158). New York, NY: Cambridge University Press.

Schön, D. A. (1983). *The reflective practitioner: How professionals think in action*. New York, NY: Basic Books.

Schreiber, D.A. (1998). Organizational technology and its impact on distance training. In D.A. Schreiber and Z.L. Berge (Eds.), *Distance Training: How Innovative Organizations are Using Technology to Maximize Learning and Meet Business Objectives*. San Francisco, CA: Jossey-Bass Inc., Publishers. pp. 3-18.

Schulman, A.H. & Sims, R.L. (1999). Learning in an online format versus an in-class format: An experimental study. *THE Journal, 26*(11), 54-56.

Schumacher, E. (1977). *A guide for the perplexed*. USA: Harper & Row.

Schutz, A. (1932/1967). *The phenomenology of the social world*. Evanston, IL: Northwestern University Press.

Schwartzman, R. (2006). Virtual group problem solving in the basic communication course: Lessons for online learning. *Journal of Instructional Psychology, 33*(1), 3-14.

Seigenthaler, J. (2005). A false Wikipedia biography. *USA Today*, Nov 29.

Seitzinger, J. (2006). Be constructive: blogs, podcasts, and wikis as constructivist learning tools. *Learning Solutions*, July.

Selwyn, N. (2000). Creating a "connected" community? Teachers' use of an electronic discussion group. *Teachers College Record, 102*(4), 750-778.

Seng, J. and Lin, S. (2004). A mobility and knowledge-centric e-learning application design method. *International Journal of Innovation and Learning 2004, 1*(3) 293-311.

Seng, L. C., & Al-Hawamdeh, S. (2001). New mode of course delivery for virtual classroom. *Aslib Proceedings, 53*(6), 238-242.

Senge, P. M. (1990). *The fifth discipline: The art and practice of the learning organization*. New York: Doubleday/Currency.

Senge, P.M. (1992). *The fifth discipline: The art and practice of the learning organization*. Adelaide: Random House Australia.

Seymour, E. (1999). Therole of socialization in shaping the career-related choices of undergraduate women in science, mathematics, and engineering majors. In C.C. Selby (Ed.), *Women in science and engineering: Choices for success* (pp. 118-126). New York: The New York Academy of Sciences.

Sharan, S. and Shaulov, A. (1990). Cooperative learning, motivation to learn and academic achievement, *Co-operative Learning: Theory and research*, S. Sharan (ed.), 173-202.

Sharma, S (2007). From Chaos to Clarity: Using the Research Portfolio to Teach and Assess Information Literacy Skills. *The Journal of Academic Librarianship*. 33 (1). Retrieved Dec 5, 2006, from http://www.sciencedirect.com/science/article/B6W50-4MK0HTH-4/2/de225b2c2d9329a19623c1fe1365d0db.

Shelley, D. J. (2005). Developing, implementing and assessing a university level online U.S. history course. *Proc. 8th IASTED International Conf. on Computing and Advanced Technology in Education*, Oranjestad, Aruba, 2005, pp. 312-314.

Shelley, D.J., Swartz, L. B. & Cole, M.T. (2007). A comparative analysis of online and traditional undergraduate business law classes. *International Journal of Information and Communication Technology Education, 3*(1), 10-21.

Shelley, D.J., Swartz, L. B. & Cole, M.T. (2008). Learning business law online vs. onland: A mixed methods analysis. *International Journal of Information and Communication Technology Education*. In press.

Shelly, G. B., Cashman, T. J., & Rosenblatt, H. J. (2006). *Systems Analysis and Design* (6th ed.). Boston: Course Technology.

Shepherd, C. (2006). Objects of Interest. Retrieved from http://www.fastrak- consulting.co.uk/tactix/Features/objects/objects.htm.

Shulman, L. (1986). Those who understand: Knowledge growth in teaching. *Educational Researcher, 15*(2), 4-14.

Sidman, M. (1960). *Tactics of scientific research*. New York: Basic Books.

Signorielli, N., & Morgan, M. (1990). Cultivation analysis: New directions in media effects research. Newbury Park, CA: Sage Publications.

Simon, H.A. (1976). *Administrative Behaviour: A Study of Decision Making Processes in Administrative Organization 3rd ed.* New York NY: Free Press.

Simonson, M., Smaldino, S., Albright, M., & Zvacek, S. (2000). *Teaching and learning at a distance: Foundation of distance education.* Upper Saddle River, NJ: Merrill.

Simpson, O. (2003). *Student Retention in Online, Open and Distance Learning*. Sterling, VA: Kogan Page.

Smith, P. L., & Ragan, T. J. (1996). *Impact of R. M. Gagne's work on instructional theory*. Paper presented at the 1996 National Convention of the Association for Educational Communications and Technology, Indianapolis, IN. (ERIC Document Reproduction Service No. ED397841)

Snyder, L., Caccamise, D. and Wise, B. (2005). The Assessment of *Reading* Comprehension: Considerations and Cautions. *Topics in Language Disorders, 25*(1), 33.

Sørebø, A. M. (2004). A replication of the Post-Acceptance Model in the context of E-Learning, *Proceedings of 2004 IRMA International Conference*. Louisiana, USA.

SPSS. (2004). *SPSS 13.0 for Windows*. Chicago, IL: Apache Software Foundation.

Stager, G.S. (2001, July 7-25). *Constructionism as a high-tech intervention strategy for at-risk learners*. National Educational Computing Conference, Building on the Future 1. Chicago, IL

Stahl, G. (2006). *Group Cognition*. Cambridge MA: MIT Press.

Stansberry, S and Kymes, A. (2007). Transformative learning through "Teaching With Technology" electronic portfolios. *The Journal of Adolescent & Adult Literacy. Vol.* 50 (1):488-497.

State-wide portfolio project for the state of Minnesota. (http://www.efoliominnesota.com) [retrieved 3/10/2006]

Steer, D., Turner, P. and Spencer, S. (2000). Issues in Adapting the Technology Acceptance Model (TAM) to Investigate Non-workplace Usage Behavior on the World-Wide-Web, School of Information Systems, Working Chapter, University of Tasmania.

Steinke, J. (1997). A portrait of a woman as a scientist: Breaking down barriers erected by gender role stereotypes. *Public Understanding of Science, 6*, 409-428.

Stephens, M. (2007). Messaging in a 2.0 World. *Library Technology Reports, 43*(5), 62-66.

Stewart, C. M., Shields, S. F., & Sen, N. (2001). Diversity in On-Line Discussions: A study of Cultural and Gender Differences in Listervs. In C. Ess & F. Sudweeks (Eds.), *Culture, Technology, Communication: Towards an Intercultural Global Village* (pp. 161-186). Albany, NJ: State University of New York Press.

Straub, D., Limayem, M. and Karahanna, E. (1995). Measuring System Usage: Implications for IS Theory Testing, *Management Science, 41*(8), 1328-1342.

Suanpang, P., Petocz, P. & Kalceff, W. (2004). Student attitudes to learning business statistics: comparison of online and traditional methods. *Educational Technology & Society, 7*(3), 9-20.

Sullivan, M. (1993). *A meta-analysis of experimental research studies based on the Dunn & Dunn learning styles model and its relationship to academic achievement and performance.* Doctoral dissertation, St. John's University, Jamaica, New York.

Sunal, W. D., Sunal, S. C., Odell, R. M. & Sundberg, A. C. (2003). Research-supported best practices for developing online learning. *The Journal of Interactive Online Learning, 2* (1), 1-40. Retrieved November 18, 2007 from http://www.ncolr.org/jiol/issues/PDF/2.1.1.pdf#search=%22Sunal%20%22Research-supported%20best%20practices%20for%20developing%20online%20learning%22%22

Swan, K. (2001). Virtual interaction: Design factors affecting student satisfaction and perceived learning in asynchronous online courses. *Distance Education, 22*(2), 306-331. Retrieved February 18, 2006, from the ProQuest database.

Szajna, B. (1996). Empirical Evaluation of the Revised Technology Acceptance Model, *Management Science, 42*(1), 85-92.

Tallent-Runnels, M. K., Thomas, J. A., Lan, W. Y., Cooper, S., Ahern, T. C., Shaw, S. M., Liu, Xiaoming. (2007). Teaching courses online: A review of the research. *Review of Educational Research, 76*(1), *93-135.*

Tannen, D. (1991). Teachers' Classroom Strategies Should Recognize that Men and Women Use Language Differently. *The Chronicle of Higher Education, June 19*, B1-B3.

Tannen, D. (1994). *You Just Don't Understand Women and Men in Conversation*. New York, NY: Ballantine Books.

Tannen, D. (1995). The Power of Talk: Who Gets Heard and Why. *Harvard Business Review, 73*(5), 138-148.

Tavangarian, D. Leypold, M. E., Nölting, K., Röser, K. M. and Voigt, D. (2004). Is E-Learning the Solution for Individual Learning?, *Electronic Journal of E-Learning, 2*(2), 273-280.

Taylor, F.W. (1967). *The principles of scientific management*. New York: Norton.

Taylor, R. (2001). Current, best instructional strategies for your gifted and highly capable students (workshop). Presented by Bureau of Education and Research.

Tham, C. M., & Werner, J. M. (2005). Designing and evaluating e-learning in higher education: A review and recommendations. *Journal of Leadership & Organizational Studies, 11*(2), 15-25.

Thompson, A. (2005). Scientifically based research: Establishing a research agenda for the technology in teacher education community. *Journal of Research on Technology in Education, 37*(4), 331-337.

Thompson, A., Bull, G., & Bell, L. (2005). A proactive approach to a research agenda: A call to action (editor's remarks). *Journal of Computing in Teacher Education, 21*(3), 74, 76.

Thompson, L., & Hing-Yu, K. (2006). A case study of collaborative online learning. *Quarterly Review of Distance Education, 7*(4), 361-375.

Tiger Leap Foundation (1997) Retrieved March 9, 2007, from http://www.tiigrihype.ee/?op=&id=

Tinto, V. (1998). Colleges as Communities: Taking Research on Student Persistence Seriously. *The Review of Higher Education, 21*(2), 167-177

Tobin, T. (2004). Assessing educators' online performance. *The Online Journal of Distance Learning Administration, 7*(2). Retrieved September 15, 2006, from http://www.westga.edu/%7Edistance/ojdla/summer72/tobin72.html

Toffler, A. & Toffler, H. (1995). *Creating a new civilization: The politics of the third wave*. Atlanta: Turner Publishing.

Tomasello, M. (1999). *The cultural origins of human cognition*. Cambridge, MA: Cambridge University Press.

Tomei, L. A. (2007). *Learning theories: A primer exercise.* Retrieved January 14, 2007, from http://www.acacemics.rmu.edu/~tomei/ed711psy/human.htm

Tomlinson, C. & Allan, S. (2000). *Leadership for Differentiating Schools and Classrooms.* ASCD.

Topper, A. (1995). *Intersubjectivity and educational computer-mediated communication*. Unpublished manuscript.

Truong, N., Bancroft, P., & Roe, P. (2005, June 27–29). Learning to program through the web. *Proceedings of the 10th annual SIGSCE conference on innovation and technology in computer science education*, ITiCSE'05. Monte de Caparica, Portugal:ACM Press.

Tsang, A.C.W., & Chan, N. (2004). An online problem-based model for the learning of java. *Journal of Electronic Commerce in Organizations, 2*(2), 55-64.

Tu, C. (2000). Strategies to increase interaction in online social learning environments. Paper presented at the Society for Information Technology and Teacher Education International Conference, San Diego, CA.

Tuchman, G. (1978). The symbolic annihilation of women by the mass media. In Tuchman, G., A. Kaplan Daniels, & J. Benet (Eds.). pp. 3-17. *Hearth and home: Images of women and the media.* New York: Oxford University Press.

Turkle, S. (1995): Life on the Screen: Identity in the age of the Internet. *London: Phoenix*

Tyack, D., & Cuban, L. (2004). Tinkering toward utopia: A century of public school reform. In J. H. Ballantine & J. Z. Spade (Eds.), *Schools and society: A sociological approach to education* (2nd Ed., pp. 459-478). Belmont, CA: Wadsworth/Thomson Learning.

Tynjala, P. (1999). Towards expert knowledge? A comparison between a constructivist and traditional learning environment in the university. *International Journal of Educational Research 31*(5), 357-442.

Van Merrienboer, J. J. G., & Paas, F. (2003). Powerful learning and the many faces of instructional design: Towards a framework for the design of powerful learning environments. In E. De Corte, L. Verschaffel, N. Enstwistle, & J. J. G. Van merrienboer Eds.), *Powerful learning environments: Unravelling basic components and dimensions*. Oxford: Elsevier Science.

Vavrus, M. D. (2002). *Postfeminist News: Political Women in Media Culture.* Albany: State University of New York Press, p.13).

Venkatesh, V. and. Davis, F. D (2000) A Theoretical Extension of the Technology Acceptance Model: Four Longitudinal Field Studies. *Management Science, 46*(2), 186-204.

Venkatesh, V., & Morris, M. G. (2000). Why Don't Men Ever Stop to Ask for Directions? Gender, Social Influence, and their Role in Technology Acceptance and Usage Behavior. *MIS Quarterly, 24*(1), 115-139.

Vessey, I., & Conger, S. (1994). Requirements Specification: Learning Object, Process, and Data Methodologies. *Communications of the ACM, 37*(5), 102-113.

Vigotsky, L. S. (1978). Mind in society: The development of the higher psychological processes(Cambridge , MA: Harvard University Press).

Von Krogh, G., Ichijo, K. & Nonaka, I. (2000). *Enabling knowledge creation.* USA: Oxford University Press.

Vonderwell, S. (2003). An examination of asynchronous communication experiences and perspectives of students in an online course: A case study. *The Internet and Higher Education, 6,* 77-90.

Vouk, M.A., Bitzer, D.L., & Klevans, R.L. (1999). Workflow and end-user quality of service issues in web-based education. *IEEE Trans. on Knowledge and Data Engineering, 11*(4), 673-687.

Vygotsky, L. (1986). *Thought and language* (A. Kozulin, Ed. & Trans.). Cambridge, MA: The MIT Press. (Original work published in 1934).

Wade, S. E., & Fauske, J. R. (2004). Dialogue online: Prospective teachers' discourse strategies in computer-mediated discussions. *Reading Research Quarterly, 39*(2), 134-160.

Wagner, C. (2004). Wiki: A Technology for Conversational Knowledge Management and Group Collaboration. *Communications of the AIS, 13*, 265-289.

Wagner, C. (2006). Breaking the Knowledge Acquisition Bottleneck through Conversational Knowledge Management. *Information Management Resources Journal, 19*(1), 70-83.

Walker, J. (2005). Weblog. Definition from the Routledge Encyclopedia of Narrative Theory.

Wallace, M. (2005). Notes towards a literacy for the digital age. Retrieved from http://uclaccc.ucla.edu/articles/article-digitalage.htm

Wallace, R. M. (2004). A framework for understanding teaching with the internet. *American Educational Research Journal, 41*(2), 447-488.

Wang, A. Y., & Newlin, M. H. (2000). Characteristics of students who enroll and succeed in psychology web-based courses. *Journal of Educational Psychology, 92*,137-143.

Wang, F., & Hannafin, M.J. (2005). Design-based research and technology-enhanced learning environments. *Educational Technology Research and Development, 55*(4), 5-24.

Wang, J. and Fang, Y. (2005). Benefits of cooperative learning in Weblog networks. Proceedings APA-MALL2005 and ROCMELIA2005, December 9-11, 2005, Kun Shan University.

Wang, Y. (2003). Assessment of learner satisfaction with asynchronous electronic learning systems. *Information & Management, 41,* 75-86.

Wännman, G. T. (2002). *Women creating knowledge on the web.* Doctoral dissertation, Umeå: Umeå University, Department of Education. (In Swedish)

Waukesha School District. (2007). Retrieved September 15, 2006 from http://www.waukesha.k12.wi.us/District/Default.asp

Weatherall, A. (1998). Re-visioning Gender and Language Research. *Women and Language, 21*(1), 1-9.

Weaver-Kaulis, A. & Crutsinger, C. (2006). Assessment of student learning outcomes in FCS programs. *Journal of Family and Consumer Sciences, 98*(1), 74-81.

Web Portfolio Commercial Site. http://www.portfolio-village.com

Webb, H. W., Gill, G., & Poe, G. (2005). Teaching with the case method online: Pure versus hybrid approaches. *Decision Sciences Journal of Innovative Education, 3*(2), 223-250.

Weick, K.E. (1978). *The social psychology of organizing*. Reading: Addison-Wesley.

Weinberg, G. (1971). *The psychology of computer programming*. New York: Van Rostrand Reinhold.

Weiss, M. (2003). A gentle introduction to agents and their applications. Online presentation at http://www.magma.ca/~mrw/agents/, 2004.

Wellins, R.S., Byham, W.C & Wilson, A.M. (1991). *Empowered teams: Creating self directed work groups that improve quality, productivity and participation*. San Francisco: Jossey-Bass.

Wenden, A. (1991). *Learning Strategies for Learner Autonomy*. New York: Prentice-Hall

Wenger, E. (1998). *Communities of practices*. Cambridge, UK: Cambridge University Press.

Wertsch, J. V. (1984). The zone of proximal development: Some conceptual issues. In B. Rogoff & J. Wertsch (Eds.), *Children's learning in the "zone of proximal development"* (pp. 7-18). San Francisco, CA: Jossey-Bass Inc., Publishers.

Wertsch, J. V. (1985). *Vygotsky and the social formation of the mind*. Cambridge, MA: Harvard University Press.

Wertsch, J. V. (1991). *Voices of the mind: A sociocultural approach to mediated action*. Cambridge, MA: Harvard University Press.

West, C., & Zimmerman, D. (1983). Small Insults: A Study of Interruptions in Cross-sex Conversations between Unacquainted Persons. In B. Thorne, H. Kramarae & N. Henley (Eds.), *Language, Gender and Society* (pp. 103-118). Rowley, MA: Newbury House.

Wickramasinghe, N. and Lichtenstein, S. (2006). Supporting knowledge creation with e-mail. *International Journal of Innovation and Learning, 3*(4), 416-426.

Wiesner, H. (2002). Die Inszenierung der Geschlechter in den Naturwissenschaften. Wissenschafts- und Geschlechterforschung im Dialog. *Campus-Verlag.*

Wiesner, H. (2004). Handlungsträgerschaft von Robotern: Robotik zur Förderung von Chancengleichheit im schulischen Bildungsbereich. *Historical Social Research, Zentrum für Historische Sozialforschung, 29*, 120-154.

Wiesner, H., Kamphans, M., Schelhowe, H., Metz-Göckel, S., Zorn, I., Drag, A., Peter, U. and Schottmüller, H. (2004). Leitfaden zur Umsetzung des Gender Mainstreaming in den "Neuen Medien in der Bildung—Förderbereich Hochschule," *Bremen—Dortmund.*

Wiesner-Steiner, A., Wiesner, H., Schelhowe, H. (2006). Technik als didaktischer Akteur: Robotik zur Förderung von Technikinteresse, *Hochschulinnovation. Gender-Initiativen in der Technik. Reihe: Gender Studies in den Angewandten Wissenschaften. Gender Studies & Applied Sciences*, Gransee, C. (ed.), LIT-Verlag Hamburg, 89-115.

Wikipedia (2006). Mobile Computing, definition. Retrieved from http://en.wikipedia.org/wiki/Mobile_computing, accessed 8/25/2006

Wilhelmson, L. (2002). On the theory of transformative learning. In A. Bron & M. Scheemann (Eds.), *Social science theories in adult education research* (pp. 180-210). Münster: LIT Verlag.

Wilhelmson, L. (2006). Dialogue meetings as non-formal adult education in a municipal context. *Journal of Transformative Education, 4*(3), 243-256.

Wilhelmson, L., & Döös, M. (2002). *Dialogue competence for development in working life*. Stockholm: National Institute for Working Life. (In Swedish)

Wilkins, H. (1991). Computer talk. *Written Communication, 8*(1), 56-78.

Williams, L.A., Wiebe, E., Yang, K., Ferzli, M., & Miller, C. (2002). In support of pair programming in the introductory computer science course. *Computer Science Education, 12*(3), 197-212.

Williams, S. W. (2001). *Experiences of web-based instruction among African-American students enrolled in training and development graduate courses.* Paper presented at the Academy of Human Resource Development 2001 Conference, Raleigh, NC. (AHRD Reference No. 059)

Wilson, B. G. (1996). *Constructivist learning environments: Case studies in instructional design.* Englewood Cliffs, NJ: Educational Technology Publications.

Wisconsin Department of Public Instruction. (2005). *Public school open enrollment: 2005-2006 timelines for public school open enrollment.* Retrieved September 15, 2006, from http://www.dpi.state.wi.us/dpi/dfm/sms/oetime05.html

Wisconsin Department of Public Instruction. (2006, March 3). Public school open enrollment: 2005-2006 timelines for public school open enrollment. Retrieved November 18, 2007 from http://www.dpi.state.wi.us/dpi/dfm/sms/oetime05.html

Wolf, K. Linkels, S. and Meinel, C. (2007). Teleteaching Anywhere Solution Kit (tele-TASK) Goes Mobile, In Proceedings of SIGUCCS'07, October 7-10, 2007, 366-371.

Wolff, T. (2006). Podcasting Made Simple. In Proceedings of SIGUCCS'06, Edmonton Alberta Canada, Nov 5-8 2006, 413-417.

Woodley, A. (2004). Conceptualizing Student Dropout in Part-Time Distance Education: Pathologizing the Normal? *Open Learning, 19*(1), 47-63.

Wu, Y. B., & Chen, X. (2005). E-learning assessment through textual analysis of class discussions. In *Proceedings of the Fifth IEEE International Conference on Advanced Learning Technologies* (pp. 388-390). Retrieved January 8, 2006, from the IEEE database.

Yamane, T. (1973). *Statistics: An introductory analysis,* (3rd Ed.). New York, NY:Harper & Row.

Yates, S. J. (2001). Gender, Language an CMC for Education. *Learning and Instruction, 11*(1), 21-34.

Yeh, H. (2005). The use of instructor's feedback and grading in enhancing students' participation in asynchronous online discussion. In *Proceedings of the Fifth IEEE International Conference on Advanced Learning Technologies.* Retrieved December 14, 2005, from the IEEE database.

Ying-Shao, H., & Fu-Kwun, H. (2002, June). *The use of multiple representations in a web-based and situated learning environment.* Paper presented at the ED-MEDIA 2002 World Conference on Educational Multimedia, Hypermedia & Telecommunications, Denver, CO. (ERIC Document Reproduction Service No. ED477029)

Yli-Renko, H., Autio, E., & Sapienza, H. J. (2001). Social capital, knowledge acquisition, and knowledge exploitation in young technology-based firms. *Strategic Management Journal, 22*(6-7), 587-613.

Yu, D., Zhang, W., & Chen, X. (2006). New ceneration of e-learning technologies. In *proceedings of the First International Multi-Symposiums on Computer and Computational Sciences (IMSCCS'06).*

Yu, E. (2001). Agent-oriented modelling: software versus the world. Agent-Oriented Software Engineering AOSE-2001 Workshop Proceedings (pp. 206-225), Montreal, Canada, May 2001. LNCS 2222. Springer-Verlag, Berlin, Germany.

Yuen, A.H.K. (2006). Learning to program through interactive simulation. *Educational Media International, 43*(3), 251-268.

Yuzer, T. V. (2007). Generating virtual eye contacts through online synchronous communications in virtual classroom applications. *Turkish Online Journal of Distance Education, 8*(1).

Zeichner, K., & Liston, D. (1996) *Reflective teaching: An introduction.* Mahwah, NJ: Lawrence Erlbaum Association.

Zhang, D, Zhou, L., Briggs, B., & Nunamaker, J. F. (2006). Instructional video in e-learning: Assessing the impact of interactive video on learning effectiveness. *Information & Management. 43*(1), pp. 15-27.

Zhang, D. (2004). Virtual mentor and the LBA system—towards building an interactive, personalized, and intelligent e-learning environment. *Journal of Computer Information Systems, XLIV*, (3), 35-43.

Zhang, D., Zhou, L., Briggs, R. O., & Nunamaker, J. F. Jr. (2005). Instructional video in e-learning: Assessing the impact of interactive video on learning effectiveness. *Information Management, 43* (1), 15-27.

Zhang, X., Prybutok, V. R., & Strutton, D. (2007). Modeling Influences on Impulse Purchasing Behaviors During Online Marketing Transactions. *Journal of Marketing Theory & Practice, 15*(1), 79-89.

Zhao, Y., & Frank, K. A. (2003). Factors affecting technology uses in schools: An ecological perspective. *American Educational Research Journal, 40*(4), 807-840.

Zhao, Y., & Rop, S. (2001). A critical review of the literature on electronic networks as reflective discourse communities for inservice teachers. (ERIC Document Reproduction Service No. ED459483)

Zimmerman, D., & West, C. (1975). Sex-roles, interruptions and silences in conversation. In B. Thorne, H. Kramarae & N. Henley (Eds.), *Language and Sex: Difference and Dominance* (pp. 89-101). Rowley, MA: Newbury House.

Zualkernan, I. A. (2006). A framework and a methodology for developing authentic constructivist e-Learning environments. *Educational Technology & Society, 9* (2), 198-212.

Zwart, W. J. (1992, June). *Instructional transaction theory applied to computer simulations.* (ERIC Document Reproduction Service No. ED352945)

About the Contributors

Lawrence A. Tomei is the associate vice president of Academic Affairs and associate professor of education at Robert Morris University. Born in Akron, Ohio, he earned a BSBA from the University of Akron (1972) and entered the US Air Force, serving until his retirement as a Lieutenant Colonel in 1994. Dr. Tomei completed his MPA and MEd at the University of Oklahoma (1975, 1978) and EdD from USC (1983). His articles and books on instructional technology include: Professional Portfolios for Teachers (1999); Teaching Digitally: Integrating Technology Into the Classroom (2001); Technology Facade (2002); Challenges of Teaching with Technology Across the Curriculum (2003); and Taxonomy for the Technology Domain (2005).

* * *

Karin Alvemark has her MA and is a lecturer in Educational Psychology within the field of working life and leadership at Högskolan Dalarna, Sweden. She is experienced in e-learning and distance education when making use of video conferencing in Marratech software. Alvemark has experience from both personnel training, and organisation and leadership development within the public sector as well as in private companies. Her main interest deals with adult learning, organisational preconditions for learning at work, and leadership.

Zane Berge is professor and former director of the training systems graduate programs at the University of Maryland System, UMBC Campus, USA. He teaches graduate courses involving training in the workplace and distance education. Prior to UMBC, Berge was founder and director of the Center for Teaching and Technology, Academic Computer Center, Georgetown University, Washington, DC. It was there that he first combined his background in business with educational technology to work in the areas of online journals, moderated online discussion lists, and online education and training. Berge's publications include work as a primary author, editor, or presenter of 10 books and over 200 book chapters, articles, conference presentations and invited speeches worldwide.

Silvia Braidic serves as an associate professor in the administrative program for principals at California University of Pennsylvania. Her research interests include principal/teacher leadership, instructional strategies/differentiation, and online teaching and professional development. In addition to her work at the university level, she has experience as a principal and assistant principal in the Mt. Lebanon School District in Pittsburgh, Pennsylvania. She also served as the district's coordinator for strategic planning. Prior to her work in administration, she taught middle school mathematics. She holds a doctorate of education in administrative and policy studies – educational administration from the University of Pittsburgh, K-12 principal certification from Carnegie Mellon University, MSEd in elementary education and BSEd in secondary education – mathematics from Duquesne University.

MarySue Cicciarelli lives in Peoria, Illinois with her husband Greg and their five children. Her current administrative work as the executive assistant to the president at Peoria Notre Dame High School involves collaborating with the school's principal, managing the school's business department, advancement and admissions department, project and alumni department, Foundation Board, and its Board of Trustees. She received a BA from the University of Iowa in 1987, a MA in curriculum and instruction from Bradley University in 2001, and her EdD in instructional technology from Duquesne University in 2007. Recent articles written by Cicciarelli and published in the *International Journal of Information and Communication Technology Education* include A Description of Online Instructors Use of Design Theory and Behavioral, Cognitive, and Humanistic Theories: Which Theories Do Online Instructors Utilize? Her work on behavioral, cognitive, and humanistic theory can also be found in the *Encyclopedia of Information Technology Curriculum Integration*. Future work written by Cicciarelli will be on implementing secondary education global learning programs and, as her high school builds a new facility, constructing high schools equipped with technology that meets the needs of twenty-first century students.

Michele T. Cole is the director of the masters program in nonprofit management and the director of the Massey Center for Business Innovation and Development at Robert Morris University in Pittsburgh, PA. She received her AB in English Literature from Wheeling Jesuit University in 1968, a master's in public and international affairs from the University of Pittsburgh in 1974, a juris doctorate from Duquesne University in 1982 and a PhD in public administration from the University of Pittsburgh in 1993. She is a former Peace Corps volunteer and is licensed to practice law in Pennsylvania. She has published in the areas of online education, accountability in the nonprofit sector, and models of vocational rehabilitation.

John DiMarco (dimarcoj@stjohns.edu) is a communications professor, consultant, designer, researcher, and writer. John is an assistant professor and the director of the undergraduate public relations program at St. John's University in New York City. DiMarco teaches courses in mass communications, public relations, media graphics, advertising, and animation. He has held faculty positions at LIU, SUNY Old Westbury, Molloy College, and Nassau Community College. His latest book, *Web Portfolio Design and Applications*, was published in 2006. In 2004, he published an edited book titled *Computer Graphics and Multimedia, Applications, Problems, and Solutions* for Idea Group Publishing. He is the founder of PortfolioVillage.com, a company and website dedicated to providing educational products and services. DiMarco is in the final stages of completing a PhD in information studies at Long Island University. He holds a master's degree in communication design from Long Island University and a bachelor's degree in communication & public relations from the University at Buffalo.

Marianne Döös, adjunct professor in educational psychology within the field of organization pedagogies at Lund University, Sweden, and affiliated to the Swedish Agency for Innovation Systems. Her research deals with the processes of experiential learning in contemporary settings, on individual, collective and organizational levels. Topical issues concern interaction as carrier of competence in relations, shared or joint leadership, conditions for competence in work-life and organizational change and development, the aims of which are to generate theory within organization pedagogies and to subject outcomes to empirical investigation. Döös is the author or co-author of many articles and books. Some recent publications in the English language: Functioning at the edge of knowledge – a study of learning processes in new product development (2004), Organizational learning. Competence-bearing relations and the breakdowns of workplace relatonics (2007) and Competent web dialogue: Thoughts linked in digital conversations (2007).

Michelle Emerson obtained her PhD in sociology from Georgia State University. She has a MS in criminal justice from Northeastern University and a BS in public and social services from Kennesaw State University. Her area of research is in the area of violence against women, and international crime. She is also involved in the scholarship of teaching. She currently teaches at Kennesaw State University in the department of sociology and criminal justice. Her courses include Foundations of Criminal Justice, Victimology, Sociology of Violence, and Research Methods.

Eva R. Fåhræus has a PhD in technology focused on IT and learning, especially electronically mediated collaborative learning at a distance. 1995-2007 she was a teacher and researcher at the department of computer and systems sciences at Stockholm University and KTH (Royal Institute of Technology in Stockholm). Before that Fåhræus worked in the industry for 25 years with systems and organizations development, economic control, leadership and education. Fåhræus is the author or co-author of many articles and book chapters. Two examples in the English language: A Triple Helix of Learning Processes – How to cultivate learning, communication and collaboration among distance-education learners (PhD dissertation, 2003), and Competent Web dialogue: Thoughts linked in digital conversations (2007). Her most recent work is Learn where you are: a handbook for distance learners and others learning via the net (in Swedish, 2008).

Michael Fedisson serves as a seventh grade language arts teacher at the Bellefonte Area Middle School. In addition, he is also the school's newspaper advisor. He recently finished his MEd from California University of Pennsylvania in the administrative program for principals. In addition, he holds a BSEd in elementary education from Lock Haven University with minors in special education and reading. He also holds a mid-level English certification.

David Gefen is associate professor of MIS at Drexel University, Philadelphia USA, where he teaches Strategic Management of IT, Database Analysis and Design, and VB.NET. He received his PhD in CIS from Georgia State University and a master of sciences in MIS from Tel-Aviv University. His research focuses on trust and culture as they apply to the psychological and rational processes involved in ERP, CMC, and e-commerce implementation management, and to outsourcing. Gefen's wide interests in IT adoption stem from his 12 years of experience in developing and managing large information systems. His research findings have been published in *MISQ, ISR, IEEE TEM, JMIS, JSIS, The DATA BASE for Advances in Information Systems, Omega: the International Journal of Management Science, JAIS, CAIS,* and elsewhere. David is an author of a textbook on VB.NET programming. David is on the Editorial Boards of *MISQ, DATABASE* and *IJeC*.

Nitza Geri is head of undergraduate management studies at the department of management and economics at The Open University of Israel and a member of the Chais Research Center for the Integration of Technology in Education. She holds a BA in accounting and economics, an MSc in management sciences and a PhD in technology and information systems management from Tel-Aviv University. She is a CPA (Israel) and prior to her academic career she had over 12 years of business experience. Her research interests and publications focus on various aspects of the value of information, and information systems adoption and implementation, which include: strategic information systems, e-business, value creation and the Theory of Constraints, managerial aspects of e-learning systems adoption and use.

Jeffrey Hsu is an associate professor of information systems at the Silberman College of Business, Fairleigh Dickinson University. He is the author of numerous papers, chapters, and books, and has previous business experience in the software, telecommunications, and financial industries. His research interests include human-computer interaction, e-commerce, IS education, and mobile/ubiquitous computing. Hsu received his PhD in information systems from Rutgers University.

Diane Hui, is a post-doctoral fellow and lecturer in the faculty of education, the University of Hong Kong. Hui, a Spencer scholar, received her PhD in education from Washington University in St. Louis, USA, and an M.Sc. in applied linguistics from the University of Edinburgh, Scotland. Her research interests have revolved around sociocultural and cognitive aspects of teacher and student learning in both formal and informal, individual and collaborative settings. Her current research involves the development of an online language diagnostic assessment tool to be used by teachers within the communities of school-based assessment within the Hong Kong education reform. The project is entitled, "Diagnostic and innovative assessment of language by oral genre with the use of engagement" (or DIALOGUE). Her doctoral dissertation examined Engagement in supporting new teachers: A role for computer-mediated communication in teacher learning within informal professional communities (2006). She has published several articles concerned with intersubjectivity and learning through technological mediation including: Understanding innovative professional development for educators through the analysis of intersubjectivity in online collaborative dialogues (2007), A new role for computer-mediated communication in engaging teacher learning within informal professional communities (2005), Managing intersubjectivity in the context of a museum learning environment (2003), and a published Review of We've got blog: How weblogs are changing our culture (2003).

Princely Ifinedo is an assistant professor at the Shannon School of Business, Cape Breton University, Canada. He earned his PhD in information systems science from the University of Jyväskylä, Finland. He also holds an MBA in international management from Royal Holloway College, University of London, the UK, a MSc in informatics from Tallinn University of Technology, Estonia and a BS. in mathematics/computer science from the University of Port-Harcourt, Nigeria. His current research interests include e-learning, e-business, e-government, ERP success measurement, organizational computing, social informatics, IT/business alignment, and the diffusion of IS/IT in transiting and developing economies. He has presented at various international IS conferences and his works have appeared (and are scheduled to appear) in such journals as *The Journal of Computer Information Systems*, *Enterprise Information Systems*, *Journal of Information Technology Management*, and *Journal of Global Information Technology Management*. Ifinedo has authored (and co-authored) 50 peer-reviewed papers. He is affiliated with AIS, ASAC, DSI, and ACM.

Hong Lin is currently an associate professor in computer science at the University of Houston-Downtown. He earned his doctoral degree from the University of Science and Technology of China in 1994. His research interests include parallel/distributed computing, multi-agent systems, and formal methods.

Petra Luck is the award director of the online BA early years management at Liverpool Hope University. Research interests are gender and professionalisation in the early years sector, gender and management and on-line learning. Luck is a fellow of the Higher Education Academy in the UK, member of the European Distance and E-learning Network (EDEN) and member of the British Academy of Management (BAM). Luck has also project managed a range of European funded initiatives aimed at work based learning and the use of online technology.

Ido Millet (www.MilletSoftware.com) is a professor of MIS at Penn State Erie. He received his PhD in decision sciences from the Wharton School at the University of Pennsylvania. His research interests include the Analytic Hierarchy Process (AHP), business intelligence, and online reverse auctions. He is also a consultant and software developer.

Solomon Negash specializes in ICT for economically developing countries, e-learning, and business intelligence. He is the 2005 recipient of the distinguished eLearning award from his department and recipient of the 2007 Distinguished Graduate Teaching Award from his university. His work is published in Information & Management, Communication of the ACM, Psychology and Marketing, Communication of AIS, and at conference proceedings in the US, Canada, Spain, Ethiopia, Kenya, and Malaysia. Negash is the program coordinator for the bachelor of science in information systems (BSIS) program at Kennesaw State University. With an engineering, management, and information systems background, his over 20 years of industry experience include consulting, entrepreneurship, management, and systems analysis. He worked as a business analyst at Cambridge Technology Partners and managed his own consulting firm.

Robert Nelson is a lecturer in MIS and computer science at Penn State Erie. He teaches courses such as Systems Analysis and Introduction to ERP & Business Processes using SAP. Prior to joining Penn State, Nelson worked as an MIS project manager for a large manufacturing corporation. Nelson received his BS in mathematics and his MEd from Edinboro University. Nelson enjoys traveling, playing golf and tennis, and spending time with his family and grandchildren.

Narasimha Paravastu is an assistant professor of MIS at the Metropolitan State University, Minneapolis, MN. He received his PhD from Drexel University, Philadelphia PA.

Donna Russell is an assistant professor at the University of Missouri-Kansas City, USA. She has a bachelor's and master's degree in education specializing in instructional design. Her PhD is in educational psychology with an emphasis on cognition and technology. She is currently developing a NSF grant to develop problem-based curriculum in high school engineering programs. She is also implementing a Kaufman Foundation grant that develops a problem-based virtual learning scenario in a 3D virtual learning environment to teach the geosciences. She has published several articles and book chapters on online learning including: Online professional development for educators: A case study analysis using cultural historical activity theory, Implementing an innovation cluster in educational settings to develop constructivist-based learning environments, Transformation in an urban school: Using systemic analysis to understand an innovative urban teacher's implementation of an online problem-based unit, and Using cultural historical activity theory to understand how group collaboration impacts an online problem-based university course.

Heidi Schelhowe is professor for "Digital Media in Education" at the Computer Science Department of the University of Bremen since 2001. Her special field of research and teaching is application of digital media in schools as well as in university teaching, and vocational training. She is head of an interdisciplinary team of researchers. She studied theology and German in Freiburg/Brsg and Muenster and worked as a teacher in Bremen. Later she earned a degree and a PhD in computing science in Bremen (1989). She worked at the "interdisciplinary research center work and technology" (artec), University of Bremen and at the computer science department, University of Hamburg (1992-1996) and at the Computer Science Institute, Humboldt University of Berlin (1996-2001).

Daniel J. Shelley earned his BS in elementary education from Penn State University in 1971. He completed a master's degree in social science with an emphasis in American history at Penn State in 1972. He earned his PhD in education at the University of Pittsburgh in 1986. Shelley is also a certified elementary principal and a curriculum program specialist. He began his teaching career as a fifth grade teacher in Connellsville Area School District in 1972. During his 14 years in Connellsville he taught fifth grade, fourth grade, elementary gifted education and secondary gifted education. In 1987, he accepted a professorship at Edinboro University of Pennsylvania, where he worked for 15 years. While a professor at Edinboro, he developed the Educational Technology Center for students and faculty. In addition, he served as the director of the Miller Laboratory School. The last six years at Edinboro University he served as department chair in elementary education (1995 2002). He was hired in August, 2002 by Robert Morris University to be the director of elementary education. Since his arrival he has revised and developed courses at the undergraduate and graduate level. He developed and co-taught the first course; Applications of Instructional Technology in Education (EDML 8110) in the new PhD in Educational Management and Leadership. One of his major areas of research and study has been enhancing pre-service teacher's skills and expertise in applying educational technology to their teaching. He was the co-author of a three-year, $1.7 million PT-3 (Preparing Tomorrow's Teachers to use Technology) grant (1999-2002). He currently serves as the director of the Southwestern region of the Pennsylvania Association for Educational Computing and Technology (PAECT). His classroom interests include robotics and instructional software and authoring. He has also written numerous articles and given presentations at national and international conferences on the integration of technology into classroom teaching. In recent years he has worked with professors at the University of Costa Rica, the London Institute of Education and the Institute for Pedagogical Advancement in Aruba who have similar interests in instructional technology. Shelley is also a certified online teaching instructor and has developed and taught several undergraduate/graduate courses in the online format. In recent years his publication and research agenda has focused on online instruction.

Louis B. Swartz, JD, is a full-time assistant professor at Robert Morris University, Moon Township, PA, in the department of economics, finance and legal studies. He teaches Legal Environment of Business and The Constitution and Current Legal Issues at the undergraduate level and Legal Issues of Executive Management in the MBA program. He received his bachelor's degree from the University of Wisconsin in Madison, Wisconsin (1966) and his juris doctorate from Duquesne University in Pittsburgh, PA (1969). He is the coordinator of the Robert Morris University Pre-Law Advisory Program and a member of the Northeast Association of Pre-Law Advisers (NAPLA).

Chris Thompson is the director of technology for the School District of Elmbrook, a 7,700 student K-12 public school district just west of Milwaukee, WI. Thompson earned his BSBA from Georgetown University (1993) and completed his master's degree from the University of Maryland University College (2005). Thompson continues to research best practice staff development models, especially those focused on the integration of technology in the classroom. Thompson spends much of his free time with his wife, Melissa, and two daughters, Kate and Lauren.

John Vandegrieft has been associated with consulting, project management, IT and development for over 20 years. Having consulted for a couple of years, he went to work for Hewlett-Packard in 1983, working as a programmer analyst, system programmer and system manager in HP's Southern Sales Region IT. Then for 6 years, Vandegrieft was a technical consultant delivering training and consulting to HP customers. Moving back into what was now HP's America's IT, Vandegrieft was a Technology Research Engineer before taking the helm of the development arm of the Technology Solutions Lab, where he managed a group that grew to 22 people for 8 years. For the last 4 years at HP, Vandegrieft was a program manager delivering programs on a worldwide scale to HP's internal as well as external customers. Vandegrieft joined Blackstone and Cullen in 2006 as a senior consultant, bringing with him enterprise level experience. Vandegrieft is in his final semester at Kennesaw State University and will be graduating in May with a MS in information systems. John has a PMP certification in project management and is a Microsoft Certified Technology Specialist in Microsoft Office SharePoint Server 2007.

Mara H. Wasburn is an associate professor in the department of organizational leadership in the College of Technology, Purdue University. Prior to joining the faculty, her experience was in fundraising and publicity/public relations. Her research and consulting focus is on mentoring, with an emphasis on women in technology. She recently developed a team mentoring model, which is in the process of being trademarked. She holds a PhD from Purdue University.

Andreas Wieser-Steiner is a sociologist trained in the field of science and technology studies. He works as a lecturer at the University of Applied Science in Bremen and has conducted interdisciplinary work in the fields of human genome research, climate change research and digital media in education. He studied sociology and economics in Bremen and has earned a doctoral degree at the Technical University of Darmstadt (2004).

Heike Wiesner is visiting professor at the Berlin School of Economics in the Harriet Taylor-Mill-Institute with the denomination "Knowledge Management, eLearning and Gender," Germany. Her main work areas are business informatics and eLearning, science and technology studies, gender and digital media. She is the author of the book *Die Inszenierung der Geschlechter in den Naturwissenschaften.*

Lena Wilhelmson, reader in education, is a senior researcher within the field of organization pedagogies, and a university teacher at Högskolan Dalarna, Sweden, and also affiliated to the department of education, Stockholm University. Her research deals with individual and collective learning in renewal processes in working life. Another area of interest is adult education, dialogue and learning processes in adult life. Also, Wilhelmson has conducted studies concerning shared and joint leadership. Wilhelmson is the author or co-author of many articles and books. Some recent publications in the English language: Dialogue Meetings as Non-formal Adult Education in a Municipal Context (2006), Transformative Learning in Joint Leadership (2006), Sustainable Heritage in a Rapidly Changing Environment (accepted).

Index

A

ADDIE Model 136, 138
agent based laboratory 103–121
analysis phase 137
architectural design 105, 106, 112, 120
assessment 126, 128, 130, 131
asynchronous classrooms 269–282
asynchronous discussion 266, 267, 331, 350
asynchronous text meetings 226

B

behaviorism 152, 153–154, 157
blogs 234–258, 259
Bloom's At-A-Glance 305, 306, 307, 308, 309, 311, 330
Bruner's Three-Form Theory 155
business law 76–93

C

Chemical Reaction Model (CRM) 103, 105
cognitive flexibility theory 157, 161
cognitivism 153, 155–156, 157–158
collaboration 235, 236, 237, 238, 239, 242, 244, 246, 248, 249, 251, 252, 253
collaborative activity 283–302
collaborative peer tutoring xii, 185, 186, 187, 189, 194, 196, 197, 199, 332
collective learning 223, 230
comparison of e-learning outcomes 50–52
computer-supported collaborative learning (CSCL) 47
computer science 2, 13, 332
constructivist, distributed learning environment (CoDE) 47
constructivist e-learning environment (CEE) ix, 45, 46, 47–48, 50, 54
constructivist e-learning system (CES) 45–58
constructivist learning 62, 235, 236, 251, 258, 350
control agent 117
control dimension 272
conversational technologies 235, 236, 237, 251, 253
cooperative learning theory 156–157
course design 81
course management system 29

D

data flow diagram (DFD) 94–102
design phase 137
development phase 137
dialogue competence 219, 220, 221, 223–224, 227, 228, 230, 231
didactical actor 63–65
didactical agency 59–75
didactics 59–75
differentiating qualities 224, 229
digital portfolios 209
direct instruction 157
discussion 219, 221, 223, 225
discussion forums xiv, 259, 260, 261, 266, 331
distance education 142, 144, 145, 148, 149, 150, 260
distance learning 15–28
distance training 144, 150, 350
divergent thinking 308, 309
dual-coding theory 160

E

e-learning 59–75
e-meetings 224, 227, 230

e-pedagogy 76–93
effectiveness-breadth 264, 265
effectiveness-depth 264
effective questioning 303–312
elaboration theory 157, 160
Electronic Classroom of Tomorrow (ECOT) 143, 144, 145, 146, 148, 149, 332
electronic portfolios 209
enhancing Missouri's Instructional Networked Teaching Strategies (eMINTS) 289, 297
Estonia 29–44
European Enhancement of Early Years Management Skills (EEEYMS) 67, 68, 69, 71
evaluation phase 138

F

face-to-face instruction 313–323
facilitating online learning 303–312

G

Gagne's Conditions of Learning 157–158, 161
gender 15–28, 60, 61, 62, 63, 64, 65, 66, 68, 75
gender communication style differences 17–18
group dynamics 293
group work 70–71, 127–128

H

humanism 156–157, 158–159
hybrid e-learning 269–282

I

I.Q. – I QUESTION 305, 306, 308, 309
I.Q. – I Question a Picture 309, 310
implementation phase 137
information agents 115–116
information and communication technology (ICT) 30, 32, 39
information systems (IS) 30, 31, 32, 33, 34, 38, 39, 40, 42, 335, 352
initiative 262, 264, 265
instant messaging 236, 238, 239, 252, 253
instructor points 265
integrating qualities 228
Interactive Lesson 135–141
interactive lesson 135
interface agents 115
intersubjectivity 283–302
interteaching 189–190
iQ Academies (iQ) 143, 144, 145, 148, 149, 339

J

Java™ vi–vii, xii–xvi, 185–203
java program 187–188
java tutor 187–188, 189

K

knowledge management 122, 123, 128, 132, 133, 147, 236, 247, 248, 253, 340

L

lab interface 112–113
learning community 144, 319
learning design 46, 56, 341
learning dialogues 221, 223, 224
learning ePortfolio 206
learning management systems (LMS) 259, 260, 261, 262, 263, 266
learning outcomes ix–xvi, 45–58
learning processes 222–223
legal environment of business 77

M

maintenance agent 117
managing instructional tactics vi–vii, xii–xvi, 185–203
master control agent 116–117
mastery learning 154
media, and women in technology 1–14
media strategy 3–6
Merrill's Instructional Theory 158, 160
message passing interface (MPI) 109, 112, 113, 115, 117, 118
methodologies 94, 95, 96, 97, 98
Moore's Theory of Transactional Distance 155–156, 160
multi-agent systems 104, 106
multi-agent system simulation 115–120

N

National Educational Technology Standards (NETS) 168, 183, 346
New York colleges and universities 204–218

O

online class, and organization 316
online course discussions 18–19
online information security 319
online instruction 313–323

online instructors 151–165
online learning 235, 244, 252
online versus traditional instructional issues 77–78

P

parallel learning processes 222, 228
partial least squares (PLS) 29, 36, 37, 38, 39
participation vi, xiv, 259, 262, 264, 265
pedagogy 126, 132
phenomenal field theory 158–159
podcasting 259
podcasts 234–258
PowerPoint presentations 135–141, 166–184
problem based learning 68, 72
programmed instruction xii, 185, 186, 196, 197, 198, 199, 332

Q

qualitative and subjective forum assessment 262–263
quantitative and objective forum assessment 261–262

R

research model 34, 37
resistance and disagreement 283, 286, 292, 296
Roberta—girls conquer robotics 59, 60
Robert Morris University (RMU) 77, 80, 91
robotics 59, 60, 61, 62, 63, 65

S

science, technology, engineering, and mathematics (STEM) 1, 2, 3, 5, 6, 7, 8, 10, 11, 12
self-actualization theory 159
showcase ePortfolio 206
social-cultural model of learning 153–154
sociolinguistics 16, 17, 18, 19, 23, 25, 335
software self-efficacy xiii, 185, 186, 188, 189, 193, 194, 195, 198, 332
staff development efforts 144
staff training 142–150
structural equation modeling 29, 36
structured ePortfolio 206
student attitudes 166–184
student information agent 117–120
student Web portfolios 204–218
student websites 209
style 265
synchronous e-learning 269–282
synchronous technology 275
synchronous text meetings 226
systems analysis 94, 95, 97, 99, 335

T

teaching 94
team behavior 129–130
team competency principles xi–xvi, 122–134
team mindset 128–129
team work 128, 131
technological aids 166, 169, 182
technology, women in 1–14
technology acceptance model (TAM) 29–44
technology survey 166, 171
temporary suspension 283, 286, 288, 291, 292, 296
theory of immediacy and social presence 156
theory of multiple representations 155, 160
threaded discussions 15–28
time dimension 271
timeliness 264
traditional classrooms 269–282

U

undergraduate education 104
use cases 94–102

V

value 264, 265
video-casting 259
video or talking head pane 276
virtual high schools 142–150
Virtual I.D.E.A.L. (IDEAL) 145, 146, 147, 148, 149
virtual learning environments (VLEs) 270, 272, 273, 281

W

WebCT 29–44
Web dialogues 219–233
weblog 241, 242
web portfolios 209
whiteboard and Web-browsing pane 275
wikis 234–258
women 1–14

Z

zone of proximal development (ZPD) 285, 286, 287, 295

Printed in the United States
By Bookmasters